Real World Dog Training

Why **CORRECTION SUCCEEDS** and **"POSITIVE" TRAINING FAILS**

TULLY WILLIAMS

Copyright © 2025 by Tully Williams

First printed 2025

All rights reserved. This book may not be reproduced in whole or in part, stored in a retrieval system, or transmitted in any form or by any means—electronic, mechanical, or other—without prior written permission from the copyright owner, except by a reviewer, who may quote brief passages in a review.

Title: Real world dog training – why correction succeeds and "positive" training fails.

ISBN: 978-1-7638209-0-6

Praise for *Real World Dog Training*

"Dog training today suffers from an ideology, and this book is the antidote.

Years ago, renowned behavioral psychologist B. F. Skinner claimed that punishment was ineffective for controlling behavior. He fueled a well-meaning but misguided movement against even mild correction in animal training—from pigeons to people. But research shows that Skinner was wrong: punishment is effective, natural, and necessary.

Through vivid examples and careful explanations, experienced dog trainer Tully Williams shows how punishment is part of life and, when properly applied makes a better, happier dog (and owner!) than reward-only methods. A 'balance-trained' dog not only learns faster but winds up happier than those unfortunate animals trained by less effective methods based not on science, but on ideology.

A great and necessary book."

- Professor John Staddon
James B. Duke Distinguished Professor Emeritus of Psychology and Neuroscience.

Author of more than 200 research papers and numerous books, including co-editor of the *Handbook of Operant Behavior* and author of *Adaptive Dynamics: The Theoretical Analysis of Behavior; Adaptive Behavior and Learning; Scientific Method: How science works, fails to work and pretends to work; Science in an age of unreason;* and *The New Behaviorism: Foundations of behavioral science.*

"A must-read for veterinary professionals—and indeed all dog trainers—who value science- and evidence-based methods. My own dogs are proof that what Tully says and teaches, works."

- Dr Deborah Maxwell BVSc
Veterinarian and Working Dog Trainer

"It only takes a little reflection on our own lives to see how much of our behaviour involves avoiding consequences that will hurt us. We avoid touching a hot stove, driving through red lights, and making jokes in immigration queues at airports. We do these things not because somebody has rewarded us for doing so, but because the world in which we've evolved naturally arranges aversive consequences. And we learn very quickly and efficiently to avoid them.

We share this evolutionary history with other animals, including dogs. So it's both surprising and misguided that the dog-training profession has been so strongly influenced by an ideology that demonizes the use of punishment. We should, apparently, never even say 'no' to our dogs. This philosophy of only allowing our dogs to use half of the learning system they've evolved to have not only results in training methods that are time-consuming and often ineffective, but betrays a serious misunderstanding of learning principles researched through 100 years of behavioural psychology, including unjustified appeals to research that has been debunked long ago.

Tully Williams' important book draws on both his extensive experience as a dog trainer and on his confident understanding of learning principles. It is a much-needed and bracing corrective to the misguided 'reward only' approach, and it's also an engaging and entertaining read. This book should be required reading for all dog trainers, and probably for all dog owners as well."

- Professor Douglas Elliffe
Former Head of Psychology and Deputy Dean of Science,
University of Auckland

"The dog training world is in crisis! More dogs are ending up in shelters due to unresolved behavioural issues than ever before. At the root of this growing problem is the rise of the 'Force Free' cult—an ideology that dismisses tried and tested training methods, leaving dogs and owners without real solutions.

In *Real World Dog Training*, Tully Williams boldly challenges this ideology. Drawing on decades of experience across multiple training disciplines, he delivers a powerful critique of the force-free approach and supports it with evidence of what truly works—and what doesn't.

An essential read for professional dog trainers and dedicated dog owners alike."

- Alan Brooks
International Award Winning Guide Dog Specialist

About the author

Tully Williams has been training dogs since he was twelve years old (over thirty-five years) and started competing in sheep dog trials when he was fifteen. He taught his first obedience classes and private one-on-one problem-solving consults in 1993. He has personally owned and trained many hundreds of working dogs, from various breeders and bloodlines, mostly Australian border collies but including many kelpies. He generally has between 15 to 20 of his own working dogs in his kennels at any time, plus a variety of Turkish Kangal and Maremma livestock guardian dogs.

For the past 18 years Tully has operated his highly successful dog training and behavioural problem-solving business *Bendigo Dog Training* (*www.bendigodogtraining.com.au*). In this time, he has assisted many thousands of dog owners and their dogs with all types of general training and also every behavioural problem imaginable. This includes all types and levels of aggression, which many trainers, particularly positive reinforcement trainers, refuse to work with.

Tully is also one of the most experienced breeders of working sheepdogs in the world—over the past 30 years he has bred over 150 litters (an average of five litters per year) of his highly sought after *Campaspe Working Dogs* (*www.campaspeworkingdogs.com*). This has included personally rearing almost every single pup to working age (when the pup's working instinct switches on, at between four and seven months of age, when the pups can be assessed for suitability for potential buyers or be retained for breeding or training).

Tully served 7 years on the governing board of the *Victorian Working Sheep Dog Association* (VWSDA). He was the youngest member ever to be elected to the board (as a teenager) and was on the Open judging panel.

More recently he was a founding member, Vice President, and judge, of the *Australian Utility Stock Dog Society* (AUSDS), a society established with the aim of "fostering excellence in breeding practical utility stock dogs", through the running of better designed working dog trials and a superior judging system. In 2020 he was awarded a Life Membership of the AUSDS.

In 2007, Tully's ground-breaking book, *"Working Sheepdogs – A practical guide to breeding, training and handling,"* was published by Landlinks Publishing (a division of the CSIRO—*Commonwealth Scientific and Industrial Research Organisation*—Australia's peak scientific body). It has been described as "the finest book written on the selection, breeding, training and handling of working dogs… a must read for every working dog breeder, trainer and farmer," and "destined to become a classic in its field".

He has also produced an online video streaming series (with more on the way) of sheepdog training instructional videos (*www.workingsheepdogtraining.com*).

Prior to a bad fall from a horse while out mustering sheep in steep, rocky, mountainous terrain, which resulted in a serious back injury, he was also an experienced horse trainer, including retraining problem horses.

Tully is also an experienced trainer of humans. Besides training thousands of clients in the art of dog training, he has also trained thousands of students (both adults and children) through his and his wife's successful ballroom dancing studios, including coaching DanceSport competitors.

He currently resides on his farm in Victoria, Australia, with his wife Bronwyn, along with sheep, cattle, horses, goats, chooks, and dogs.

Figure 1 - The author with sheepdogs Campaspe Ring and Campaspe King

Praise for

WORKING SHEEP DOGS
A Practical Guide to Breeding, Training and Handling - *By Tully Williams*

"…the finest book written on the selection, breeding, training and handling of working dogs… a must read for every working dog breeder, trainer and farmer."

"A truly exceptional book on breeding…the painstaking obsession of brilliance."

"… a whole new approach to breeding, training and handling farm dogs."

"THE book I've been needing for years… my BOOK OF BOOKS!"

"…one of the best training books I have."

"'Working Sheep Dogs is a seminal book on the subject."

"One of the best, most insightful books on working stock dogs"

"…an outstanding piece of work…. destined to become a classic in its field."

Buy it now at

www.realworlddogtraining.com.au
www.campaspeworkingdogs.com

Client Testimonials

The following testimonials are just a small sample of the thousands of success stories from people who have applied the principles described in this book. These examples provide a glimpse into the transformative impact this approach has on dogs and their owners. See the Appendix for more.

Couldn't believe my eyes!

So grateful! After months of training with positive reinforcement with another trainer without any result, Max our one-year-old puppy walked on the lead within 2 minutes without any pulling thanks to Bendigo Dog Training. Couldn't believe my eyes. He is a different puppy on the lead. Our walks are now enjoyable instead of frustrating. Worth every dollar! Thanks.

Christa Goodall

A completely different dog since working with you

Hi Tully. I just wanted to thank you for coming out the other day! Jed has been a completely different dog since working with you. No jumping and no scratching at the door, and now we are using the same techniques for other little issues that arise! He was even super calm and gentle with my friends 8-month-old baby! Cannot thank you enough!

Tanika Love

Can't thank Tully enough...

We can't thank Tully enough, we were on the verge of having to re-home our family pet, and after three sessions we have a dog who will calmly interact with our family and is quickly learning to do everything he's told.

Nikki Anthony

Cured in 45 minutes

Fantastic and fast dog training. Our Ridgeback Sam had a phobia of bridges (refused to cross the Mitchell St train bridge, or any others), and was hard to get into the car. Cured in 45 Minutes. Highly Recommended.

Andrew Wilkinson

Amazing outcome

I had problems with my dog Kia jumping on everyone she saw whilst out walking. Since my lessons with Tully we now walk up the street with no jumping, no chasing cats or other dogs, and best of all NO LEAD! Thankyou Tully—an amazing outcome.

Jenny Round

Amazing... tried other avenues—they didn't work.

Amazing. Our staffy x has stopped jumping, nipping, rushing through doors and we have had calm walks that everyone enjoyed. Tully explained what we should be doing clearly and the reasons why. Tully was always prompt and easy to organise sessions with. I'd recommend him 100%, we had tried other avenues—they didn't work. Thanks Tully—one happy dog and very happy humans

Lucy Schepisi

Tried multiple dog trainers, but none of them compare to Tully

Tully is a highly professional and knowledgeable dog trainer. He took incredible pride in training our two border collies. We have tried multiple dog trainers, but none of them compare to Tully. Highly recommend.

Daniel & Morgan Leist

See Appendix for more

Contents

Introduction — 15

Part 1 – The demonisation of punishment — 19

Chapter 1 – The demonisation of punishment — 21

    Dr Ian Dunbar 23
    Karen Pryor 27
    Zak George 28
    Other influential figures 29
    Conclusion – the demonisation of punishment 29

Part 2 – Some terms explained — 31

Chapter 2 – Some terms explained — 33

    Reward 33
    Reward-only trainers 34
    Punishment 35
    Correction 36
    Classical conditioning 36
    Operant conditioning 37
    Continuous or variable schedules of reinforcement 41

Chapter 3 – Some problems with operant conditioning — 43

    Overlap between the quadrants 43
    Confusion 48
    Problems with definitions 49

Part 3 – Answering the case against punishment — 53

Chapter 4 – Is punishment effective when training dogs? — 55

"Positive" trainers and the effectiveness of punishment............ 56
Punishment in nature... 66
A mother with her pups.. 69
Dog to dog interactions.. 72
On wolves and dogs.. 74
Dominance... 80
How can these trainers all get it so wrong?........................... 87
Conclusion – Is punishment effective when training dogs?......... 89

Chapter 5 – Is punishment effective when training people? — 91

Debutante problems.. 93
Misbehaving students.. 93
Competition training... 95
Self-injurious behaviour... 98
Speeding fines... 100
Differences between people and dogs................................. 100
Conclusion – Is punishment effective when training people?...... 103

Chapter 6 – Does punishment cause undesirable side effects? — 105

When might punishment cause problematic side effects?............ 105
Aggression... 111
Fear and anxiety.. 121
Learned helplessness... 131
On being "operant", and spoilt brats................................... 136
My dog won't like me.. 138
Conclusion – Does punishment cause undesirable side effects.... 145

Chapter 7 – Can punishment *create* behaviours? — 146

Conclusion – Answering the case against punishment............... 151

Part 4 – Investigating reward — 153

Chapter 8 – Does reward work? — 155

Teaching the meaning of commands..................................... 156
Gaining obedience to commands.. 158

Recall..	159
The "Really Reliable Recall"...	172
Walking on the lead..	174
Competing rewards—and Premack.....................................	182
Stopping unwanted behaviour with reward.........................	188
Bin diving...	189
Rewarding alternative behaviours......................................	191
On extinction—and consistency...	198
Confinement...	203
But it worked for my dog!...	204
Conclusion – does reward work?.......................................	210

Chapter 9 – More downsides and side effects of "positive" training   213

Positive reinforcement methods contain punishment.............	215
Aggression..	218
Learned helplessness..	220
Fear / anxiety..	222
Hunger..	226
Addiction..	232
Stringent and extended restriction of freedom.....................	235
Manipulation and dishonesty...	239
Anger and abuse..	240
Rehoming and euthanasia..	243
Conclusion – downsides and side effects of "positive" training...	248

Part 5 – The scientific basis—fact and fallacy   251

Chapter 10 – The scientific basis—fact and fallacy   253

Why observational evidence is so often superior to "science"......	257
What is wrong with science – the people............................	259
What is wrong with science – surveys................................	260

Chapter 11 – Murray Sidman's "Coercion and its fallout"   265

Punishment doesn't work?...	266
The inductive fallacy – wild, fanciful extrapolation.............	270
Punishment is coercive and reward is not?.........................	272
Deprivation – positive reinforcement IS negative reinforcement...	275
Destroying the bond..	277
Conclusion – "Coercion and its fallout".............................	281

Chapter 12 – Some real-life junk "science"   283

    Aggression survey – Herron, Shofer & Reisner 2009............... 283
    Livestock chasing – E-collar verses positive reinforcement........ 291
    Some real science... 300

Chapter 13 – LIMA, LIFE and PROTECT   303

    PROTECT.. 306
    Prioritize Results.. 306
    Optimal Training.. 308
    Ethical Transformation... 309
    Conclusion – LIMA, LIFE and PROTECT........................... 310

Part 6 – Myths about principles of punishment   311

Chapter 14 – Myths about principles of punishment..................... 313

    Punishment must be consistent?.. 314
    Punishment must be immediate?... 317
    Punishment must be instructive?... 322
    Punishment must have a warning?... 325
    Conclusion – myths about principles of punishment................ 328

Chapter 15 – A model of obedience—balance scales................... 329

Part 7 – The morality of punishment   339

Chapter 16 – The morality of punishment........................... 341

    Can we train without punishment?.. 342
    Is there anything morally wrong with using punishment?......... 343
    Is punishment really that bad?... 344
    Do punishment's positives outweigh its negatives?................... 345
    Conclusion – the morality of punishment................................. 347

Conclusion   349

    Appendix – Client testimonials... 363
    Index.. 373

Introduction

Dogs, so they say, are man's best friend. And, indeed, a well-trained and inherently good-natured dog can certainly be a great companion. But for many owners their dogs can also be a source of great frustration and annoyance, many sadly ending up in the pound or at the vet surgery waiting to be euthanized.

For me, one of the most gratifying things about teaching people to better handle their pets is seeing the great changes that occur in the dog, usually in a *very* short space of time (when sound training methods are used). Equally gratifying is seeing the load that this lifts from the owner's shoulders, and the large number of dogs that consequently are saved from being rehomed or put to death. Once dogs have some good manners and general obedience, there is often a marked improvement in quality of life for both dog and owner.

In my work as a dog behavioural and training consultant, I often arrive for the first consultation to find owners at their wit's end. They're often frustrated and desperate, believing there's just no hope for their pet. The professional dog trainer is usually their last resort. And yet, when I return a week or two later after that initial one-and-a-half-hour session (if a second visit is even required), a great weight has been lifted from their shoulders. They can see light at the end of the tunnel!

Comments often centre on amazement that such a dramatic change has occurred in such a short space of time: "He's like a new dog!" Walking on the lead issues are cured in a week, sometimes even in minutes:

Introduction

> *"So grateful! After months of training with positive reinforcement with another trainer without any result, Max our one-year-old puppy walked on the lead within 2 minutes without any pulling thanks to Bendigo Dog Training.*
>
> *Couldn't believe my eyes. He is a different puppy on the lead. Our walks are now enjoyable instead of frustrating. Worth every dollar!!! Thanks."*
>
> *- Christa Goodall*

Certain long-standing phobias can be cured in minutes. Recall (come when called) sorted in a week. And most other issues likewise:

> *"Fantastic and fast dog training. Our Ridgeback Sam had a phobia of bridges (refused to cross the Mitchell St train bridge, or any others), and was hard to get into the car. Cured in 45 Minutes. Highly Recommended."*
>
> *- Andrew Wilkinson*

Yet there really isn't anything amazing or miraculous in these turn-arounds. Or, at least, *there shouldn't be*. They are simply the result of a sound understanding of the principles of dog training and psychology, and of their application. The only really amazing thing is that most professional dog trainers nowadays (who are often of the so-called "positive" or "force-free" variety) do not really understand (or simply refuse to acknowledge) *the most fundamental aspects* of dog (or human) psychology and training.

This book has come about largely in response to this relatively recent proliferation of so-called "positive" dog training. Many books and dog trainers have surfaced in recent years advocating this extreme "reward-only" policy, and condemning outright any use of correction or punishment. The results have been *disastrous*.

These extreme, reward-only views are not driven by "science" (as they would have us believe), or by results, but by *ideology*. Those subscribing to these views believe that all correction or punishment is bad, and so they ignore the *reality of the real world*. They attempt to train in a way that is contrary to nature, contrary to experience, contrary to logic, contrary to common sense, and contrary to *true* science.

The "positive" trainers have done an impressive job of vilifying the concept of "punishment", giving it a strongly negative connotation. The word often conjures up images in people's minds of what can only be described as *abuse*, not punishment. I will get into the full definition of punishment shortly, however, just know that when you read "punishment" throughout this book, it means *anything*

Introduction

the dog (or person) finds even *faintly* unpleasant—such as a "stern look", telling your dog off, a time-out, a tug on a leash, or, as we go up the scale, a smack, and so on. A speeding fine is punishment for us. A lunch-time detention is punishment for kids. As you keep reading you will discover what a great job the "positive" trainers have done at demonising something that is perfectly normal and natural, extremely effective, *entirely necessary*, and which happens all around us, and to us, and to our dogs, all the time.

In my practice as a dog behavioural and training consultant dealing with problem dogs of all breeds, I see far greater problems *caused by reward* than by punishment—a fact which will be demonstrated throughout this book. It is rare that I come across problems that have been caused by punishment, **and there has never been even a *single case* of problems or side-effects caused by a *correct* usage of correction or punishment.**

Far more problems are caused by a *lack of effective discipline* than are caused by too much, or ill-used, correction. Furthermore, *the vast majority* of the dogs ending up at the pound with behavioural problems, to await adoption to become someone else's problem, or to be put to death, are there largely *as a result* of this *"avoid correction at all costs"* policy. And so, the "positive" trainers must accept the lion's share of the blame for this state of affairs. The truth of this will become evident as we continue.

> *Far* more problems are caused by a *lack of effective discipline* than are caused by too much, or ill-used, correction.

Unfortunately, I don't get to help all the pets with behavioural problems. And so, many do sadly end up being rehomed or euthanised. Or many are simply left to languish in the back yard (what the reward-only crowd, as exemplified by the *American Veterinary Society of Animal Behaviour* (AVSAB), euphemistically refers to as "management").

In this book, I will outline and discuss the main arguments levelled against punishment and correction by such "reward-only" advocates. I will explain why their arguments—which are often deceptive or even outright lies—do not withstand scrutiny. I will also explain why such an extreme avoidance of correction or punishment is not only *unnecessary*, but is also often much *slower*, far less *effective*, generally *ill-advised*, and very often *harmful*. I will also show that such methods are contradicted by the *sound, credible* scientific research.

I will outline a more balanced, quicker, easier, longer lasting, and far more effective approach. Just like it always has, *the judicious mix of both reward and correction, with occasional resort to punishment,* provides the fastest and most

Introduction

effective results in producing a happy, well-adjusted member of your family "pack".

That phrase, *"the judicious mix of reward and correction, with occasional resort to punishment,"* encapsulates in a nutshell my approach to dog training. I use reward, and rely on it heavily (though usually not in the way the treat trainers do). I also use correction and punishment, just like in nature, in *the real world*. After all, as everyone knows, that's the way nature trains—by reward *and* punishment. That's the way bitches train their pups, or dogs deal with each other, or pups learn about their environment. And they do so with *none* of the often-claimed adverse side-effects (as we will see).

> Good Dog Training in a Nutshell
>
> The judicious mix of both reward and correction, with occasional resort to punishment.

Why spend months or even *years* attempting to teach something, or attempting to cure problems, by long, time consuming, difficult, convoluted, and ineffective "positive" methods, when the same result (and usually far superior) could be accomplished by a more suitable method in *a tiny fraction* of the time? And especially when we consider that simultaneously we achieve *far superior* welfare outcomes?

So, let's get started with some commonsense *real world* dog training. Let's debunk the myths and lies of the ineffective, ideologically and financially driven utopian fantasy that is "positive" training.

Part 1

THE DEMONISATION OF PUNISHMENT

Chapter 1 – The demonisation of punishment

1

The demonisation of punishment

As mentioned in the introduction, there has been quite an attack on the use of correction and punishment in dog training (and also in child rearing, school teaching, and even in the judicial system) in recent years. And, conversely, a push to encourage—and even *legislate*—the use of "reward-only" methods of training.

The school of "reward good behaviour and ignore bad behaviour" argues for the absolute avoidance of *all* correction or punishment, regardless of the situation. This is quite an extreme view, but unfortunately it has been widely promoted. Some of these trainers even go so far as to say that you should never even give your dog a "stern look" or a "sigh", because these constitute punishment (and by definition, according to them, *all* punishment is bad):

> "Have you ever thought of a stern look as punishing?... my trainer pointed out to me the effect those sighs were having on my Jessie."[1]

The reward-only ideologues, like this *Karen Pryor Certified Training Partner* just quoted, tell us that this should be avoided at all costs. In fact, Kay Lawrence CPDT-KA (Certified Professional Dog Trainer—Knowledge Assessed) tells us we should not even use what is known as "negative punishment", such as turning our back on our dog if it jumps on us.[2]

---

[1] Benson, Cindy. *Livestock Guardian Dog Training Manual.* Self-published, 2021.
[2] Pryor, Karen. *On My Mind – Reflections on Animal Behavior and Learning.* Waltham, MA. Sunshine Books. 2014.

PART 1 – The Demonisation of Punishment

Trainers who utilise punishment as one of their tools are often demonised. Dr. Ian Dunbar talks about "Trainers from The Dark Side", and much more from various people along the same vein.

The problem is that this fear of utilising correction or punishment in any form is largely the reason for the very many uncontrollable dogs ending up in the pound (and for the drastically increasing numbers of children and teenagers ending up in various sorts of trouble). It is also responsible for a large percentage of the dogs that are euthanised due to behavioural problems.

Up until the late 1980s, or even the 1990s, dog training was fairly uncontroversial. Trainers nearly all used a combination of reward and punishment, as they have done since time immemorial.

Then along came *a small number of influential figures* who were to greatly alter the dog training landscape. And it wasn't as though these small number of influential figures were even particularly qualified for the role, as we will soon see. It is quite amazing how a tiny number of people can greatly influence the world, and not for the better. And, as we will also soon learn, not necessarily because they knew (or know) what they were (or are) talking about.

> It is quite amazing how a small number of people can greatly influence the world, and not for the better.
>
> And not necessarily because they know what they are talking about.

This small number of influential figures started what has been called a "revolution" in dog training—the revolution of demonising the use of correction or punishment, and of pushing the ideology of ineffective "positive" training. The results have been devasting by every measure.

Throughout this book, I will be fairly critical of a number of these major figures. Perhaps some will think this is unfair and/or unnecessarily harsh (those in the know will realize that I haven't been nearly harsh enough). So, I'd like to explain why I believe this is necessary, and why I chose these particular figures to focus on. If I had just written a book expressing my ideas and opinions, backed up perhaps by examples, or by some science, that would be one thing. But the reward-only advocates would simply dismiss it as just another author who is "living in the dark ages", who "doesn't understand the science", and so on.

However, what I have chosen to do is to use the examples and words of some of the best-known, big-name, so-called "positive" trainers, to prove the case against their approach. I have chosen to attack the myths head on, and to deal with them directly. Only in that way, I believe, is there any chance of people realizing the truth. Maybe then we can prevent family's lives being ruined by out-of-control

CHAPTER 1 – The Demonisation of Punishment

canines, and maybe then we can prevent so many dogs from being destroyed for behavioural problems that are easily remedied. The "positive training" crowd have a lot of blood on their hands.

> The "positive training" crowd have a lot of blood on their hands.

A variety of sources will be cited throughout this book, however a number of names will keep surfacing. These main figures will get a brief introduction here, before we get into debunking the myths and lies about punishment, proving that "correction succeeds and positive training fails", and outlining what actually does work in *the real world*.

Dr. Ian Dunbar

One of the main figures in the "positive" world has been Dr. Ian Dunbar. He has been called "the godfather of modern dog training."[3] On his website, Dogstar Daily, we read:

> "He received his veterinary degree and a Special Honors degree in Physiology & Biochemistry from the Royal Veterinary College (London University) and a doctorate in animal behavior from the Psychology Department at the University of California in Berkeley, where he spent ten years researching olfactory communication, the development of hierarchical social behavior, and aggression in domestic dogs.
> "Dr. Dunbar is a member of the Royal College of Veterinary Surgeons, the International Society for Applied Ethology, the American Veterinary Society of Animal Behavior, the California Veterinary Medical Association, the Sierra Veterinary Medical Association, and the Association of Pet Dog Trainers (which he founded)."[4]

Not only did Dunbar found the *Association of Pet Dog Trainers* (renamed in 2013 as the *Association of Professional Dog Trainers*) which claims it is "the world's largest professional dog training community", but the *Certification Council of Professional Dog Trainers* is an offshoot of APDT, and so also linked to Dunbar. He was also involved in establishing the *American Veterinary Society of Animal*

---

[3] Dunbar, Ian, Dr. *Barking Up The Right Tree - The science and practice of positive dog training*. Novato, CA: New World Library, 2023.
[4] Dunbar, Ian, Dr. *Dr. Ian Dunbar*, DogStarDaily, accessed December 11, 2024, https://www.dogstardaily.com/blogs/dr-ian-dunbar.

PART 1 – The Demonisation of Punishment

Behaviour (AVSAB)[5], of which you will read more about in this book. The deeper you dig the more obvious it becomes that it has been a very small number of people at the root of the reward-only push, with Dunbar perhaps at the epicentre.

He was also the first to really popularise puppy classes throughout the world on a large scale. He, and now his son Jamie and ex-wife Kelly, continue to be major figures in the "positive" training field. He has often referred to trainers who use punishment as "Trainers from the Dark Side."[6]

Evidently a highly savvy businessman, Dunbar has been enormously influential. However, like so many others, he became highly influential despite having *virtually no real experience*. By his own admission he started teaching dog training courses with *no real experience*, and also started his puppy classes in much the same manner. Most people, I suspect, would think of a highly experienced dog trainer deliberately setting out to teach his extensively tried and proven techniques. That was, in fact, far from the case.

> By Dunbar's own admission he became highly influential despite having *virtually no real experience*.
>
> He started teaching dog training courses with *no real experience*, and also started his puppy classes in much the same manner.

In Dunbar's own words:

> "…when I taught the ten-week course on dog behaviour at the UC Berkeley Extension, I included one-week on dog training. *In preparation, I read two books* by Dr. Leon Whitney—The Natural Method of Dog Training and Dog Psychology: The Basis of Dog Training. I loved these books because they talked in detail about using food lures and reward for off-leash training… for beagles!... *I just gave the technique a name for the lecture.*"[7] (emphasis added)

And elsewhere:

> "The ten-week extension course was primarily about behaviour, but by unanimous request, I included a week on dog training—about which I thought *I knew little to nothing.* So I just chatted about how, when I was

---

[5] *Ian Dunbar*. Wikipedia, Wikimedia Foundation, 7 May 2024, https://en.wikipedia.org/wiki/Ian_Dunbar.

[6] Dunbar, Ian. *The Good Little Dog Book*. 3rd ed. Berkeley, CA: James & Kenneth Publishers, 2003.

[7] Dunbar, Ian, Dr. *Barking Up The Right Tree - The science and practice of positive dog training.* Novato, CA: New World Library, 2023.

CHAPTER 1 – The Demonisation of Punishment

growing up on the farm, we lure trained all animals off-leash, dogs included."[8] (emphasis added)

There are at least two problems here. Firstly, I believe Dr. Dunbar misrepresents Whitney's work. If you have never read Whitney's books you would get the impression that Whitney trained his dogs with treats off leash (like Dunbar's SIRIUS puppy classes), and you would also likely assume that Whitney was a reward-only trainer. Both would be completely false assumptions. In relation to training off-leash (of which Dunbar makes a big deal), here is what Whitney *actually* advises in the book Dunbar mentions:

"You will see that most of the training sessions were conducted *on a table*... I suggest that you train your dog on a similar *raised platform*. In the wall above the table, you will notice a hook, or eye-bolt... the dog's *leash is fastened* to it..."[9] (emphasis added)

So not only *on* leash, but *tied to the wall*, and *on a raised table*! I bet that isn't the impression you got from Dunbar's statement about "using food lures and rewards for *off-leash* training", is it? In fact, when I first read Whitney's book perhaps thirty years ago, I experimented with some of his methods. Training using treats with the dog on a raised platform tied to the wall was exactly what I did.

And, in relation to the impression that Whitney was a reward-only trainer, this is simply not the case. Yes, he used treats, but he also used punishment, as he knew its necessity. To give just one example from his book:

"When it's necessary, give your pup a sharp smack beside the cheek. And I do mean smack, not tap. Except with tiny dogs, to be effective, a smack must make your hand sting. Your hand will hurt more than the dog's cheek will hurt him... the rolled-up newspaper is effective, but only when used properly. It does have one special advantage. The sound it makes when you whack adds to the effect..."[10]

So there is really very little in common between the methods of Whitney's and Dunbar's. Whitney was a highly experienced trainer and breeder who actually knew how to train dogs in *the real world*.

---

[8] Dunbar, Ian, Dr. *Barking Up The Right Tree - The science and practice of positive dog training.* Novato, CA: New World Library, 2023.
[9] Whitney, Leon F., Dr. *The Natural Method of Dog Training.* New York: Tandem Books, 1969.
[10] As above.

PART 1 – The Demonisation of Punishment

> Dr. Dunbar admits he *knew very little* about dog training when he started teaching it.

Secondly, Dr. Dunbar admits he *knew very little* about dog training when he started teaching it. This is a common story for so many "dog trainers", academics, and teachers. They've owned a pet or two, or have some theoretical learning, and think that teaching people to train dogs for a living sounds nice, and so away they go. Or, they've done some research in a lab or trained a few dolphins, and think this qualifies them to tell others how they should train *in the real world*.

This reminds me of a similar story of this lack of experience in academia. There was a geneticist professor I corresponded with a few years back from the University of Sydney here in Australia, discussing dog breeding. Like many academics, she had strong opinions, but I suspected very little experience. When I pressed her on her actual experience breeding dogs, she admitted she had *never bred a single litter*. That story is typical of academics and researchers who think they can inform people in the *real world* how they should be doing things (and even worse, influence *legislation* on how they should be doing things), when they have *absolutely no experience*.

So, after teaching about dog training in his behaviour classes, *about which he admits he knew "little to nothing"*, Dunbar moved on and started teaching puppy classes. Again, the case was the same:

> "The whole time *I just followed my nose*. I planned very little. In the early 1980's, as I was starting SIRIUS, I remember driving across the Richmond Bridge to teach what was then the world's very first off-leash puppy socialization and training class and thinking, apart from a few dogs on the farm, *I have never actually trained a dog myself. Hope it works*. It did." [11] (emphasis added)

And yet, he also claims:

> "I first introduced science-based training techniques to dog trainers way back in 1971." [12]

---

[11] Dunbar, Ian, Dr. *Barking Up The Right Tree - The science and practice of positive dog training*. Novato, CA: New World Library, 2023.
[12] Dunbar, Ian. *Barking up the wrong tree – for 110 years?* Dog Star Daily, May 7, 2015. https://www.dogstardaily.com/blogs/dr-ian-dunbar/barking-wrong-tree-110-years.

So, by his own admission, he had *never actually trained a dog himself* when he started his highly influential puppy classes in the 1980's. And yet, he claims to have "introduced 'science-based' training techniques to dog trainers" at least a decade earlier!

> By his own admission, Dunbar had *never actually trained a dog himself* when he started his highly influential puppy classes in the 1980's.

Karen Pryor

Perhaps one of the next most influential figures in the "positive" field is Karen Pryor, who has been called "the mother of modern dog training."[13] Her book, *Don't Shoot the Dog—The New Art of Teaching and Training* was at the forefront of the "positive" push (even though, as we will see, she did use punishment very effectively at the same time as claiming it "almost never really works"[14]).

In the 1960s Pryor started working as a dolphin trainer at "Sea Life Park" in Hawaii, her husband's marine park. After half-a-dozen or more wild dolphins were captured straight out of the ocean, she set about training them to perform *in captivity*. The reader might already recognize just a hint of hypocrisy here, as Pryor—and her students—often talk about the deleterious side-effects of punishment—including such mild punishment as "stern looks" and "sighs". Yet, she was working with dolphins captured straight from the wild with what was definitely far greater trauma than any sound methods of punishment training ever produce. (If you'd like to learn more about this, watch the documentary *"Blackfish"* about Tilikum the killer whale.)

In 1984, following on from her experience training *dolphins in captivity*, her book, *Don't Shoot the Dog*, was published. It had a huge impact on the dog training world, even though Pryor—like Dunbar—had very little actual experience *training dogs*. Her experience came from training *captive dolphins* in a completely controlled environment, using techniques developed training rats or pigeons *in cages*. Unfortunately, these are all very different from training dogs out in the *real world* like general dog owners experience.

> *Don't Shoot the Dog* had a huge impact on the dog training world, even though Pryor—like Dunbar—had very little actual experience *training dogs*.

---

[13] Donaldson, Jean. *The Culture Clash*. 2nd ed. Berkeley, CA: James & Kenneth Publishers, 2005.
[14] Pryor, Karen. *Don't Shoot the Dog!: The New Art of Teaching and Training*. Rev. ed. New York: Bantam Books, 1999.

PART 1 – The Demonisation of Punishment

Those with an ideological bent towards the avoidance of punishment jumped on the bandwagon, and so the revolution of "reward-only" training was further strengthened. Ever since, Pryor has been highly influential in the world of "positive" dog training, nowadays particularly through her *Karen Pryor Academy* where people can become certified in her methods. She is particularly well-known for popularising "clicker" training (although Whitney amongst others had written about its use previously).

So here we have Karen Pryor (the "mother" of modern dog training), a *dolphin* trainer, and Dr. Dunbar (the "godfather" of modern dog training), who, even worse, had *never actually trained a dog or even a dolphin himself*, making huge waves in the dog training world. No tried, tested or proven methods (unlike Dr. Whitney, who actually had a lifetime of real experience), but just "follow your nose" and make it up as you go along. Or 'here's how I trained *dolphins* in *captivity*, let's train *dogs* (or people) in the *real world* the same way'.

Zak George

Zak George is possibly the most modern incarnation of the reward-only ideologue. He was a real estate agent who, after training his first dog, decided to quit real estate and make a living training dogs. So, yet another "trainer" with very limited experience.

George is a YouTube influencer, having amassed a huge following (in the millions) and claims to be the "#1 most subscribed dog trainer on YouTube" [15]. He has also published two books (with free-lance journalist Dina Roth Port) titled, *Zak George's Guide to a well-behaved dog*, and *Zak George's Dog training revolution*. In fact, there is nothing revolutionary about any of it. It is simply parroting the same old "positive" talking points. In fact, there is a huge amount that is wrong with the majority of the methods he teaches (as there is for all "positive" trainers). Many of the techniques simply *will not work*. Many will even *make things worse* (you'll learn more about this as we continue).

Part of the reason I have included many mentions of Zak George in this book (apart from the fact that his advice is typical of most of the "positive" trainers), is that he is a vocal advocate for the introduction of legislation *forcing* trainers to adopt ineffective "positive" methods like the ones he advocates.

As mentioned earlier, you will also hear a bit about the *American Veterinary Society of Animal Behaviour* (of which Dunbar was involved in establishing, and is

---

[15] George, Zak, and Dina Roth Port. *Dog Training Revolution: The Complete Guide to Raising the Perfect Pet with Love*. New York: Ten Speed Press, 2016.

CHAPTER 1 – The Demonisation of Punishment

a member), because this society also attempts to push these reward-only methods on all dog trainers. In fact, a past president of AVSAB, Dr. John Ciribassi, claims that the methods George outlines are "spot-on". Dr. Ciribassi couldn't be further from the truth, as we will discover.

Other influential figures

A number of other "positive" influencers also get fairly frequent mentions, again due to their popularity and influence.

Jean Donaldson, author of *"The Culture Clash"* (1996), is highly critical of anyone who uses punishment. However, as you will see as we progress, Donaldson admits that punishment can be indispensable at times, and at times uses it while claiming to be a "positive" trainer. Her book, described by Dr. Dunbar as "quite simply the very best dog book I have ever read,"[16] has been quite influential.

Pat Miller, author of *"The power of positive dog training"* (2008), and a past board member of the *Association of Pet Dog Trainers* (the association founded by Dunbar) has been another influential figure.

Victoria Stilwell, the actress who became a television dog training star, who describes herself as "one of the world's most vocal proponents of force-free dog training methods" (whilst herself using force), gets various mentions.

World champion agility trainer and author Susan Garrett will get a few mentions, as will influential New Zealander Daniel Abdelnoor (aka "Doggy Dan").

Murray Sidman, author of *"Coercion and Its Fallout"*, perhaps the king of the malignant reward-only ideology in the scientific and human realms, gets a whole chapter devoted to debunking his massively flawed ideas. The well-known B.F. Skinner will also get a few mentions.

Conclusion – The Demonisation of Punishment

So that's a quick introduction to the major figures responsible for the demonisation of sound training methods. Before we get into debunking the myths, the deceptions, and even the outright lies, it's important to clarify some terms that are used throughout this book. Let's do that now.

---

[16] Donaldson, Jean. *The Culture Clash*. 2nd ed. Berkeley, CA: James & Kenneth Publishers, 2005.

PART 1 – The Demonisation of Punishment

Part 2

SOME TERMS EXPLAINED

Chapter 2 – Some terms explained

Chapter 3 – Some problems with operant conditioning

2

Some terms explained

Throughout this book, I will use a variety of terminology. Many believe that we should stick solely to the behavioural scientist B. F. Skinner's terms of "positive reinforcement", "positive punishment", and so on. I don't accept that premise for a number of reasons, as will be explained here. I generally prefer the simple, intuitive terms "reward" and "punishment", and also "correction", and some other intuitive terms like "warning" and "praise" and so on.

First, I will give the simple definitions of various terms you will come across. These are the terms I usually use when explaining concepts to clients (depending on their level of understanding). Then I will give a bit more in-depth and technical assessment of operant conditioning and its terminology, and some of the issues I have with it.

Reward

I define reward as basically *anything the dog likes*. For example, food, or a pat, or praise, or toys.

Some dogs value these things more than others, and so they function better or worse as rewards. The more rewarding a reward is to that particular dog in that particular circumstance, the more it will reward behaviour! The less rewarding a reward is, the less it will reward behaviour. To a highly food-driven dog (and some dogs are *highly* food driven), or a starving dog, food is more effective than to a dog with low food drive, or one who has just eaten. This is why, when captive dolphins or killer whales (and dogs, as you will discover) are being trained, *food deprivation* is an essential aspect of their so-called "positive" training and performing.

Praise and patting are the same, as are toys, and games, and so on, if the dog likes those things. They are more rewarding for some dogs, and less for others. And more rewarding in some situations than others, depending on whether the dog wants them at the time or not.

So, reward is basically anything the dog likes or wants at the time. If the dog is scratching at the back door and you open the door to let it in, you have rewarded it for scratching the door. If your dog growls at you and you back off, you have rewarded the aggression. These behaviours will be maintained or will increase.

Reward-only trainers

Throughout this book I often refer to trainers, and/or their ideology, as "reward-only". By that, I mean they never use punishment (or at least they claim they don't).

This can vary from those who claim they never use "positive punishment" (see definition below), or "negative reinforcement", to even those who also don't use "negative punishment" (or think or claim they don't). This ideology is also sometimes referred to as "positive" or "force free", etc., which are all complete misnomers. So, when I refer to reward-only trainers, or their ideology, this is what I am referring to—the extreme (but now common) position that demonises basically all use of correction or punishment.

When I refer to "treat training", I am generally using this as shorthand for any type of "positive reinforcement" training, whether that involves food, or toys, or what some refer to as "life rewards".

On the other side of the coin are what are sometimes referred to as "balanced trainers". This doesn't mean that they use 50/50 reward and punishment, but that they use all the tools at their disposal, where each is applicable.

Also, note that there is no such thing as a "punishment-only" trainer. There is no "punishment-only" opposite to the extreme "reward-only" approach. There are just normal trainers who use all the tools of training depending on which is appropriate in any given situation, including positive reinforcement.

CHAPTER 2 – Some terms explained

And when I say "positive training fails" I am not saying that positive reinforcement doesn't work in *some* situations at *some* times for *some* things—what I am saying is that as an *overall strategy* for having a well-trained dog, when punishment is excluded, training fails.

> As an *overall strategy* for having a well-trained dog, when punishment is excluded, "positive" training fails.

Punishment

Punishment, the way I use the term throughout this book, is the opposite to reward—*anything the dog doesn't like,* or wants to avoid. And this is consistent with the way the term is used by most "positive" trainers. It doesn't have to be *physical* punishment, i.e. a smack. It may be, but it often isn't. Growling at your dog, "ahh," when it goes to jump in the wrong door of your car, can be punishment.

So-called "positive" trainers are generally against *any* and *all* forms of punishment. They generally claim they are all *ineffective*, will even make the behaviour *worse*, and will definitely cause *terrible, problematic* side effects. The following two quotes are typical. The first is from psychologist Jillian Enright:

> "Punishing Unwanted Behaviour *Just Makes it Worse*…when I refer to punishment, I don't mean only physical punishment, I am referring to *anything* that a child may experience as *unpleasant*. This can mean *time-out*, yelling, *removal of privileges, grounding, suspension* from school, and spanking."[17](emphasis added)

Or as we've heard from Cindy Benson, a "Karen Pryor Certified Training Partner":

> "Have you ever thought of a *stern look* as punishing?... my trainer pointed out to me the effect those *sighs* were having on my Jessie."[18](emphasis added)

Punishment could be a "stern look" or even a "sigh". It could be telling your dog off, "uh, uh, *Ruby*, no," if she tries to jump in the front seat of the car instead of the back seat. Take a human example—speeding fines. That is punishment, and it reduces our behaviour to drive above the speed limit (humans can be punished a fair time after the fact and it still influences behaviour—this is not true in dogs).

---

[17] Neurodiversified. *Punishing Unwanted Behaviour Just Makes it Worse*. Medium, February 2024. https://medium.com/neurodiversified/how-punishing-unwanted-behaviour-just-makes-it-worse-baf22793d07b.
[18] Benson, Cindy. *Livestock Guardian Dog Training Manual*. Self-published, 2021.

If your dog jumps on the kitchen table to eat your dinner, and gets a smack for his trouble, he is less likely to do it again in the future (and no, you aren't going to cause all sorts of terrible side effects—that myth will be debunked later).

So, when you read "punishment" throughout this book, it generally describes this *whole range* of consequences.

Punishment is ANY consequence for a behaviour that the dog (or person) doesn't like.

Correction

Correction is a term I use to denote something akin to a warning or low-level punishment. For example, if I growl "ahhhh" at a dog for doing something wrong, or about to do something wrong, that is what I mean by correction. The dog knows that if it ignores this there will be a consequence to follow (punishment).

Correction can also technically be termed "positive punishment" (in the operant conditioning terminology), or in some cases as a "conditioned punisher". However, it is very low-level punishment, if at all, and the dog views it as such.

Take a human example: If a child is thinking of bopping her brother over the head with a toy, and mum warns her, "Don't you even think about it…" in a stern voice, I call that correction. Does the child feel punished? Not yet! Or maybe very mildly. But if she ignores mum, and gets a smack, that's punishment.

Or the boss calls you over and says: "If you do that again, I will dock your pay". I would not call that punishment (although it can certainly act to reduce behaviour, so it could technically be termed punishment in the Skinnerian sense). I call it correction—it *corrects* a behaviour without requiring punishment, *provided it is known that the punishment will follow as a consequence if it is ignored.*

We know that being told not to do something, or being given a warning to stop doing something, isn't the same as being punished for doing it. It is the same for dog training. Common sense.

Classical Conditioning

Classical conditioning is the idea that behaviours or associations are learned by connecting a neutral stimulus with a primary reinforcer (eg. food). The term was coined by Russian scientist, Ivan Pavlov, in the 1890's, after he noticed that when

CHAPTER 2 – Some terms explained

dogs repeatedly heard a bell ring prior to being fed they soon came to salivate to the sound of the bell, or just to the presence of the assistant who fed them. Like when your doorbell rings and the dog goes berserk. Or when we feed a litter of pups and they come to associate us with good things.

Operant Conditioning

Operant conditioning theory consists of four "quadrants". This simply means there are four aspects to it—positive and negative reinforcement, and positive and negative punishment. (In Skinner's original work he only had three—he lumped positive and negative punishment together. And, as will be discussed, there is a sound scientific argument that there is really no meaningful distinction between positive and negative reinforcement—so we return to simply reward and punishment.)

	Reinforcement (reward)	Punishment
Positive + plus (add something)	Positive Reinforcement (reward)	Positive Punishment
Negative - minus (take something away)	Negative Reinforcement (reward)	Negative Punishment

Figure 2 - The four quadrants of operant conditioning.

In the scientific jargon, positive means we add something, while negative means we take something away. Reinforcement means, technically, something occurring after the behaviour that strengthens (reinforces) the behaviour. Punishment, by this definition, is something that decreases the likelihood of a behaviour, or weakens it.

37

PART 2 – Some terms explained

Operant Conditioning is the term B. F. Skinner coined in the 1950s to describe these theories on behaviour modification. Note the word "theories". There are various different theories and models of learning—Skinner's is just the best known. Many people, including many dog trainers, consider Skinner's theories to be something akin to fact or gospel—they are not. There have been many criticisms related to his theories from various quarters, including scientific critique, and even from some of his students.

For example, two of Skinner's most well-known students and proteges, Keller and Marian Breland, found various flaws in his theories. I won't go into great detail, because it won't benefit the vast majority of readers. However, just be aware that operant conditioning is not the be-all and end-all of animal training. Consider just a couple of quick quotes from the Breland's from an article they wrote titled *"The Misbehaviour of Organisms"* (a play on the title of Skinner's book *"The Behaviour of Organisms"*):

> "…we have ventured further and further from the security of the Skinner box. However, in this *cavalier extrapolation,* we have run afoul of a persistent pattern of discomforting *failures.* These failures, although disconcertingly *frequent* and seemingly diverse, fall into a very interesting pattern. They all represent *breakdowns of conditioned operant behavior.*"[19] (emphasis added)

So, with these *disconcertingly frequent discomforting failures,* all isn't necessarily rosy with Skinner's views on behaviour.

Here is a short overview of Skinner's terminology.

Positive reinforcement

Positive reinforcement means we *add* something the dog likes in order to *strengthen* a behaviour. So we give the dog a treat when it sits. Or open the door when it scratches on it (a bad idea!). Or pat the dog when it jumps on us (another bad idea). Or back off when our pup growls at us (also bad). Technically, positive reinforcement "increases the likelihood of the behaviour occurring in the future", or strengthens (thus the term "reinforces") the behaviour.

---

[19] Breland, K., and Breland, M. *The Misbehavior of Organisms*. American Psychologist 16, (1961): 681-684.

CHAPTER 2 – Some terms explained

Negative reinforcement

Negative reinforcement means we *remove* (like minus in maths) something which the dog *doesn't* like, in order to reward (reinforce) a behaviour. Why do we eat when we are hungry? Because eating removes the hunger, and thus eating is negative reinforcement. Or our pup goes to the toilet because of an increasingly uncomfortable bladder, which removes the discomfort. Or a bitch feeds her pups because of her increasingly uncomfortable udder. The reward is that the discomfort goes away. A fly landing on us is annoying—we shoo it away and it stops annoying us. These are all examples of negative reinforcement.

So when picking up a pup, we might hold it more tightly when it struggles, until it relaxes, and then we relax our hold. This corrects (punishes) the pup for struggling, and rewards it for relaxing and holding still. The reward-only crowd tells us that negative reinforcement, like positive punishment, is *bad*.

Figure 3 - The author holding a kelpie pup

PART 2 – Some terms explained

Positive punishment

Positive punishment is where we *add* (positive) something which the dog doesn't like (such as a smack when our dog jumps on us), in order to weaken or reduce the behaviour (or to teach or increase behaviour—the reward-only ideologues claim this cannot happen, but later you will learn that it certainly can—see *Chapter Seven – Can punishment create behaviours?*). We give the dog a smack when it is chewing the electrical power cord. Or an old dog bites the young pup when it tries to take her bone. Technically, it "reduces the likelihood of the behaviour occurring in the future".

Negative punishment

Negative punishment is where we take something away in order to weaken a behaviour. For example, we are teaching our pup to sit with treats, and as we go to give the pup the treat it jumps up for it. We immediately pull the treat away. This punishes the pup for jumping up instead of staying seated. Or, some trainers (such as Victoria Stilwell, and many, many others) advise that when your pup jumps on you, you should turn your back on it (don't bother, it doesn't work). However, there are even well-known "positive" trainers today who are so extreme that they say we shouldn't even use *negative* punishment like this (and they argue this not on the grounds of it not working, but simply because it is punishment, and punishment is *bad*). Karen Pryor approvingly quotes her friend Kay Lawrence (author of "*Learning About Dogs*") as saying:

> "If I am teaching a dog, *I avoid every atom of punishment or removal of something good* to get the behavior."[20] (emphasis added)

In other words, supposedly Kay doesn't use negative reinforcement, positive punishment, or even negative punishment ("removal of something good"). So, this rules out even "positive" trainer Victoria Stilwell's advice to "turn your back on your dog"[21] when it jumps on you, or pulling the treat away if the dog tries to jump up to get it. Kay obviously fails to realize that when she withholds reward during a variable schedule of reinforcement or shaping procedure (see below) that this also constitutes punishment, and so she is, in reality, not at all avoiding *every atom* of punishment as she claims (but more on that later).

So these are the four "quadrants" or tools of Skinner's "operant conditioning"— positive reinforcement, negative reinforcement, positive punishment, and negative

---

[20] Pryor, Karen. *On My Mind – Reflections on Animal Behavior and Learning.* Waltham, MA. Sunshine Books. 2014.
[21] Stilwell, Victoria. *Train Your Dog Positively.* USA. Ten Speed Press. 2013.

punishment. We *apply or remove something the dog likes or doesn't like*, in order to reward or punish it.

Continuous or Variable schedules of reinforcement

A reinforcement schedule, or reward schedule, is simply the frequency and pattern with which behaviours are rewarded. There are two main types of reinforcement schedules in operant conditioning, which also occur in the real world. These are important to understand. A "continuous" reward schedule is when we reward the pup *every time* it sits. A "variable" schedule is when we only reward a certain percentage of the sits, in some variable manner.

So, if training with treats, when we train a new behaviour, we will use continuous reinforcement by rewarding *every* sit, until the pup understands the command. And then we stop rewarding *every* sit, and instead do so *intermittently* or *variably*. Again, different people can use the words in slightly different ways, and for the vast majority of people it isn't important. Just think rewarding *every* behaviour (continuous), or slowly fading out the treats in a random manner (variable). In the process we generally reward the better behaviours and not the poorer ones—in that way we can "shape" the behaviour to improve it.

Take a human example. If we go up to a coke vending machine and put our money in, out comes the bottle of coke. *Every time*. That is "continuous reinforcement". We are rewarded with a bottle of coke *every time* we put our money in. However, if we put our money in and the bottle of coke *doesn't* come out, we don't keep putting money in. We kick the machine, or we walk away (yes—reward schedules can cause aggression when the reward isn't forthcoming—more on this is *Chapter Nine – More downsides and side-effects of reward-based training*).

Now, instead of the drink vending machine, consider gambling or pokies machines (known as slot machines internationally). In this situation, instead of kicking the machine (although gamblers can still get angry) or walking away, people *keep putting money in* even though they often get *no* reward. Why? Because they know that *at some point* they will get a win. This is variable or random reinforcement. In this case, unlike with the coke machine, they don't quit as soon as they don't get the reward—they keep trying.

This is the paradox that once a reward-trained pup knows the behaviour (created using continuous reinforcement), and the trainer begins reducing the reward randomly (or at least, it *seems* random to the pup), the *behaviour actually gets stronger and more resistant to extinction* (fading away). Whereas, if we have only ever rewarded our pup *every* time it sat, and then when we have no treats with us,

PART 2 – Some terms explained

the pup might sit once or twice and then figure out we have no treats, and walk away (or maybe start barking at us in frustration)—just like we do with the coke machine. But if we *sometimes* reward the pup and sometimes we don't, the pup is more likely to keep sitting, hoping the treat isn't far away.

Think vending machines, slot machines, gambling and addiction. These examples summarize continuous and variable schedules of reinforcement. Addiction (like food deprivation) is a major and essential element of so-called "positive" training. You will read a lot more about this in *Chapter Nine – More downsides and side-effects of "positive" training.*

So, these are just some quick explanations of the common terms I use throughout this book. Let's now move on and have a look at some of the problems with Skinner's "operant conditioning" theories and terminology. If the science isn't particularly your thing, feel free to skip the next chapter and get right into debunking the "positive training" myths.

3

Some problems with operant conditioning

(NOTE: If you aren't much interested in the deeper science, feel free to skip this chapter and move on. You can always come back and read it later.)

There are a number of problems I have with Skinner's "operant conditioning" model:

1. There is considerable overlap between the four quadrants.
2. Even many influential trainers who should know better often confuse them.
3. The definitions often don't make much sense.

There are other problems also, but they are beyond the scope of what I consider pertinent to this book. Let's consider these points in turn.

Overlap between the quadrants

The four aspects of operant conditioning—positive reinforcement, negative reinforcement, positive punishment and negative punishment—although they may appear quite neat and tidy and distinct, are actually quite messy. In fact, there is considerable *overlap* between these four areas.

PART 2 – Some terms explained

Positive reinforcement

For example, positive reinforcement (reward) contains punishment, *every time the reward is withheld*. Imagine I ask you to "paint me a picture, and if I like it I will buy it from you". You paint the picture, and I say, "No, sorry, it's not good enough, I'm not going to buy it". It is unpleasant (aversive) and you feel punished. You certainly don't feel rewarded, and you don't feel neutral. It's an unpleasant and punishing experience. Steven R. Lindsay (author of the three volume *"Handbook of Applied Dog Behaviour and Training"*) says:

> "For example, if a hungry dog fails to obtain a piece of food for sitting because it misses a signal or fails to sit in a timely fashion, the *dog is punished*—not indirectly as a result of the withdrawal of the appetitive opportunity—but *directly* as the result of its failure to control the opportunity to obtain food."[22] (emphasis added)

Or, when a gambler bets on a horse race and loses. He feels dejected and unhappy, maybe even angry (yes, as we saw before, and will learn a lot more about, positive reinforcement schedules can cause aggression), because he *failed to gain the reward*. Positive-reinforcement trainer Jane Killion notes:

> "The operant dog looks for reinforcers as confirmation that he is doing the right thing. *If a reinforcer is not forthcoming, he assumes he is doing something "wrong"*. Think about this statement carefully, for it is of profound significance. The ***absence of reinforcement is actually punishment*** to the operant dog. In contrast, a dog that has been trained to heel using leash checks thinks he is in good shape so long as he is not getting leash checks."[23] (emphasis added)

> The absence of reinforcement is punishment to the positive reinforcement trained dog.

So, there we have a "positive" trainer recognizing that positive reinforcement methods include punishment by their very nature—contrary to Kay Lawrence's claims about avoiding "every atom of punishment".

Dr. Michael Perone (Professor of Psychology at West Virginia University, who specialized in the experimental analysis of operant behaviour for over 30 years) provides a good example:

---

[22] Lindsay, Steven R. *Handbook of Applied Dog Behavior and Training, Volume 1: Adaptation and Learning*. Ames, IA: Blackwell Science, 2000.
[23] Killion, Jane. *When Pigs Fly!: Training Success with Impossible Dogs*. Wenatchee, WA: Dogwise Publishing, 2007.

CHAPTER 3 - Some problems with operant conditioning

> "Outside the laboratory, I cannot help but be impressed with the *propensity of people to respond to the negative side of positive contingencies*. Consider college students. In my large undergraduate courses, I have tried a variety of contingencies to encourage class attendance. Early on, I simply scored attendance and gave the score a weighting of 10% of the course grade. There were lots of complaints. The students clearly saw this system as punitive: Each absence represents a loss of points towards the course grade. So I switched to a system to positively reinforce attendance. When students come to class on time, they earn a point above and beyond the points needed to earn a perfect score in the course. Thus, a student with perfect attendance, and a perfect course performance, would earn 103% of the so-called maximum. A student who never came to class, but otherwise performed flawlessly, would earn 100%. If course points function as reinforcers, then this surely is a positive contingency. But the students reacted pretty much the same as before. *They saw this as another form of punishment:* With each absence, I was *denying them a bonus point. Of course, the students are right. Whenever a reinforcer is contingent on behavior, it must be denied in the absence of that behavior.*"[24] (emphasis added)

And yet, *withholding reward* is an intrinsic part of so-called "positive" training methods. This occurs as soon as the trainer moves into a variable schedule of reinforcement (and even before that point), which is exactly what gambling is. Variable reinforcement schedules simply turn your dog into a gambling addict (see *Chapter Nine – More downsides and side effects of "positive" training*).

Perone also discusses the problems with the distinction between positive and negative reinforcement. And he correctly talks about the aversive effects of positive reinforcement:

> "Theoretically, the distinction between positive and negative reinforcement has proven difficult (some would say the distinction is untenable). When the distinction is made purely in operational terms, **experiments reveal that positive reinforcement has aversive functions.**"[25] (emphasis added)

So positive reinforcement can also be considered to be negative reinforcement a majority of the time, and it has aversive effects. Steven Lindsay writes:

> "Thus, from this perspective, *working for food* may be interpreted as escape-avoidance behaviour aimed at *reducing or terminating the aversive condition of starvation*. Unfortunately, the terms positive reinforcement

---

[24] Perone, Michael. "Negative Effects of Positive Reinforcement." *The Behavior Analyst* 26, no. 1 (Spring 2003): 1-14.
[25] As above.

PART 2 – Some terms explained

and negative reinforcement—although of some practical value in the everyday control of behaviour—are highly subjective and appear to depend on the experimenter's point of view and bias."[26] (emphasis added)

As you will learn later, "positive" trainers do indeed often use a number of strategies to make sure their animals are *hungry*, in an effort to make them more receptive to treat training (like Pryor with her dolphins, or the trainers of Tilikum the killer whale). Dr. Dunbar goes so far as to say you should *never feed your dog a meal* (more on this in *Chapter Nine – More downsides and side-effects of "positive" training*). Thus, we are clearly now in negative reinforcement territory, and can no longer claim to be using positive reinforcement.

So, what many consider *positive* reinforcement training is more often *negative* reinforcement, and positive reinforcement methods are also aversive when those rewards are withheld—as they are in *all* "positive" methods. And keeping your dog hungry so your treats work better is also aversive.

So there is a lot of overlap and confusion even just in the basic operant behaviour terminology.

Negative Reinforcement

As discussed, negative reinforcement (just think negative *reward*) means we remove something the dog doesn't like in order to reward it. The dog doesn't like getting wet when it is raining, so it goes into its kennel, and is rewarded because the rain stops wetting it. If we are cold, we put a jumper on, and are rewarded because the cold goes away.

However, when using negative reinforcement in a training sense, we have to first *create* the unpleasant situation. This is why it is often called *pressure/release* training. This generally doesn't have to be painful (although in some situations, such as high-level aggression, it might need to be). It can often simply be *annoying*. I often think of a fly landing on us—it isn't painful, but it *is* annoying, and we want it to go away (just like a low-level tingle on an e-collar).

So, we might put tension on the lead for our dog to follow us, creating an unpleasant sensation (pressure), and then when the dog starts to follow us, the lead

---

[26] Lindsay, Steven R. *Handbook of Applied Dog Behavior and Training, Volume 1: Adaptation and Learning.* Ames, IA: Blackwell Science, 2000.

goes slack (release). And so the dog is rewarded because the unpleasant sensation of a tight lead ceases. That is negative reinforcement. However, by creating the unpleasant situation in the first place, we are punishing our dog for not following us. So negative reinforcement can contain positive punishment.

Karen Pryor says:

> "it's important to remember that each instance of negative reinforcement also contains a punisher."[27]

And yet, how many so called "positive" trainers have ever pulled on their dog's lead when it isn't following them, or pulled it away when it is reacting to another dog? The answer is 100%. They all have, and do. If so, these so-called "positive" or "force free" trainers are using punishment and/or negative reinforcement and force.

So, positive reinforcement is often negative reinforcement, and negative reinforcement usually contains positive punishment. As I mentioned earlier, it all gets a bit messy.

Positive punishment

When we consider positive punishment, we find that it also contains either negative or positive reinforcement—every time the punishment is avoided. Just like positive reinforcement contains punishment when the reward isn't forthcoming, so punishment contains reward when the punishment isn't forthcoming.

For example, if I am about to lean on a fence on the farm and then remember it is electrified, and catch myself in time, I feel pretty relieved! I am reinforced for *not* touching it. One of the reasons punishment is so effective is that after it has occurred it is then *self-reinforcing when we avoid it*.

Negative punishment

Negative punishment, the *removal* of something the dog *likes* in order to punish it, can often simply also be considered positive punishment. For example, imagine we are teaching our dog to "sit" using treats. As we go to give it a treat for sitting, it jumps up to get it. We immediately pull the treat away—that is definitely negative punishment for jumping up. But is it really any different from positive punishment? Something the dog doesn't like occurs (the food gets further away) as a consequence of its behaviour.

---

[27] Pryor, Karen. *Don't Shoot the Dog!: The New Art of Teaching and Training*. Rev. ed. New York: Bantam Books, 1999.

PART 2 – Some terms explained

Also, consider time-outs (useful in humans, practically useless in dogs) such as putting a dog outside or in the laundry or in its crate when it misbehaves. Timeouts are often considered to be *negative* punishment. However, we didn't really *remove* something—we *added* something the dog or person doesn't like. If you say to your child, "if you do that again, you are going to your room," I would consider that to be *positive* punishment rather than negative.

Again, it's all a bit messy and there is a lot of overlap and confusion. Which, again, is why I generally prefer the simple terms of reward and punishment.

Confusion

The second problem with "operant conditioning" as a model for dog training is that many people, even some of the most influential and most respected people in the reward-only world, often get things mixed up (it's not hard to do). For example, Karen Pryor frequently confuses negative reinforcement and positive punishment, as do many others. Here are examples Pryor gives of what she says are negative reinforcement:

> "a derisive glance from a friend when you make a poor joke, or a slight draught from an air conditioner that causes you to move to another chair."[28]

Did you spot the problem? Moving to another chair to get away from the air conditioner is indeed a good example of negative reinforcement. There is something unpleasant, and we are rewarded when we move away because it stops annoying us. However, a "derisive glance from a friend" when we make a poor joke is *not* negative reinforcement. It is positive punishment. It happened after the event. There was no rewarding aspect. Pryor also claims that the following scenarios are all examples of negative reinforcement:

- A cow touching an electric fence.
- A coach correcting your faulty tennis swing by saying "No!".
- Rebuking a shirking or lazy employee in each instance in which their work falls below par.[29]

These are, however, all examples of positive punishment, not negative reinforcement. Mark Plonsky, Ph.D., makes the same observation about this confusion:

---

[28] Pryor, Karen. *Don't Shoot the Dog!: The New Art of Teaching and Training*. Rev. ed. New York: Bantam Books, 1999.
[29] As above.

CHAPTER 3 - Some problems with operant conditioning

> "I have found that many people confuse negative reinforcement and punishment. These people include dog trainers, college students, and surprisingly, even many college professors and Introduction to Psychology textbooks. In fact, I recently read the book Don't Shoot the Dog! by Karen Pryor and was a bit dismayed to find this confusion there as well. Note that this is not an attempt to flame the book or criticize it unduly. I like the book and learned a lot from it. I recommend it highly to anyone interested in training animals (and sometimes even those interested in training their children). Nonetheless, I do believe that the book suffers from the problem of confusing negative reinforcement and punishment. As such it propagates this confusion because many dog trainers look to this book as a definitive source on the science of behavior."[30]

So, operant conditioning terminology can be quite confusing.

Problems with definitions

Dr. Ian Dunbar likes to use a definition of punishment that is quite different to that of most people:

> "Strictly speaking, anything (even if enjoyable) that decreases behavioral frequency is a "punishment," for example, using the command Sit to decrease or stop jumping up."[31]

So now "even something enjoyable" like telling your dog to "Sit" is punishment, because the definition of punishment is something that makes the behaviour decrease. Dunbar makes a distinction between "aversives" (something the dog doesn't like) and "punishment", which can even be something it *does* like! I suspect there are very few people anywhere in any practical sense who view that as the meaning of punishment, although an argument could be made that he is correct in some semantic, technical way (which is simply further evidence for my point about the problems with operant conditioning theories). Here are two other definitions of punishment, first from the "Medical News Today" website:

> "Positive punishment involves adding an aversive stimulus after an unwanted behavior to discourage a person from repeating the behavior. Spanking and chores are examples of this."[32]

---

[30] Plonsky, M. (1996). *Confusing Consequences: A Brief Introduction to Operant Conditioning.* Retrieved from https://www4.uwsp.edu/psych/dog/LA/DrP2.htm

[31] Dunbar, Ian, Dr. *Barking Up The Right Tree - The science and practice of positive dog training.* Novato, CA: New World Library, 2023.

[32] Medical News Today. *Positive punishment: Examples, definition, and risks.* Accessed 21/2/2024, https://www.medicalnewstoday.com/articles/positive-punishment.

PART 2 – Some terms explained

Or a "derisive glance from a friend when you make a bad joke" to quote Karen Pryor's example. Note that this definition includes "aversive stimulus" in its definition. And the following is similar, from the "Simply Psychology" website:

> "Positive punishment involves adding an aversive stimulus or something unpleasant immediately following a behavior to decrease the likelihood of that behavior happening in the future."[33]

Again, the definition includes adding an "aversive" stimulus.

If we use Dunbar's definition, then all trainers use punishment. However, that is not how the vast majority of trainers understand the term. Again, this is why I generally prefer just to talk about the commonsense meanings of the terms "reward" and "punishment". It is simply more accurate and works better. (Throughout this book, when I specifically reference "positive reinforcement" or "positive punishment," etc., then I am using those terms in the generally used sense, not in Dunbar's.)

There is another problem with the technical definition of punishment, that it *reduces* the frequency of a behaviour. In fact, punishment can also be used to *increase* the frequency of behaviour. I discuss this further in *Chapter Seven – Can punishment create behaviours?* If a dog is asked to "sit", and it doesn't, then a punishment (whether a tug on the leash or a smack on the rump), will in fact *increase* the likelihood of the behaviour (sitting) occurring next time the command is given. So now, according to Dunbar's definitions and logic, we would have to call this "positive reinforcement", not "punishment", because it increases the likelihood of the behaviour!

Perhaps now you are starting to see some of the problems.

Conclusion – Some terms explained

So that is a breakdown of some of the terms used in this book.

I believe in common sense—the good thing about most people, unlike dogs (despite claims to the contrary, as we will see) is that *we* understand language. We understand that words have a "semantic range" (each word has a range of different meanings), and that the meaning of words vary depending on *context*. For example, I could say that something is "awfully pretty" or I could say it is "pretty awful," and you would get my meaning.

---

[33] Simply Psychology. *Operant Conditioning In Psychology: B.F. Skinner Theory.* Accessed 21/2/2024. https://www.simplypsychology.org/operant-conditioning.html

CHAPTER 3 - Some problems with operant conditioning

Or consider the word "bark"—the noise a dog makes, or the outer covering on a tree. We understand language largely through context. We don't need to get too hung up in always using words in the picture-perfect scientific definition (as we have seen, that can be messy and confusing anyway). As long as you understand what I am getting at, I have done my job. However, I do try to use the technical terms consistent with their commonly understood definitions where applicable.

So, I've now introduced some of the main figures involved in the push for reward-only training and in the demonisation of punishment, including Ian Dunbar, Karen Pryor, Zak George, Victoria Stilwell and others. I've briefly defined some of the main terms used throughout this book. And I've shown some of the problems with "operant conditioning", such as the confusing overlap. Now, finally, let's get stuck in and discuss the common arguments levelled against the use of punishment, and discover just how false they are.

Part 3

ANSWERING THE CASE AGAINST PUNISHMENT

Chapter 4 – Is punishment effective when training dogs?

Chapter 5 – Is punishment effective when training people?

Chapter 6 – Does punishment cause undesirable side effects?

Chapter 7 – Can punishment create behaviours?

4

Is punishment effective when training dogs?

The "positive" or "force free" advocates generally argue against the use of correction and punishment on three broad fronts:

1. They argue that correction and/or punishment is ineffective (and, conversely, that reward is more effective).

2. That correction and/or punishment inevitably cause undesirable side effects, and that these are so terrible (often likened to "nuclear fallout"[34]) that it should therefore be avoided at all costs, and

3. That punishment doesn't create behaviours, it only stops them.

These are *all* fundamentally untrue. In fact, just about the exact opposite applies. So, let's discover the truth about these three claims made by the "reward-only" advocates. To begin with, let's address the first of the three—that *punishment simply doesn't work*. For example, Karen Pryor says:

> "Method 2: Punishment. (Everybody's favourite, in spite of the fact that it *almost never really works*.)"[35] (emphasis added)

---

[34] Sidman, Murray. *Coercion and Its Fallout*. 2000. United States. Author's Cooperative.
[35] Pryor, Karen. *Don't Shoot the Dog!: The New Art of Teaching and Training*. Rev. ed. New York: Bantam Books, 1999.

PART 3 – Answering the case against punishment

Dr. Ian Dunbar likewise says:

> "Most punishments are notoriously ineffective..."[36]

And Dunbar again:

> "...most aversive punishments *don't work*, since they seldom reduce or eliminate undesired behaviors."[37]

Psychologist Jillian Enright goes even further:

> "Punishing Unwanted Behaviour Just Makes it Worse."[38]

These statements are all completely untrue. All we need in order to see the error of these statements is a tiny bit of common sense, some *real world* experience, and some statements and examples from these very same authors themselves. In addition, if we need more, we'll look at the *real* science. The truth is, that when used correctly, punishment is *extremely* effective, entirely *natural*, and very *long lasting* (often being completely *permanent*), as demonstrated by the following examples.

> Punishment is *extremely* effective, entirely *natural*, and very *long lasting* (often being completely *permanent*).

"Positive" trainers and the effectiveness of punishment

Karen Pryor (despite, as just quoted, claiming that "punishment *almost never really works*"), actually gives various examples of the great effectiveness of punishment in other parts of her very same book (and in her other books). In fact, all *within a few paragraphs* she says:

---

[36] Dunbar, Ian. *Six Simple Steps to Solve Your Dog's Behavior Problems*. Dunbar Academy. Accessed Feb 2024. https://www.dunbaracademy.com/courses/six-simple-steps.
[37] Dunbar, Ian, Dr. *Barking Up The Right Tree - The science and practice of positive dog training*. Novato, CA: New World Library, 2023.
[38] Neurodiversified. *Punishing Unwanted Behaviour Just Makes it Worse*. Medium, https://medium.com/neurodiversified/how-punishing-unwanted-behaviour-just-makes-it-worse-baf22793d07b.

CHAPTER 4 – Is punishment effective when training dogs?

> "The most the punisher can hope for is that the child's motivation will change: The child *will try to alter future behavior in order to avoid future punishment.*"[39] (emphasis added)

That, I would have thought, is the whole point! And yet, a few paragraphs later, Pryor then claims:

> "…punishment or the threat of it doesn't help the subject learn how to modify the behaviour involved…"[40]

But then, in the *very next paragraph* she again contradicts herself:

> "A child who is scolded severely the first time he or she crayons on the wall may very well stop defacing the house. A citizen who cheats on his income tax and gets fined for it may not try it again."[41]

And two paragraphs later:

> "My parents punished me exactly twice in my whole upbringing… both behaviours stopped instantly."[42]

So, all within a few paragraphs, Pryor says a child will try to alter future behaviour to avoid future punishment (which is the whole point), then claims punishment actually doesn't help to modify behaviour, and then gives two examples of it working perfectly to do just that! Pryor also claims:

> "They [cats] really can't be trained by punitive methods…"[43]

And yet elsewhere she says:

> "… a friend of mine quite accidently **cured her cat of clawing the couch by establishing "No!" as a conditioned aversive stimulus**. One day in the kitchen she happened to drop a large brass tray, which fell right next to the cat. She cried "No!" as the tray fell, just before it landed with a loud clatter. The cat, dreadfully startled, jumped into the air with all its fur on end. The next time the cat clawed the couch, the owner exclaimed "No!"

---

[39] Pryor, Karen. *Don't Shoot the Dog!: The New Art of Teaching and Training.* Rev. ed. New York: Bantam Books, 1999.
[40] As above.
[41] As above.
[42] As above.
[43] As above.

PART 3 – Answering the case against punishment

and the cat, looking horrified, desisted immediately. ***Two more repetitions were enough to end the behaviour permanently.***"[44] (emphasis added)

So punishment was highly effective once again! I haven't personally had much to do with training cats, but a friend of mine who was involved in training cats for movies told me that most will readily respond to a correction from a water pistol. In fact, logic tells us that any animal *must* be able to respond to correction and punishment, or they simply wouldn't still be alive. Indeed, if they didn't learn from punishment their whole species would be extinct.

Talking about a squirt from a water pistol, Pryor gives another example in her book:

> "My Border terrier, as a young dog, became fond of digging in the wastebaskets and spreading the contents around. I didn't want to punish her, but I also didn't want to constantly empty the wastebaskets.
> "I filled a spray bottle with water and added a few drops of vanilla extract: a strong but pleasant scent to me. Then I gritted my teeth and sprayed the dog in the face. She was dismayed and ran. I sprayed the wastebaskets with the scented water. She stayed away from the wastebaskets from then on."[45]

So, despite her claims that punishment *almost never really works,* and that it *doesn't help to modify behaviour,* Pryor gives plenty of examples (and more besides these) where it certainly and very effectively does!

In contrast, here is Zak George's completely ineffective "positive" method to deal with the same rubbish bin problem:

> "If you catch your dog getting into the garbage, telling her "No, don't do that" in a calm tone and then removing access to the trash is very logical to me... So after you remove the garbage and get your dog to sit and focus her attention on you, then reward the positive behaviour."[46]

There is absolutely *nothing* in this process that will teach the dog *anything whatsoever* about not getting into the garbage. It demonstrates an enormous lack of knowledge of—and real experience in—dog training.

---

[44] Pryor, Karen. *Don't Shoot the Dog!: The New Art of Teaching and Training.* Rev. ed. New York: Bantam Books, 1999.
[45] As above.
[46] George, Zak, and Dina Roth Port. *Dog Training Revolution: The Complete Guide to Raising the Perfect Pet with Love.* New York: Ten Speed Press, 2016.

CHAPTER 4 – Is punishment effective when training dogs?

No matter how "logical" it may seem to him, it will not work *in the slightest*. This applies to many of the other approaches Zak George teaches also.

Here is yet another example of the effectiveness of punishment from Karen Pryor:

> No matter how "logical" it is to Zak George, his approach for bin diving will *not work in the slightest*.
>
> And this applies to many of the other approaches he teaches also.

> "The same principle is at work in the Invisible Fence systems for keeping a dog on your property. A radio wire is strung around the area in which you want to confine the dog. The dog wears a collar with a receiver in it. If the dog gets too near the line, the collar shocks it. However, a few feet before that point, the collar gives a warning buzz... if the setup is properly installed, a trained dog can be effectively confined and will never receive an actual shock. I used such a fence when my terrier and I lived in a house in the woods. An actual fence would have been a perpetual invitation to try to dig under it or escape through an open gate; the conditioned warning signal and the Invisible Fence were far more secure."[47]

So once again, we see here Pryor's effective use of punishment methods (an e-collar containment system) and her lack of any effective positive reinforcement solutions. In fact, she says it is "far more secure" than a physical fence! Punishment can at times be not only the *most* effective method, but can also be (and often is) the *only* effective method, as Pryor evidently recognized when attempting to keep her dog from straying.

Pat Miller, author of "*The power of positive dog training*" (and past board member of Dunbar's *Association of Pet Dog Trainers*) who also argues strongly against the use of punishment or correction, found that she also needed to utilise punishment:

> "The most notable example of my reluctant application of positive punishment involved a clients Pit Bull mix who was very dedicated to chasing horses... I **pulled out the cap gun and waited. Just as he ducked under the wire into the pasture, I fired**. Happy wheeled in his tracks and dashed for the porch... It was **the first time we had succeeded** in stopping him in mid-charge... I left the gun with the owners, and we all crossed our fingers. **They only had to use it one more time. Twice was enough to convince Happy to leave horses alone forever.** Mission accomplished...To this day, I ask myself if there wasn't another, more positive way to accomplish our goal... I wonder if we could have

---

[47] Pryor, Karen. *Don't Shoot the Dog!: The New Art of Teaching and Training*. Rev. ed. New York: Bantam Books, 1999.

PART 3 – Answering the case against punishment

> succeeded with Leslie Nelson's "Really Reliable Recall". I am convinced that there was a positive way to do it—I just couldn't find it at the time."[48]

Once again, punishment worked perfectly effectively with *permanent* results with just a couple of applications. And this notwithstanding Pryor's "almost never really works", or Dunbar's "notoriously ineffective", or Enright's "just makes things worse". Dunbar also says:

> "…in fact, most of the time, punishment is neither quick, nor a fix." [49]

And yet, Happy is safe, horse is safe, owners are happy. And what was the terrible punishment required? A couple of bangs from a cap gun. How awful!

And as far as Leslie Nelson's positive reinforcement "*Really Reliable Recall*" that Pat Miller thought might work is concerned, I can guarantee her (and everyone else) that it would *not* have done the trick. And that's even assuming Happy's owners had been willing—and skilful enough—to put the considerable amount of time and effort into the long and tedious training required. *And* it's assuming Happy hadn't been killed by a well-directed hoof in the meantime, or the horse hadn't been chased through a barbed wire fence. (I will explain exactly *why* the so-called "Really Reliable Recall" wouldn't have worked later on—see *Chapter Eight – Does Reward Work? The "Really Reliable Recall"*).

Which approach is kinder to the dog? Two loud bangs from the cap gun on one side of the equation, or potentially getting kicked and badly injured by the horse, or chasing the horse through a fence, or just confining the dog and never letting it run free, on the other side of the equation? ***The answer is simple and obvious.*** (I discuss the morality of using punishment later in this book—see *Chapter Sixteen – The Morality of Punishment*.)

Jamie Dunbar, Dr. Ian Dunbar's son, says:

> "When you use effective dog training techniques, your dog's behavior improves. If you've been trying to train your dog, and your dog's behavior hasn't improved, it's not your fault. It means you've been using the wrong methods. And that's not your fault either. ***The world is full of self-proclaimed dog trainers who promote methods that sound reasonable, but they just don't work.*** Again—the proof is in the pudding. If your dog's

---

[48] Miller, Pat. *The Power of Positive Dog Training*. 2nd ed. Hoboken, NJ: Howell Book House, 2008.

[49] Dunbar, Ian, Dr. *Barking Up The Right Tree - The science and practice of positive dog training*. Novato, CA: New World Library, 2023.

CHAPTER 4 – Is punishment effective when training dogs?

behavior improves, the training methods are good. If your dog's behavior doesn't improve, the methods are not good."[50] (emphasis added)

I couldn't agree more—the proof is very definitely in the pudding. However, it is actually a great many of the methods advocated by the "positive" trainers like Jamie Dunbar, and his father, and Zak George, and so on, that fit his description of the "wrong methods" perfectly. And his description of "self-proclaimed dog trainers who promote methods that sound reasonable, but they just don't work" also fits these trainers perfectly. Like Pat Miller's positive reinforcement methods which didn't work, while her punishment approach worked perfectly, and instantly.

Zak George's rubbish bin "solution" also falls perfectly into this category. As does the "stop and go" method of teaching a dog to walk on the lead so commonly parroted by the "positive" crowd. Or Victoria Stilwell's turn your back on your dog method. And so many others. No matter how many times, or for how long, you followed Zak George's rubbish bin advice, the behaviour would not improve. Why? **Because you are *using the wrong methods* taught by *self-proclaimed dog trainers who promote methods that sound reasonable, but which just don't work*.** Karen Pryor's application of the water bottle as punishment solved the issue virtually instantly, even though she claims punishment *almost never works*.

By Jamie Dunbar's own reasoning, methods that improve behaviour almost instantly and easily, in only one or two applications, and that are long lasting (often permanent), would have to qualify as "good methods". So Pat Miller's use of punishment to cure horse chasing is a good method. Her methods using positive reinforcement were not. To give just one example amongst very many of my clients:

> "Highly Recommend! Training with my two dogs went beautifully with instant results after the first session. Second session was an overwhelmingly positive outcome that showed that a perfect walk exists. We are excited for more training with Tully."
>
> - Matt Richardson

My client's two highly reactive dogs were walking perfectly in just two sessions. So according to Jamie Dunbar, the methods are good. So why do people like Jamie, and his father, and Pryor, and so many various others, keep insisting that methods involving punishment don't work, while themselves giving many

---

[50] Dunbar, Ian. *Six Simple Steps to Solve Your Dog's Behavior Problems*. Dunbar Academy. Accessed Feb 2024. https://www.dunbaracademy.com/courses/six-simple-steps.

PART 3 – Answering the case against punishment

examples of punishment working brilliantly? And of their own positive-reinforcement methods failing?

It is extremely rare that I ever need more than two sessions, a week apart, with a client to fix *any* lead-based issues, and to have the dogs walking on a completely loose lead under any distraction. Quite often it is faster than that. And one or two sessions to get a reliable off-lead recall under any distraction. So, it works. Therefore, by their own definition, the positive punishment and negative reinforcement methods are good.

From the Mastiff that pulled his owner over and smashed her face into the concrete footpath, breaking her nose and teeth and dragging her along the ground as it tried to attack another dog; to the dog whose owner had had both shoulders operated on from damage from walking it; to the 92 year old woman attempting to walk a Siberian Husky and a Pit Bull simultaneously; all remedied rapidly and easily.

Or the Great Dane that its fit, athletic young owners could no longer walk as it just dragged them down the street while trying to attack practically every dog it saw. Two sessions and cured, and they sent me a video shortly thereafter of them taking it for a walk perfectly calmly on a completely loose lead around a crowded lake in the middle of the city. However, Dunbar claims:

> "On-leash, punishment-based training is often time-consuming and relatively ineffective." [51]

In fact, it is the *treat training methods* that are pushed by what Jamie Dunbar describes as "self-proclaimed dog trainers who promote methods that sound reasonable but just don't work", that are actually *time-consuming* and *ineffective*, and *just don't work*.

I have helped so many people who have tried all the treat training methods they have read about, or watched on YouTube, or learnt from their "positive" dog trainer. Their dogs still won't stop pulling on the lead, still won't come back when called, still won't stop attacking their cat or other dogs or people. I have helped people who have literally worked with "positive" trainers for in excess of a year with basically no improvement. I admire their persistence! And then someone recommends me, and the problems are all sorted out in a few short weeks' time.

---

[51] Dunbar, Ian, Dr. *Barking Up The Right Tree - The science and practice of positive dog training.* Novato, CA: New World Library, 2023.

CHAPTER 4 – Is punishment effective when training dogs?

And what about sheepdog training? Virtually all sheepdog trainers use correction and punishment as part of the process. Of course, there is reward also, but rarely in the treat sense. None of my sheepdogs are ever given a treat during training.

Figure 4 - The author riding Madness with well-trained sheepdogs Bounce and Kiara

Yet, somehow, despite what the "experts" like Dunbar or Pryor or Enright say about punishment never really working (or "just making things worse"), my sheepdogs (and those of many, many others) somehow magically manage to become highly trained animals at very complex tasks, and in a very short space of time. Go and watch some sheepdog trials on YouTube, and then come back and tell me that "punishment doesn't work".

(You can see videos of some of my dogs on my *Campaspe Working Dogs* website, or YouTube channel.)

What about guide dogs? Up until the last few years they were very effectively trained using punishment-based methods (such as choker chains). No one can argue the methods didn't work—guide dogs used to be some of the best trained dogs around. However, since succumbing to pressure from "welfare" groups to adopt the insane "positive" ideology (such as the STEP program— "Standard Training for Effective Partnerships"—in the UK guide dogs), which was developed largely by a

PART 3 – Answering the case against punishment

former Sea Lion trainer, the results have been disastrous (particularly for the blind people themselves).

Alan Brooks, an expert guide dog trainer from the UK with over 50 years' experience, is highly critical of the STEP program.

Amongst many other accomplishments, Brooks is the only person in the world to have been awarded both the Ken Lord Award (in 2012) by the *International Guide Dog Federation* for *"exceptional service to the international guide dog movement"*, and the "Suterko-Cory Award" *"in recognition of international leadership, dedication to the profession of Orientation and Mobility and distinguished efforts to improve services to individuals with a visual impairment. This significant honour indicates that the recipient serves as an international ambassador and role model to orientation and mobility specialists throughout the world."*

He has also worked assessing, advising, training and presenting papers on guide dogs and blind mobility in 33 countries at more than 60 guide dog programmes worldwide. His wife is also a guide dog trainer of over 35 years' experience. In an open letter to the chairman and trustees of Guide Dogs UK, Alan Brooks noted:

> "The new programme isn't working. Basically, the system is a non-punishment system, so you're not even allowed to say no to a dog in a firm voice if it does something wrong—as a result dogs aren't getting through the training programme and the population of guide dog owners has dropped by 25%"[52]

In fact, I recently come across a local case here in Australia of a lady who has long been on the waiting list to get herself a guide dog, without success. She has now been partnered with three different dogs in succession, and all three have had to be returned as not sufficiently well-trained. So the "positive" ideology is preventing many sight-impaired people from living a richer, fuller life.

Brooks also notes that many of the training staff and puppy-walking volunteers had become "demotivated and demoralised":

> "'Standard Training for Effective Partnerships' (S.T.E.P.) was inflicted upon training staff by managers inexperienced and unqualified in this specialist area. Many knowledgeable training staff and puppy walking volunteers can clearly see that S.T.E.P. is ineffective and their efforts are

---

[52] Hall, Rachel. *Rewards-Led Training Regime Behind Long Waits for UK Guide Dogs, Says Campaigner.* The Guardian, April 24, 2023. https://www.theguardian.com/society/2023/apr/23/rewards-led-training-regime-behind-long-waits-for-uk-guide-dogs-says-campaigner.

CHAPTER 4 – Is punishment effective when training dogs?

wasted. Their views are ignored, dissenting staff are threatened with disciplinary action, so they have become demotivated and demoralised. Many have since left the organisation. The success rate of dogs entering the program has plummeted but the exact figures on this are hidden as are the figures on departing staff and experienced volunteers.

"We now see the BBC 'In Touch' informing us of blind people attempting to train their own dogs, others in despair at waiting years for a replacement dog. Experienced volunteers are now following the staff and leaving in their droves."[53]

This is much like many school teachers (and preschool staff) I speak to who have likewise become demotivated and demoralised due to the same ineffective "positive" ideology that has been forced upon them. Many have left, or are seriously considering leaving, the education system. The so-called "positive" ideology is not only ineffective, but is negatively impacting a huge number of human lives as well as negatively impacting animal welfare outcomes.

Brooks also notes:

"There is an urgent need to return to ***effective, tried and tested training programmes*** which have a much higher success rate than S.T.E.P. This ill-conceived S.T.E.P. training programme has ***proved ineffective and must cease*** if we are to prevent further decline."[54] (emphasis added)

So, guide dogs (just like sheepdogs), when trained utilizing balanced methods were very well-trained animals (as police and military dogs generally continue to be). No one can seriously (or honestly) claim that "punishment doesn't work". Not only does it work, but it works *significantly* better. In fact, in many cases it is the difference between positive-reinforcement training not working at all, and tried and tested traditional methods working highly effectively.

I have tried treats. I have tested the methods. I have studied the subject. I have taught advanced trick training using clicker training (I still do). I still use treats with many clients for early training, teaching the meaning of sit, drop, stay, come here, etc., and for some other aspects. It works in these early stages, up to a point. And it's fun for kids.

---

[53] Brooks, Alan. 2023. *An Open Letter to the Chairman and Trustees of Guide Dogs.*
[54] As above.

PART 3 – Answering the case against punishment

But I rarely use these methods myself when training my own dogs (or horses). If these methods were easier, or faster, or more effective, I would swap in a heartbeat. They aren't. They are generally slower, an enormous amount more work, far more complicated and counter-intuitive, far less effective, and in many of the most important areas, fail completely. They certainly have their uses in certain cases (provided they are used in conjunction with the intelligent use of punishment), but there are often superior methods.

> If "positive" methods were easier, or faster, or more effective, I would swap in a heartbeat. They aren't.
>
> They are generally slower, an enormous amount more work, far more complicated and counter-intuitive, far less effective, and in many of the most important areas, fail completely.

If you're looking for *real world* training where your dog actually comes back when called, walks nicely on the lead, or to cure aggression or a variety of other undesirable behaviours, then if you're relying on positive reinforcement you are simply using completely the wrong tool for the job.

So, we've seen from the mouths of some of the most influential so-called "positive" trainers in the world that punishment is actually *highly* effective, even while they claim it doesn't work. And these represent only a small selection of examples—I could have added many more from these trainer's own mouths.

Punishment in nature

Quite often, reward-only training is described as "natural", with the obvious implication that everything else is "unnatural". The only problem is, like so many of the reward-only myths I outline throughout this book, it *simply isn't true*. In fact, again like the many reward-only propagated myths, the *exact opposite* is true.

There is nothing natural, either by human standards or by canine standards, about figuring out how to attempt to teach your dog to stop barking at the front door by getting out the treats, or by using other convoluted, reward-only, time-consuming techniques. Nor is there anything natural about training dogs as if they are dolphins, rats, or pigeons in captivity, and then downplaying dog's relationship to wolves, as the "positive" ideologues do (as we will see soon in the section *On wolves and dogs*).

CHAPTER 4 – Is punishment effective when training dogs?

Consider some examples of how dogs learn in nature:

Example 1 – Heights

- If a pup falls from a height and hurts himself, he will be more careful in the future. Every pup goes through this sooner or later. Pups learn by the consequences of their actions. *If punishment didn't work, all dogs (and humans) would be extinct by now.*

Example 2 – The campfire

- If a pup pokes his nose too close to the campfire, he will be punished by the campfire, and won't do it again. Ineffective? No, highly effective. *Almost never really works?* Hardly. Once is enough to cure the pup of ever putting his nose near a fire again. Quick, effective, and long lasting.

Example 3 – Chewing a bone

- If a pup gets too close to an older dog chewing a bone, the older dog will growl at the pup (correction). If the pup takes no notice, the old dog will bite him (punishment), often hard enough to make him yelp and run away. Next time a dog growls at him, the pup will take notice and remove himself. Usually, one application and the problem is cured. The pup has learnt two things—don't take liberties with other dogs that may not be appreciated, and if a dog growls at you, that growl means stop, or there is likely to be a consequence.

Example 4 – Cats

- Many cats are great puppy trainers. If the pup is annoying them, they hiss at it (correction). If the pup ignores the hiss, they smack it across the nose (punishment). Many pups soon learn to leave the cat alone, or at least to stop once the cat has had enough. And once the pup learns the boundaries, they often get along like a house on fire, with no terrible long-term side effects.

These natural ways in which dogs learn are extremely effective. If they weren't, or if punishment *just made things worse* as psychologists like Jillian Enright claim, then the pup would continue sticking his nose in the fire and getting burnt (or, in fact, do it more!), or falling from heights and getting injured, or pestering older dogs and getting bitten. You don't need scientific studies to see the truth of this. You just need a bit of common sense and logic. Punishment works.

In none of the above cases will the pup become aggressive (in fact, the cat might well cure him of his early predatory aggression towards cats), or fearful, or exhibit

PART 3 – Answering the case against punishment

"learned helplessness". Nor will he develop any of the other side-effects that the "positive" crowd claim will inevitably result from any correction or punishment. The pup simply learns to avoid crashing into trees, leave older dogs alone when they are chewing bones or if they growl at him, and so on.

In fact, even Dr. Dunbar, while on the one hand often stating that punishment is largely ineffective, on the other hand freely admits its effectiveness in the natural world:

> "In the wild, when young animals investigate the real world, the consequences of their actions range from exceedingly pleasant to downright unpleasant. Nature can be very rewarding or a very strict teacher... behaviors that have unpleasant consequences *are strongly inhibited*, such as ploughing through rosebushes or eating a wasp... dangerous, unpleasant, environmentally inappropriate behavior *quickly decrease in frequency*" [55] (emphasis added).

So, Dr. Dunbar admits that, in response to punishment, behaviours are *"strongly inhibited"* and *"quickly decrease in frequency"*. This is in stark opposition to his frequent claims to the contrary.

When we talk about science, many studies agree on the effectiveness of punishment. Consider this from scientists Azrin and Holz:

> "One of the most dramatic characteristics of punishment is the *virtual irreversibility or permanence of the response reduction* once the behaviour has become completely suppressed. Investigators have noted that the punished response does not recover for a long period of time even after the punishment contingency has been removed... How quickly does punishment reduce behaviour? Virtually all studies of punishment have been in complete agreement that *the reduction of responses by punishment is immediate* if the punishment is at all effective. When the data has been presented in terms of number of responses per day, the responses have been *drastically reduced or eliminated on the very first day* in which the punishment was administered."[56] (emphasis added)

---

[55] Dunbar, Ian, Dr. *Barking Up The Right Tree - The science and practice of positive dog training*. Novato, CA: New World Library, 2023.
[56] Azrin, N. H., & Holz, W. C. (1966). *Punishment*. In W. K. Honig (Ed.), Operant Behavior: Areas of Research and Application (pp. 213-270). New York: Appleton-Century-Crofts.

CHAPTER 4 – Is punishment effective when training dogs?

Karen Pryor's, Ian Dunbar's, and most other high-profile "positive" advocates statements (and general tone) about the ineffectiveness of punishment are looking mighty untrustworthy. And the case against their claims is only going to get stronger as we progress. In fact, in Pryor's push to demonise punishment, she tries to suggest that dogs actually *don't* learn anything from these environmental punishers! She says:

> "All dogs experience aversives in their daily lives, such as tripping over a log… Do dogs learn from these episodes? Not so you'd notice."[57]

This is, of course, absolute nonsense. If dogs *didn't* learn from these environmental punishments, they'd all be pretty banged up by now. In fact, as I mentioned previously, they'd all be dead. Pryor is letting her *ideology* get in the way of *reality,* as do all "positive" trainers. Even Murray Sidman (perhaps the king of the reward-only ideologues) acknowledges this fact:

> Karen Pryor's and Ian Dunbar's statements (and general tone) about the ineffectiveness of punishment are looking mighty untrustworthy.
>
> And it is only going to get worse for them as we go along.
>
> As all reward-only trainers do, they are letting their *ideology* get in the way of *reality*.

> "Certainly, if a species could not make use of environmental cues for reinforcement and punishment, it would not survive for long."[58]

Pups aren't born knowing about fire, or heights, or other dogs. They must learn all of these life lessons. In many cases, what a pup learns about his environment he learns through correction or punishment. These are perfectly natural ways in which dogs, and indeed all animals and humans, learn. And despite claims to the contrary, everyone can see that they generally work highly effectively.

A mother with her pups

As mentioned, there is nothing natural about long convoluted schemes to try to train dogs without recourse to correction or punishment. Nothing. It never happens in nature. As Dr. Whitney explains, talking about the ways dogs learn in the *real world*:

> "He learns, by painful reactions what he must not do. From early puppyhood he learns that his mother's growl means, "No, stay away." How

---

[57] Pryor, Karen. *On my mind – Reflections on Animal Behavior and Learning.* Waltham, MA., USA. 2014.
[58] Sidman, Murray. *Coercion and Its Fallout.* 2000. United States. Author's Cooperative.

PART 3 – Answering the case against punishment

> does he learn? By hearing the growl and when he continues doing what he should not do, feeling his mother's fangs which hurt him and frighten him terribly. It is as though there were a fence made by the growl. The mother is eating and growling. If the pup stays his distance he is safe, but woe betide him if crosses that invisible fence." [59]

Zak George disagrees. However, his lack of experience is evident when he writes:

> "… while it is true that mothers may resort to physical corrections like that, as far as I know it's a myth that this is a primary way that they communicate with their puppies. If a mother dog goes beyond those quick subtle corrections, for example if she's causing harm or *actually biting her puppies*, this is considered a major red flag amongst breeders and those who raise puppies. Typically it means there's some external issue like something medical, or possibly something in the mother dogs environment that's causing her to feel uncomfortable."[60] (emphasis added)

This is simply not true. Going beyond "quick subtle corrections," or biting their pups, is *not* considered a "major red flag amongst breeders" (certainly not *experienced* breeders), at all. Nor is it any sign of something "medical" or "external", or that she is "feeling uncomfortable". Maybe some *in*experienced breeders object to their bitches biting their pups and think it is abnormal, but that's simply because they don't understand how dogs work, and it offends their sensibilities. They, like George, simply lack the experience to know that this is how bitches operate. It is perfectly normal, everyday behaviour. It is the *rule,* not the exception.

> The mother isn't being mean, she is just training her pups the way dogs do in *the real world.*

"How could she be so mean to her puppies?!!" The mother isn't being mean, she is just training her pups *effectively*, the way dogs do in *the real world*. It is simply the way dogs work, just the same as the way wolves work. It is *reality,* not some utopian fantasy world.

Again, from Dr. Whitney (and remember, he was the author of the two books Dunbar studied up on before teaching his first dog training lecture, and just giving it a different name):

> "The mother dog doesn't gentle her puppy along by tapping him lightly with her tail. She may bite him so hard that he bleeds… I'm not advising

---

[59] Whitney, Leon, Dr. *Dog Psychology – The Basis of Dog Training*. New York: Howell Book House, 1971.
[60] George, Zak, and Dina Roth Port. *Dog Training Revolution: The Complete Guide to Raising the Perfect Pet with Love*. New York: Ten Speed Press, 2016.

CHAPTER 4 – Is punishment effective when training dogs?

that you should be so harsh. I'm simply reminding you how the natural dog is taught."[61]

Unlike Zak George, Dr. Whitney had enormous practical experience as a dog breeder (and geneticist) over a very long span. He knew what he was talking about. He was described as:

"for many years, America's best known and most widely read writer on dogs. Based on over five decades of unmatched experience—internationally saluted as geneticist, researcher, practising veterinarian and trainer of hunting and trailing dogs—he wrote more the 50 books on pets."[62]

Like Whitney, I also have had a large amount of experience breeding dogs. I have bred many litters of my Campaspe sheepdogs (*www.campaspeworkingdogs.com*) over more than thirty years. I have also reared almost every single one of those pups to between four and six months of age when they start working livestock. I generally have fifteen to twenty adult sheepdogs at a time plus numerous Turkish Kangal and Maremma livestock guardian dogs. I know how dogs interact with each other, and I know how bitches rear their pups. And I concur completely with Dr. Whitney. Zak George is simply wrong.

Figure 5 – One of the author's bitches feeding her pups

---

[61] Whitney, Leon F., Dr. *The Natural Method of Dog Training*. New York: Tandem Books, 1969.
[62] As above.

If you find the above anecdotal evidence from highly experienced breeders to be insufficient, then consider the following study:

> "…this period [weaning] is associated with increasing irritability on the mother's part toward her young, coinciding with the decline of lactation and a growing disinterest in nursing. This disinterest is not shared by her puppies, whose appetites are as sharp as their teeth. Not surprisingly, *maternal punishing* activity peaks at around this time (Rheinhold, 1963; Wilsson, 1984/85)."[63] (emphasis added)

So, given the reality, why is there this emphasis on trying to pretend that bitches *don't* punish their pups? They certainly do as weaning time approaches, and after. If a pup tries to eat his mother's dinner, he will generally get bitten. This is the way dogs work—they regularly use punishment.

> The natural way dogs interact and train their young is through punishment.

Of course, the reward-only idealogues like Zak George would like to convince everyone that dogs don't use punishment on each other as their 'go to' approach when training their pups, or interacting with each other, simply in order to bolster their case that *you* should never punish *your* dog. It is completely untrue. The natural way dogs interact and train their young is through punishment.

Dog to dog interactions

The same natural use of punishment applies to general dog interactions, as nearly anyone with multiple dogs will have experienced. This experience itself puts to rest the deception coming from those in the "positive" camp. If you feed your dogs, you will often see that one dog is the boss, and unless you intervene it will take what it wants.

If one dog is chewing a bone and the other dog comes too close, the first dog will growl (correction, or warning). If the other dog comes any closer, he will get bitten and chased away. Or, if the dog approaching is the more *dominant* one (the reward-only ideologues hate that term, but is that not exactly what is occurring?), he will growl at the dog chewing the bone, and if the bone isn't relinquished he will fall on him snapping and snarling and chase him off.

---

[63] Lindsay, Steven R. *Handbook of Applied Dog Behavior and Training, Volume 1: Adaptation and Learning*. Ames, IA: Blackwell Science, 2000.

CHAPTER 4 – Is punishment effective when training dogs?

Or, if they are playing and one dog is getting sick of the game, or a pup won't stop annoying him, he will generally nip at them, growl at them, and bite them. As anyone with the slightest experience and common sense knows, this is how dogs work in *the real world*. However, Zak George again distorts reality when he says:

> "…many older, well-adapted dogs are quite good at letting younger dogs know that they need to back off. This may mean a quick *harmless air snap* or *loud squeal*. Often a few experiences is all it takes for dogs to learn not to hump other dogs."[64] (emphasis added)

A "loud squeal"? Is that really what dogs do? No, it isn't. No "well-adapted" dog is going to make a "loud squeal" if another dog humps it!

And a harmless air snap? Maybe, sometimes. But in plenty of other cases the dog will definitely make contact with its teeth, particularly if the other dog ignores the warning that is the "air snap". It might even spin around and pin the other dog to the ground with teeth bared and stand over it to make its point. That is *punishment*, and dogs use it very effectively all the time. It is perfectly natural and normal. In fact, as George noted there:

> "Often a few experiences is all it takes…"

Once again, we see the "positive" trainer admitting that punishment is highly effective. And yet, the strategy George advises (if the other dog doesn't take care of it itself in this manner) is a pointless exercise using "time-outs" which will *never* fix the problem.

> Once again, we see a "positive" trainer admit that punishment is highly effective.

Do you want an effective strategy if your dog is humping another dog (or your leg), which follows the way dogs naturally operate? Simply warn your dog "No" (this takes the place of the warning growl). If it doesn't stop, give it a firm smack on the nose or head (this takes the place of the bite). When it stops, tell it "Good dog". If it doesn't stop, you weren't firm enough—get firmer. Simple, quick, and basically instantly effective with only a "few experiences", just the same as dogs work with each other.

Other species of animals operate in the same manner. Watch lion cubs playing with an older lion. When the cub's playfulness gets a bit too painful, when they bite the older lion's ear or tail a bit too forcefully, a growl or snap of teeth lets them know that they have crossed the line. If they still don't get the message, the

---

[64] George, Zak, and Dina Roth Port. *Zak George's Guide to a Well-Behaved Dog: Proven Solutions to the Most Common Training Problems for All Ages, Breeds, and Mixes*. New York: Ten Speed Press, 2019.

PART 3 – Answering the case against punishment

punishment will escalate. Such correction or punishment is entirely effective, and entirely natural. Watch a lamb trying to get a drink when mum says no, or from a ewe which isn't its mother—it will quickly get a solid head butt for its efforts, probably even being knocked off its feet.

And yet, influential world agility champion Susan Garrett claims that:

> "…there is *always* fallout with *any form* of punishment–even something as benign as a time-out."[65] (emphasis added)

And yet, that is exactly how dogs (and other species) interact with each other all the time—with correction and punishment. And all without any of this claimed terrible "fallout". (There is much greater detail regarding "fallout" in *Chapter Six – Does punishment cause undesirable side-effects?* and in *Chapter Eleven – Murray Sidman's "Coercion and its Fallout".*)

On wolves and dogs

One of the arguments sometimes levelled at "balanced training" (or the use of punishment in dog training), is that it is based on faulty thinking that dogs are *somehow like wolves*. I mean, how utterly preposterous to compare dogs with wolves! Dr. Dunbar says:

> "Basing dog training on a misunderstanding of wolf behavior is as useful as basing human education on a misunderstanding of chimpanzee behavior."[66]

And Zak George in like manner:

> "…no one is arguing that dogs aren't descendants of ancient wolves—they certainly are. However, dogs are not wolves, but unique animals *predisposed to learn very advanced concepts from human beings.*"[67] (emphasis added)

George's use of the term "ancient wolves" is interesting, and somewhat misleading. Wolves still exist, and they can, and do, interbreed with domestic dogs.

---

[65] Garrett, Susan. *Punishment: Pros and Cons*. Susan Garrett's Dog Training Blog, 2010, https://susangarrettdogagility.com/2010/04/punishment-pros-and-cons/
[66] Dunbar, I. (2010). *Let's Just Be Humans Training Dogs*. Dog Star Daily. Retrieved from https://www.dogstardaily.com/blogs/lets-just-be-humans-training-dogs
[67] George, Zak, and Dina Roth Port. *Dog Training Revolution: The Complete Guide to Raising the Perfect Pet with Love*. New York: Ten Speed Press, 2016.

CHAPTER 4 – Is punishment effective when training dogs?

And they have had a genetic influence on some dogs much, much closer than "ancient wolves", for example sled dogs in North America:

> "… observations of domesticated wolves, of those in sled teams (many sled teams are composed entirely of Artic Wolves) and those in zoos lead to the same conclusion so ably voice by one of the wolves' most avid students, Ernest Thompson Seton: "First, and in all ways, he is simply a big wild dog…"[68]

> "…one naturalist who adopted a wolf pup and kept it at a park says, "The day I left the park I paid a last visit to the most friendly "dog" I have ever known."[69]

> "Timber wolf puppies (some call them cubs) when reared on domestic bitches often become trustworthy human companions…"[70]

For Zak George or Dunbar to try to claim dogs are smarter than "ancient" wolves, and are somehow "predisposed to learn very advanced concepts from human beings," and use that as a justification that they should be trained using different principles, is absolute hogwash (to put it politely). *All the same behavioural principles apply.*

Consider an example of the application of this notion of George's that dogs are somehow "predisposed to learn very advanced concepts from human beings". He states that, rather than using a short, simple command like "Quiet", you should:

> "Instead, use proper grammar by saying "Stop barking please," and teach your dog your language as you would teach a young child."[71]

Seriously? *Use proper grammar* and *teach your dog your language as you would teach a young child*?! Unfortunately, he isn't alone in this madness. Dunbar also frequently talks about teaching your dog ESL—English as a Second Language. In fact, it appears to have become the mainstay of his "method". He claims:

---

[68] Whitney, Leon, Dr. *Dog Psychology – The Basis of Dog Training*. New York: Howell Book House, 1971.
[69] As above.
[70] As above.
[71] George, Zak, and Dina Roth Port. *Dog Training Revolution: The Complete Guide to Raising the Perfect Pet with Love*. New York: Ten Speed Press, 2016.

PART 3 – Answering the case against punishment

> "Lucky for us, we have language, and we can use it in ways that are far more effective than any aversive procedure… words quickly resolve most dog behavior and training problems." [72]

"Words quickly resolve most dog behaviour problems"? Really?! *We* have language, that is true, but Dunbar fails to realize that *dogs don't*. He claims:

> "Dog training is almost entirely about *communication*, specifically teaching dogs *our language*, that is, ESL." [73]

For example, here's his ESL—English as a Second Language—advice if your dog is getting a bit rough when playing with other dogs:

> "…I usually say, "Rover, Sit!" and then have a calm chat with the dog: "What on *earth* do you think you're doing? You know better than that. Now let's bring it down a notch. Alright, Rover, Go play. There's a gooood dog. Keep chill. Gooood boy." [74]

This all just makes me think about how a five-year-old child might attempt to train a dog. "Hey, Snoogums, you really shouldn't wee on the carpet, you know that makes mummy mad; perhaps you could consider weeing outside next time?"! It's quite astonishing to realize that dog training has devolved to the level where one of the world's most influential dog trainers is advocating this madness. I am truly staggered that I even feel the need to address it.

> "ESL", "talking to our dog like it is a young child", "canine consent", and being our dog's "life coach" are all simply anthropomorphism taken to absurd extremes.

(And don't even get me started on so-called "canine consent" championed by the likes of Marc Bekoff—who has nothing but glowing praise for Dunbar's methods—and Erin Jones and Chirag Patel; or Professor Paul McGreevy's claims that we should be our dog's "life coach" rather than its trainer or "parent"; all of which are nothing more than anthropomorphism taken to absurd extremes.)

Another time Dunbar says:

> "I'll turn my back on the dog and laugh…" [75]

---

[72] Dunbar, Ian, Dr. *Barking Up The Right Tree - The science and practice of positive dog training.* Novato, CA: New World Library, 2023.
[73] As above.
[74] As above.
[75] As above.

CHAPTER 4 – Is punishment effective when training dogs?

And here's another example, this time of Dunbar's advice on curing play-biting in your pup. He starts off by advising us to:

> "…verbally express your displeasure, "Ouch!!! That hurt! You miserable worm." [76]

Then he continues on:

> "For added effect, you may accentuate your injured feelings by sobbing."[77]

Sobbing?!!! But wait, there's more.

> "If you feel your puppy still ignores your feedback, call it a "Jerk!", leave the room, and shut the door."

Of course, Zak George (like Marc Bekoff) endorses Dunbar's methods:

> "Dr. Dunbar's latest book is a testament to his deep-rooted expertise… offering a blend of time-tested wisdom and innovative approaches." [78]

"Time-tested wisdom" is certainly not one of the descriptions that springs to my mind, but "innovative" certainly is one way of describing them!

Dunbar's next piece of advice, if sobbing and calling your pup a jerk fails (I would say *when* that fails), is to "schedule a home consultation with a Certified Pet Dog Trainer." The problem is that these trainers are certified in Dunbar's methods, and this is what they've been taught. So, what more will they have to offer?

Interestingly, one of Dunbar's criticisms of using punishment for play biting is that it might be *too effective* and actually stop it. He says, that if you use punishment:

> "At worst, *your puppy will stop play-biting altogether*, and hence will not develop bite inhibition."[79] (emphasis added)

---

[76] Dunbar, Ian. *The Good Little Dog Book*. 3rd ed. Berkeley, CA: James & Kenneth Publishers, 2003.
[77] As above.
[78] Dunbar, Ian, Dr. *Barking Up The Right Tree - The science and practice of positive dog training*. Novato, CA: New World Library, 2023.
[79] Dunbar, Ian. *The Good Little Dog Book*. 3rd ed. Berkeley, CA: James & Kenneth Publishers, 2003.

PART 3 – Answering the case against punishment

He's wrong about your pup not developing bite inhibition if you effectively teach it not to bite. But he's right that punishment (if applied correctly) will stop play biting quickly and effectively. His own strategies of crying and sobbing and calling your pup a jerk definitely will not help (not unless you scream at it at the top of your lungs inches from its face like Dunbar can sometimes be seen doing—which is then pure and simple punishment).

> Dunbar is right that punishment will stop play biting quickly and effectively.
>
> On the other hand, his strategies of crying and sobbing and calling your pup a jerk definitely will not help!

So, Zak George advocates using "proper grammar" and "teaching your dog your language as you would a young child"; and Ian Dunbar advocates teaching dogs ESL—English as a Second Language—using phrases like "Ouch, that hurt, you miserable worm", sobbing, and calling it a "jerk"!

In reality, dogs don't understand language (or grammar!). Trying to pretend otherwise is only going to set dog training back considerably. Despite these ludicrous claims, clear, simple commands are far superior. Dogs can learn to associate *sounds*, whether words or even collections of words, or whistles, or the doorbell—they do not understand language.

Dunbar also makes some equally inane remarks such as "all breeds are equally smart and begging for an education."[80] No, Dr. Dunbar, they certainly are not. There is a huge variation in their intelligence. And none of them are "begging for an education". That is something we want to give them in order to make them great companions or workers, and to keep them safe while giving them plenty of freedom, not something they are "begging for".

Getting back to the subject of wolves, they and dogs certainly exhibit subtle differences in their characteristics, much like the differences observed between different breeds of dogs, or between individual dogs within breeds. However, there are really *no* significant distinctions between the two species. There is no significant difference in the way dogs or wolves learn or interact with one another. There may be a difference of *degree*, like there is from breed to breed, or individuals within breeds, but that is all. Like the difference between a chihuaha and a crossbred pig hunting dog, or a pit bull terrier. Do you train them differently? Not in principle—the principles are identical.

So, while these reward-only ideologues attempt to distance dogs from wolves, what they instead advocate is basing dog training not on wolves (heaven forbid!), but on

---

[80] Dunbar, Ian, Dr. *Barking Up The Right Tree - The science and practice of positive dog training*. Novato, CA: New World Library, 2023.

CHAPTER 4 – Is punishment effective when training dogs?

dolphins trained in captivity (Karen Pryor), or *rats* and *pigeons* trained in cages in a laboratory (Skinner), and even on *humans* (teaching them English and using proper grammar).

And yet, in contrast to these attempts of George and Dunbar to distance dogs from wolves, some comments of two of Skinner's most well-known students and proteges are worth considering—Keller and Marian Breland have pointed out much the opposite in some of their work. For example, in their article "*The Misbehaviour of Organisms*," (which discusses various failings of the philosophy of operant conditioning, mostly centering around the different instincts of various species) they state:

> "There seems to be a continuing realization by psychologists that *perhaps the white rat* cannot reveal everything there is to know about behaviour."[81] (emphasis added)

And:

> "…it is our reluctant conclusion that the behaviour of any species cannot be adequately understood, predicted, or controlled without knowledge of its *instinctive patterns, evolutionary history, and ecological niche*."[82] (emphasis added)

So the Breland's suggest we need to look further than the laboratory animals, and consider the animals "instinctive patterns, evolutionary history, and ecological niche." I wonder where on earth we would look to for that?! They also state:

> "In spite of our early successes with the application of behaviouristically oriented conditioning theory, we readily admit now that *ethological facts and attitudes* in recent years have done more to advance our practical control of animal behaviour than recent reports from American "learning labs". Moreover, as we have recently discovered, if one begins with evolution and instinct as the basic format for science, a very illuminating viewpoint can be developed…"[83] (emphasis added)

Yet many of the reward-only ideologues tell us that we shouldn't compare dogs to wolves. The problem is that some of the most influential figures in the "positive" training world based their ideas not on *real world* experience with dogs at all. Take Karen Pryor as an obvious example—she was a *dolphin* trainer, who based her *dolphin* training methods on Skinner's work with *rats* and *pigeons*, who then wrote

---

[81] Breland, K., and Breland, M. *The Misbehavior of Organisms*. American Psychologist 16, (1961): 681-684.
[82] As above.
[83] As above.

PART 3 – Answering the case against punishment

a book that has become something of a bible to reward-only idealogues (both human and animal).

Like Dunbar, Pryor was *not* a professional dog trainer, and had precious little experience in the field. And yet the book she wrote at that time was a major link in the chain of the downward spiral of the "positive" based training "revolution", and she is sometimes referred to as "the mother of modern dog training". That, as far as I am concerned, is nothing to be proud of.

So basing dog training on a misunderstanding of dolphin, rat and pigeon behaviour, and arguing against the relevance of wolves, is simply absurd.

> Basing dog training on a misunderstanding of dolphin, rat and pigeon behaviour, and arguing against the relevance of wolves, is simply absurd.

In actual fact, of course, I have no problems with using relevant, properly conducted research on these other species to apply to training dogs. All animals (and humans) learn in basically the same way (with some notable differences for humans)—by the consequences of their actions both good and bad.

As long as we recognize that training dolphins *in captivity*, or rats or pigeons *in a cage*, are not the same thing as training dogs in *the real world*, or at least that there are other variables at play, everything is fine. Let those dolphins back out into the ocean, or the rats back out into the rubbish tip, and see how much control Pryor or Skinner would maintain!

> Let those dolphins back out into the ocean, or the rats back out into the rubbish tip, and see how much control Pryor or Skinner would maintain!

But for these trainers to decry the comparison of dogs to wolves because they don't like the conclusions (that wolves use punishment in their interactions), and then use methods developed on dolphins, rats, and pigeons (or in their own imaginations—such as ESL, using proper grammar, or "canine consent") is, frankly, hilarious. It also shows the lengths they will go to in their attempts to promote their reward-only ideology.

How wolves operate (which is the same way dogs operate) is, in fact, a *great way* to train dogs. It is the *natural* way to train dogs, consistent with the way they themselves operate.

Dominance

However, all this is not to say that I agree with some of the "dominance" type approaches to dog training. There are some ridiculous and harmful ideas out there. The problem is not that there isn't such a thing as "dominance"—there certainly is.

CHAPTER 4 – Is punishment effective when training dogs?

The problem is with the conclusions that have been drawn from that, or the ideas that have been developed from that. I have no problems with dogs going through the door before you, or sleeping on your bed, or being higher than you, or eating before you, and so on, if you are happy with those things. None of these are even faintly "dominance" issues, despite what is sometimes claimed.

Where you have a problem is when your dog growls at you when you try to move it *off* the bed, or off the couch, or when you go near its food. And even then, we don't need to deal with some abstract notion of "dominance"—we are generally better off dealing directly with the behaviour at hand (punishing it, just as a more dominant dog or wolf would do). The "dominance" takes care of itself in the process, if we use the right training methods.

> The "dominance" takes care of itself in that process, if we use the right training methods.

The problem is not the comparison of dogs to wolves—that is a perfectly valid comparison. Nor is the problem understanding that there is a hierarchical type of structure amongst wolves, as there is amongst dogs, chooks, sheep, horses, cattle, and people. Some animals are bigger, or stronger, or usually simply have stronger personalities. These animals generally take the first go at food and anything else they want. They are, very definitely, more "dominant".

As I am writing, I am looking out the window watching the chooks pottering around. A younger chicken is pecking at something in the grass, and an older chook has noticed. It races over with wings outstretched and chases the younger chicken away, before taking the tasty morsel for itself. That is dominance, and it is real. Or, sometimes rams will fight over something, particularly when first introduced to each other. They will back up a good maybe fifty or a hundred metres apart, and then charge! They will hit head on with a tremendous impact that can be heard from a *long* way away. It is an impressive sight and sound. Then they will back up, and repeat; until, eventually, one ram admits defeat and gives way. The next time the more dominant ram wants the best food or the female in season, he takes it without argument. *This time there is no fight*. The other ram submits to the dominant ram. This actually results in *less* ongoing fights (as noted repeatedly, when we use punishment properly, we don't use it much). Consider the following statement (and the erroneous conclusions) from Victoria Stilwell:

> That is dominance. And it is real.

> "Researchers... found that although dominant members of certain animal groups were more likely than others to display threatening or aggressive

PART 3 – Answering the case against punishment

behaviour if required, they would more often assert their influence (dominance) *without the use of force.*"[84] (Stilwell's emphasis)

She recognizes that the dominant members were more likely to "display threatening or aggressive behaviour if required" (which is their use of punishment), while emphasising that they "more often assert their influence (dominance) *without the use of force.*" The reason they can assert their influence (dominance) without the use of force, is because *previously* they have *used threatening or aggressive behaviour* and thereby *created the hierarchy*—now the other animals submit to them because they have learnt that they will be punished if they do not. And they have learnt that they will not win a confrontation. Stilwell continues:

> The reason they can assert their influence (dominance) without the use of force, is because previously they have used force to create the hierarchy.

"Other more submissive members of the animal group maintained the status quo by *offering* deference and appeasement behaviours to the more dominant members. *In other words, dominance is not so much a character trait as it is a term that describes the relationship between animals, a relationship that is usually exerted without the use of force, thereby reducing the potential for conflict and ensuring safety and survival.*"[85] (Stilwell's emphasis)

She also says:

"In a natural pack, harmony is created because, as we now know, deference behaviours are *offered freely* by the younger wolves *rather than being forced onto them by their parents.*"[86] (emphasis added)

The main reason those other members *offer* the submissive and appeasement behaviours is *because* of past experience being punished for something or other (perhaps trying to eat the other animal's food). So, Stilwell claims that this *offer* of deference and appeasement is something *they freely give*, rather than something that has been created by the more dominant animal previously. She is wrong.

And, despite her claim, dominance is *both* a character trait *and* a relationship between animals—some animals are definitely more dominant (stronger, tougher mentally, more aggressive, more motivated)—and some animals are dominant relative to others in the hierarchy.

---

[84] Stilwell, V. *Train your dog positively*. New York. Ten Speed Press, 2013.
[85] As above.
[86] As above.

CHAPTER 4 – Is punishment effective when training dogs?

Dunbar, while on the one hand criticising the idea of "dominance", also on the other hand recognizes it. In one place he says:

> "It's not that dogs physically fight and physically create this dominance hierarchy. It's totally the other way around. The hierarchy is created, if you like, by psychological experience."[87]

Created by "psychological experience"? Really? But then, in another lecture he gave, talking about the results of his *own* research:

> "What we found was *loads* of fighting within litters. *Loads* of it. They never stopped fighting, to establish a hierarchy. Loads of fighting and physical domination within litters. This is how they establish their hierarchies, and it's nonstop."[88]

So, which is it? "Loads of fighting and physical domination…this is how they establish their hierarchies", or "It's not that dogs physically fight and physically create this dominance hierarchy. It's totally the other way around"? You can't have it both ways.

Another argument often levelled at dominance by the "positive" crowd, is that Dr. L. David Mech, the scientist who is generally recognized as the world's foremost wolf expert, recanted his views on dominance. This is only partly true. He did modify his use of the term "alpha", but those who claim he rescinded his views on dominance are simply not correct. Here is some of what Victoria Stilwell had to say:

> "These captive-wolf studies have since been renounced by the very scientists who performed them and reached their original conclusions. Dr. David Mech, the wolf expert…now explains that the behavioral differences between captive wolves like those used in his study and wolves in the wild are pronounced enough to call the results of his original research into question." [89]

I decided to go straight to the source, and so I asked Mech directly about this. He provided the following to me in writing:

> "I did not state nor mean to imply that wolves do not dominate each other and have even posted a you tube video showing a wolf being dominated by another. My main point about not using the term "alpha" with wolves is

---

[87] Dog Training by K9-1.com. *Is Dr. Ian Dunbar a Fraud? Watch This Preview (k9-1.com)*. YouTube. May 27, 2024. https://www.youtube.com/watch?v=I3HSddizJc4.
[88] As above.
[89] Stilwell, V. *Train your dog positively*. New York. Ten Speed Press, 2013.

PART 3 – Answering the case against punishment

> that in a natural, free-ranging wolf pack, the individual we used to call "alpha" is merely a parent wolf. In most cases, the parent becomes a parent, not by fighting to obtain the top position in a dominance hierarchy, but rather by merely finding a mate and producing offspring with it."[90]

After speaking with Dr. Mech, I also came across this statement of his in an Ivan Balabanov (a highly regarded American dog trainer) podcast:

> "They [dog trainers] misunderstood. I didn't say there was no such thing as dominance, but a lot of the dog people thought that's what I was saying."[91]

So dominance is real, and Dr. Mech never said it wasn't, and wolves do dominate each other. Nobody who has been involved with animals for any length of time can reasonably deny it.

However, personally, I will still continue to use the term "alpha" (I disagree with Mech on this point) as it is perfectly descriptive of the most dominant animal. Even though it may be the parent, that dominant position is established while the pups are very young—if the pups are too annoying or trying to eat the parent's food, or during weaning, they will get punished. The parent's dominance isn't just a given (as Mech is incorrectly claiming), rather it is created at a young age, and then maintained (or sometimes not) if any conflicts arise.

Mech's reasoning was that the results were incorrect because the wolves being studied were in captivity and not roaming free in the wild. There are a number of things to consider in relation to this.

The first is that wolves living in captivity is actually much more applicable to domestic dogs than a free roaming wolf pack. Even if Mech is right about them behaving differently in the wild (I don't believe he is), that's not the situation our dogs are living in. So even if he's right, his observations of *captive* wolves are actually much more applicable to domestic dogs.

> Mech's observations of captive wolves are much more applicable to domestic dogs.

Secondly, when studying wolves in captivity, it is much easier to observe them closely and continuously. Whereas, when attempting to observe wolves in the wild there is only an occasional glimpse here and there, and so the data points are isolated and rare. Therefore, the observations are more likely to be accurate in the captive situation.

---

[90] Mech, L. David. *Personal correspondence*. 27th May, 2024.
[91] Balabanov, Ivan. *Training Without Conflict® Podcast Episode Thirty-Three: Dr. L. David Mech.* YouTube video. June 11, 2023. https://www.youtube.com/watch?v=YSsp8aHlOXU&ab_channel=IvanBalabanov.

CHAPTER 4 – Is punishment effective when training dogs?

As mentioned, I have personally reared over one hundred and fifty litters of pups over the last thirty plus years, nearly every single pup to sheep working age (approximately 4-6 months of age) and retained many pups past that age. My observations are that not all parents are dominant over their pups. Some bitches are too soft in nature, and some dominant natured pups will even dominate their own mother from a young age (we see this in children also). I have even seen three-month-old pups growling at their mother and keeping her away from her own food—not commonly, but it can happen. So the parents aren't just magically dominant simply by virtue of being the parents—they aren't just automatically at the top of the pecking order. The vast majority establish that dominance very early on as I described. If you aren't observing them multiple times every single day, you might miss these occurrences, *because it only needs to occur once or twice to establish the pecking order*. From then on you see the signs of submission and appeasement behaviour, and only occasionally do you see much more punishing from the mother (a warning growl usually suffices). Punishment, used properly, doesn't have to be used very often.

And then, as the pups get older, an increasing percentage are likely to end up dominant over either, or both, parents. The biggest changes occur as the pups reach adulthood at around 9 to 18 months of age—earlier in general for smaller dogs, and later for larger breeds. This is a problem I quite frequently get called out to help with. The story generally follows a very familiar pattern—the owners had an adult (usually female), and then brought a new female pup into the family "pack". Things were fine while the pup was young and content to be submissive to the older female. However, as the pup ages it becomes more confident, more independent, and less submissive; its level of dominance rises (including the various factors than can contribute to this such as its strength of character, confidence, toughness, level of aggression, size, etc.). As its level of dominance starts to roughly equal that of the older dog, you can start to see increasing levels of animosity and fights between the two. When two dogs have a similar dominance level, this is when fights tend to occur—if one dog or the other is obviously more dominant and one submissive, all is good.

At this point, I advise the owners of three possible outcomes. One is that the younger dog's level of dominance continues to rise as it ages over the next few months—if it ends up clearly the more dominant dog (i.e. it ends up the "alpha"), the roles are reversed, and peace will again reign. If its level of dominance plateaus about equal to that of the older dog, then we will have continuing problems. In that case, there are two possible outcomes—training can help the problem and, in most cases, solve it. The third possible outcome is that the problem is so intractable that one dog will have to be rehomed.

A very similar situation often occurs with dog-to-dog interactions outside the home—say, for example, at the dog park (which I am not in favour of, for a number of reasons). A young pup might be fine at the dog park, submissive, and no fights have occurred. Then, as the pup reaches adulthood, one day a fight breaks out between it and another dog it has previously been fine with. This is the same situation—the pup's level of dominance is rising as it nears adulthood, and as it comes up somewhere near equal to the older dog, fights can occur.

And this applies whether with dogs, wolves, sheep, goats, chooks, cows, horses, and so on. As Dunbar said, "fighting and physical domination… is how they establish their hierarchies" (that is, when he wasn't claiming that "It's not that dogs physically fight and physically create this dominance hierarchy"!).

And thirdly, I don't believe anything changes in principle in wolves in captivity from when they are roaming free. The interactions are simply *exaggerated* by the closer confines, as they are for example for farm animals. Take cattle for example. On one property we would sometimes have over two hundred young bulls together in the cattle yards. There was a lot of pushing and shoving and argy-bargy going on amongst the bulls. Yet when they were out in a large paddock this would only be seen much more sporadically. The close confines simply exaggerated the interactions. In principle there is no difference—it is simply one of degree. I am certain that exactly the same applies with wolves.

So Mech is absolutely correct that dominance is real, however it was the earlier observations of how dominance functions that were more correct, not his later alterations, and hence "alpha" remains an accurate term.

However, does this mean we should use an "alpha roll" (pushing the dog to the ground and rolling it onto its back) to assert our dominance? No, not as a general rule (although very occasionally it can be a good option). And does it mean we should eat out of our dog's dinner bowl before we feed it, like Daniel Abdelnoor (aka "Doggy Dan") and others recommend?! No. It is 100% pointless. There is absolutely no need to "assert" dominance in abstract ways like these. It happens simply as a *natural consequence of good training*.

To use an "alpha roll" as some trainers recommend as a punishment during training *can* work for some things at some times. However, it is often too slow by the time you catch the dog and put it on its back, and so is more disconnected (or even completely disconnected) from the behaviour. Therefore, it will be much less effective than other types of punishment that can be delivered in a timelier manner. And it is more difficult to vary the level of the consequence to suit the situation. There are simply better, more instant, more direct, less confrontational types of punishment to use in most cases.

So, when I let my fifteen or twenty sheepdogs out into their exercise paddock, I do not care if they all rush out the gate before me. And when I put their food into their dinner bowl, they may wolf it down. Do I pretend to eat out of each dog's dinner bowl first at every meal as in Daniel Abdelnoor's recommendations so I can maintain my alpha status?! Not on your life.

Does that mean my dogs are dominant, and I am not? Not at all. I just don't consider it worth bothering to teach them to go quietly out the gate. But they certainly know who the boss is—if I want to take the bone they are chewing away, I do so. There are no arguments. They know their position in the hierarchy.

So dominance is a real thing, but it comes about simply as a matter of course during normal training (provided this includes punishment). It is not something you need to give much thought to, and it is something I rarely think about or talk about (other than perhaps "dominance aggression"). However, if you are relying solely on positive reinforcement, then with some dogs dominance can indeed become a very real problem for you.

> If you rely solely on positive reinforcement then dominance can become a very real problem for you.

We reward good behaviour, and we correct or punish undesirable behaviour—the dominance takes care of itself in that process. It is really rather simple (and it actually is—unlike the false claims about the ease of "positive" training).

How can these trainers all get it so wrong?

So how can Pryor, and Dunbar, and all these other so-called "positive" trainers—or psychologists parroting the same nonsense—possibly claim that punishment "almost never really works"? It obviously does, and we've seen that *straight from their own mouths*. I suspect there are three possible reasons for these quite evidently false statements.

Firstly, perhaps these trainers *want it to be true,* because they have an awful lot vested in it being true (often their life's work). For example, Dunbar, Stilwell, Pryor, George, etc., have made a lot of money out of their training empires built largely on these fallacies. If the wider public, or their students, realized just how brilliantly effective the correct use of correction and punishment is, it may put all this at risk.

Secondly, many have an *ideological, utopian wish* that it is true—they simply don't like the *idea* of punishment. That's understandable I suppose. It's a nice sentiment. The only problem is that it *isn't based on reality*. It has nothing to do with the *real world.* As the saying goes, "Facts don't care about your feelings." You can wish

> You can wish you didn't have to use correction or punishment in training, or that it doesn't work, but that won't change the facts.

PART 3 – Answering the case against punishment

you didn't have to use correction or punishment in training, but that won't change the facts.

As Ayn Rand said:

> "You can avoid reality. But you can't avoid the consequences of avoiding reality."

From a marketing perspective, so-called "force free" or "positive" training is an *easy sell*. It is much easier to sell the idea to the general public, particularly in this day and age when most people are so disconnected from the natural world, that you don't need to punish or correct your dog; and that if you just be nice to it all the time and feed it treats it will magically turn into a well-trained dog that is a pleasure to have around. Unfortunately, no, it won't (and nor will your kids or your students).

Perhaps in some *utopian fantasy land* we can avoid all punishment. However, whether the ideologues like it or not, we inhabit the *real world*. And in the *real world* punishment not only works, and is highly effective, but is indeed entirely *necessary*.

And the third reason "positive" trainers get it so wrong is that most of them have only actually owned and personally trained a small number of dogs themselves. Then they decide they want to make a living out of it, so they get a quick "qualification". Then they set about teaching everyone else how to train dogs, and in some cases promoting themselves as experts. And because they have so little *real world* dog training experience themselves, they simply repeat what they have heard others say, and regurgitate what they have been taught.

> Unfortunately, the lack of experience and perpetuation of these myths and lies is reducing the quality of life for so many dogs out in *the real world*, and for their owners.

Unfortunately, their lack of experience and their perpetuation of these myths and lies is reducing the quality of life for so many dogs out in *the real world*, and for their owners.

So, these are the three possible reasons why I believe the "positive" trainers get it so wrong when it comes to claiming punishment doesn't work: it is in their own deeply invested self-interest; they have an ideological wish that it is true; and/or they simply lack the experience and understanding to know better.

Conclusion – Is punishment effective when training dogs

So, we can safely and convincingly debunk the first argument levelled at punishment—that it doesn't work.

We have done so not only from the clear examples of nature (a pup won't put his nose in the campfire twice, and soon learns not to jump from too high a height, or learns to leave the old dog alone), but also from the "positive" advocates own mouths. Karen Pryor gives us plenty of examples of the brilliant effectiveness of punishment, as do Pat Miller, Dunbar, Stilwell, Donaldson, George, and various others.

Here is a quote from James Mazur, from the Sixth Edition of the textbook *"Learning and Behaviour"*. Mazur is Professor Emeritus at the Southern Connecticut State University. He has taught and conducted research on learning for over 40 years and was formerly the editor of the *Journal of the Experimental Analysis of Behavior*:

> "The question of whether punishment can be an effective way of controlling behaviour *has been settled*. This chapter has presented numerous studies, from both inside and outside the laboratory, that have demonstrated that punishment can change behaviour and, in many cases, *change it permanently*."[92] (emphasis added)

Anyone with actual *real world* animal training experience has realized this all along. Yet, as we have seen, so many try to claim otherwise. When used correctly, punishment not only works, but is *highly* effective. In fact, it is actually *far* more effective than reward in many situations (which often doesn't work at all). And the results achieved by it are usually much faster than reward-based methods, often by a *massive* amount. Zak George, while continually advocating methods that take a huge amount of time and effort (and probably still won't get results), constantly warns people to stay away from the "quick fix" solutions of trainers who utilise correction and punishment. However, in reality, good training—training that is actually effective—*is* usually quick.

> Good training—training that is actually effective—*is* usually quick.

By Jamie Dunbar's own logic, if the behaviour improves, the methods are good.

---

[92] Mazur, James E. *Learning and Behavior, 6th Edition.* New Jersey. Pearson Prentice Hall, 2006.

PART 3 – Answering the case against punishment

> "Again—the proof is in the pudding. If your dog's behavior improves, the training methods are good. If your dog's behavior doesn't improve, the methods are not good."[93]

Think of the two applications of the cap gun by Pat Miller which cured "Happy" from chasing the horse; one or two repetitions cured Pryor's friend's cat of clawing the couch; one application of punishment to each of Pryor's two childhood misbehaviour's and they "stopped instantly"; one application of the spray gun cured Karen Pryor's bin diving border terrier; and so on. Or think of any highly trained sheepdog, or police or military dog; or, until recently, well-trained guide dogs.

So that's Myth #1—that punishment doesn't work. Punishment is, in fact, *highly* effective in dog training. Let's now look at how it works with people.

---

[93] Dunbar, Ian. *Six Simple Steps to Solve Your Dog's Behavior Problems*. Dunbar Academy. Accessed Feb 2024. https://www.dunbaracademy.com/courses/six-simple-steps.

5

Is punishment effective when training people?

Karen Pryor, Dr. Dunbar, and others, give examples in their dog training books and elsewhere of many of these training principles involving people. Karen Pryor gives examples of what she identifies as "positive reinforcement" which are, in fact, quite obvious examples of "negative reinforcement". Therefore, by her own admission, they include punishment. She says:

> "Here is the Karen Pryor system of effecting a change in a hard case:
>
> *Karen Pryor* (Seeing a young visitor's wet bathing suit and towel on the living-room couch): Please take your wet things off the couch and put them in the dryer.
> *Young Visitor*: Okay, in a minute.
> *K.P.* (Physically goes to the young visitor and stands there, saying nothing.)
> *Y.V.* What's the matter with you?
> *K.P.* Please take your wet bathing suit off the couch and put it in the dryer. (N.B. Without adding "Now!" or "Right this minute," or "I mean it," or anything else. I am training this person to obey requests the first time, not to wait until the signal has been heightened with further details or threats).
> *Y.V.* Well, jeez, if you're in such a hurry, why don't you do it yourself?
> *K.P.* (Pleasant smile, no comment. I am waiting to reinforce the behaviour I want. Giving me an argument is not the behaviour I want, so I ignore it.)

PART 3 – Answering the case against punishment

> Y.V. Okay, okay. (Gets up, goes to couch, picks up stuff, tosses it at the laundry room.)
> K.P. In the dryer.
> Y.V. (Grumbling, puts stuff in the dryer.)
> K.P. (Big smile, sincere, no sarcasm): Thank you!"[94]

Was this an example of positive reinforcement in practice? No, it was a perfect example of the effectiveness of *negative* reinforcement and punishment. It seems that Pryor simply doesn't understand the methods she is using. And as she points out elsewhere:

> "While negative reinforcement is a useful process, it's important to remember that each instance of negative reinforcement also contains a punisher."[95]

Pryor going and standing imposingly in the girl's personal space—when her young visitor didn't do as she was asked—was punishment for not doing as she was told; it made her young visitor feel distinctly uncomfortable. Remaining there in her space, looking at her (regardless of the fake "pleasant smile") until her commands were obeyed, was negative reinforcement. The main reward was not the "thankyou" and the big fake smile (what Pryor elsewhere might describe as a "known manipulator"[96] as she suggests in her criticisms of using praise—see *Chapter Nine – More downsides and side-effects of reward-based training— Manipulation and dishonesty*). Rather, the main reward was being left alone once the visitor had done what was asked. So, Pryor's example of positive reinforcement was in fact negative reinforcement (sometimes called escape training or pressure/release training) and it did indeed include punishment.

So punishment and negative reinforcement both worked well with a person in what Pryor described as "effecting a change in a hard case". And don't forget Pryor's own statement that:

> "My parents punished me exactly twice in my whole upbringing… both behaviours stopped instantly."[97]

---

[94] Pryor, Karen. *Don't Shoot the Dog!: The New Art of Teaching and Training*. Rev. ed. New York: Bantam Books, 1999.
[95] As above.
[96] Pryor, Karen. *On My Mind – Reflections on Animal Behavior and Learning*. Waltham, MA. Sunshine Books. 2014.
[97] Pryor, Karen. *Don't Shoot the Dog!: The New Art of Teaching and Training*. Rev. ed. New York: Bantam Books, 1999.

So Pryor obviously realizes the effectiveness of negative reinforcement and punishment with people (as well as—as we have already seen—with dogs and animals).

Debutante problems

A few years back, my wife and I owned and operated a ballroom dancing studio. As part of that, we generally trained half-a-dozen debutante groups or so every year. Often, we would have the students (sometimes over 70 students in a single class) dancing in a big rectangle around the outside edge of the dance floor. One of the problems we faced when training them was the students constantly cutting the corners and encroaching on the centre of the dance floor. They would start out dancing nicely around the outer edge of the dance floor, and gradually bunch up in a tighter and tighter circle in the middle of the floor.

Despite our constant pleadings (nagging), they would not improve. Why? Because they had competing priorities. They were more interested in remembering their steps, or making a good impression on their dance partner by not stepping on their toes. Just like your dog is weighing up your treat versus the environmental rewards (much more on this later), so too were the dance students simply weighing up what was more important to them at the time. And keeping out on the wall wasn't high enough on the list.

So, how to fix it? Simple, add a consequence they won't like, that will outweigh their other considerations. In other words, a punishment. Or, at least, the threat of a punishment. We simply said, "This time, whoever is closest to the centre of the dance floor when the music stops will have to dance a solo performance." And, hey, presto! They all stuck to the walls like glue. Problem solved by the threat of a punishment which the students wanted to avoid.

And did it create all sorts of terrible "fallout" as Susan Garrett claims *always* accompanies *any form* of punishment including "time-outs"? Not at all. It simply fixed the problem quickly and easily.

Misbehaving students

Here is another example from one particular debutante group. In this group, we had three girls who were being quite disruptive. Not knowing what we were permitted to use in terms of punishment (school teachers have their hands tied behind their backs nowadays thanks to the same crazy, reward-only ideology), we contacted the school. They assured us we could expel them from the group if we felt it necessary.

PART 3 – Answering the case against punishment

So, the next class, when they began to misbehave, I stopped the class and told them the rules. "If you do that again, you will sit the rest of the class out. If you repeat it after that, you will sit the next class out also. If you do it a third time, you will be out of the debutante ball altogether."

Exactly as I expected, they again misbehaved. No doubt they were expecting the ineffective nagging that they got from many of their teachers. Instead, they instantly found themselves sitting the rest of the class out. They got quite a shock when we did exactly what we said we would do, with no further chances. They never gave us any trouble after that, no doubt because they now fully believed that the *promised escalating punishments* would occur. So, once again, a great example of the effectiveness of punishment. Problem solved instantly and permanently. And yet, like many others, Australian counsellor Silvia Cataudo-Williams says:

> "No matter what the behaviour is, responding with punishment *simply will not work*."[98] (emphasis added)

Or psychologist Jillian Enright:

> "Punishing Unwanted Behaviour Just Makes it Worse... when I refer to punishment, I don't mean only physical punishment, I am referring to anything that a child may experience as unpleasant. This can mean time-out, yelling, removal of privileges, grounding, suspension from school, and spanking."[99]

So, according to these child training "experts", "no matter what the behaviour is punishment simply will not work", and it will "just make it worse". Really? It worked perfectly. It certainly didn't make anything worse—it fixed the behaviour basically *instantly and permanently*, as it so often does when used correctly.

Also consider the following study:

> "For example, because fourth-grade children in one elementary school were constantly unruly and disruptive during gym class, their teachers set up a time-out contingency. Any child who behaved in a disruptive way was immediately told to stop playing and to go and sit on the side of the room, where he or she had to remain until all the sand flowed through a large hourglass (which took about 3 minutes). Children who repeatedly

---

[98] Lemon Drops Kids Therapy. *Why Punishing Our Children Doesn't Work.* Lemon Drops Kids Therapy, www.lemondropskidstherapy.com.au/blog/why-punishing-our-children-doesnt-work.

[99] Neurodiversified. *Punishing Unwanted Behaviour Just Makes it Worse.* Medium, https://medium.com/neurodiversified/how-punishing-unwanted-behaviour-just-makes-it-worse-baf22793d07b.

CHAPTER 5 – Is punishment effective when training people?

misbehaved also lost free play time and other desirable activities (a response-cost contingency). This omission procedure was very effective, and disruptive behavior during gym class soon dropped by 95% (White and Bailey, 1990)."[100]

And yet, the "positive" crowd say "punishment doesn't work and only makes things worse."

Figure 6 – The failure of "positive" training

How can "qualified" people make these claims? Ignorance? Deceit? Ideology? All of the above? But unfortunately, many parents, dog trainers, educators, school principals, and legislators listen to them. And today's dogs, owners, youth, students, teachers, and parents are suffering because of it.

Competition training

A technique I often used with dance students—particularly DanceSport competition couples who were training to win ballroom dancing competitions—was simply to introduce a consequence (punishment) for failing to achieve some aspect of training.

---

[100] Mazur, James E. *Learning and Behavior, 6th Edition.* New Jersey. Pearson Prentice Hall, 2006.

PART 3 – Answering the case against punishment

For example, we might be working on Quickstep, and with all the jumping around the student's arms are constantly dropping too low. So, I'll tell them, "This time, if your arms drop down even once, you have to begin again from the start. And we will keep doing that until they stay up the entire dance."

Figure 7 - The author and dance partner (2011)

CHAPTER 5 – Is punishment effective when training people?

This had to be achievable, so for lower-level students it might be, "your arms must stay up for this *part* of the dance." If their arms dropped down, I would sing out "start again", and they would have to stop, go back to the beginning, and repeat this until they achieved what I was asking. And this quickly becomes very tiring.

It was remarkable how their focus immediately sharpened, their level of effort rose, and their performance drastically improved. And also how happy they were (and therefore how rewarding it was) when they achieved it! Steven Lindsay notes:

> "The withdrawal of aversive stimulation evokes *opposing pleasurable emotional reactions*. When an aversive stimulus is terminated, the opposing *pleasurable recoil* provides a source of covert reinforcement…"[101] (emphasis added)

Of course, I also made careful and deliberate use of reward and praise, etc., during training—it is only the reward-only ideologues who hold the extreme position of, "I will only use this one training tool, everything else is evil". There is no such thing as a pure-punishment trainer. There are trainers who make use of all the tools at their disposal (often referred to as "balanced" trainers), and then there are the extremists who use only one—positive reinforcement (or claim they do).

Another similar approach I often use is what I call "the hat trick rule". For those of you who don't know the sport of cricket, a hat trick is when the bowler gets three batsmen out with three consecutive balls. So, for some aspect of the dance, or sometimes even the whole dance, I would say, "you now have to dance *three complete repetitions* of this exercise correctly (for whichever aspect we were working on), or you have to start all over again from zero, and do another three until you get it right *three times in a row*."

Once again, this punishment marvellously focused their attention and efforts, particularly as they were approaching the end of the 3rd repetition, which is when the level of dancing usually falls off as fatigue sets in and concentration falters. They didn't want to have to go back to the beginning and do another 3! I used this technique quite a bit with a young couple who won a number of Australian Championship titles, amongst various others.

---

[101] Lindsay, Steven R. *Handbook of Applied Dog Behavior and Training, Volume 1: Adaptation and Learning*. Ames, IA: Blackwell Science, 2000.

PART 3 – Answering the case against punishment

(I originally developed this "hat trick rule" many years ago when training dogs and horses, and then adopted it to humans—I believe 3 to be a significant number in training. Firstly, like in the above example, it is less likely to be random chance if the student or animal gets something right three times in a row. In other scenarios, it is the number where *confidence in a pattern* starts to occur. Often, if we just do something once, the animal can simply see it as a *random* occurrence, and hence ignores it or can be unsure or confused. The second time provides *confirmation*, and the animal can start to see a *pattern*. The third time acts as further confirmation and gives *confidence*.)

> 3 is a significant number in training.
>
> Once can appear random, and can be confusing. Twice a pattern is emerging. Three times adds further confirmation and creates confidence.

So, once again in these examples we see punishment working highly effectively. Did it make things worse? No. Did it cause terrible "fallout"? No. It was highly effective. These psychologists simply do not know what they are talking about, and many children (and adults) are suffering because of their ignorance and misguided ideology.

Self-injurious behaviour

Remember psychologist Jillian Enright's statement:

> "Punishing Unwanted Behaviour Just Makes it Worse... when I refer to punishment, I don't mean only physical punishment, I am referring to anything that a child may experience as unpleasant. This can mean time-out, yelling, removal of privileges, grounding, suspension from school, and spanking."[102]

We've already seen how utterly false this is. Now consider the following example, described in Mazur's *"Learning and Behavior"*. It involves a 9-year-old girl named Sharon who would punch herself in the face about 200 times per hour if not physically restrained.

As you read the example, take note of a few points—first, her therapists tried positive reinforcement strategies, and they failed (as they so often do). Secondly, note how quickly and completely the punishment strategies succeeded, and put a permanent end to the horrible behaviour. Thirdly, notice the greatly improved

---

[102] Neurodiversified. *Punishing Unwanted Behaviour Just Makes it Worse.* Medium, https://medium.com/neurodiversified/how-punishing-unwanted-behaviour-just-makes-it-worse-baf22793d07b.

freedom for Sharon, now freed from constant restraint (management). And fourthly, notice that contrary to Enright's claims, the punishment not only worked brilliantly but it also had other positive side-effects:

"…some children who are retarded, autistic or schizophrenic engage in self-injurious behaviors such as repeatedly slapping themselves in the face, biting deep into their skin, or banging their heads against any solid object. Because of the risk of severe injury, these children *are sometimes kept in physical restraints around the clock*, except when a therapist is in the immediate vicinity. Prochaska, Smith, Marzilli, Colby, and Donovan (1974) described the treatment of one 9-year-old girl named Sharon who was profoundly retarded. She would hit her nose and chin with her fist at a rate of about 200 blows per hour if she was not restrained. *Behavior therapists first tried to decrease her head banging with negative punishment and* **with positive reinforcement for other behaviors**, *but with Sharon,* **these procedures were ineffective**. They then began to use a shock to Sharon's leg as a punisher for head banging, and her rate of head banging *decreased dramatically*. Unfortunately, there was little generalization of this learning to Sharon's home or school environments. Evidently, Sharon had learned that head banging was punished only in the clinic, with the electrodes on her leg and with the therapist watching her.

"To reduce Sharon's head banging in her normal environments, the therapists made use of a remote-controlled unit to deliver the shocks, and her behavior was continuously monitored at school and at home. Under these new conditions, *Sharon's head banging* **dropped to zero within a week**. The disappearance of head banging generalized to times when the shock generator was removed, and eventually the punishment procedures were terminated with *no return of the behavior*. There were also **improvements in other behaviors**: Sharon's *crying spells (previously frequent) ceased*, so she could *now go with her parents to public places such as shopping malls and restaurants*…

"…the pain of several dozen half-second electric shocks seems small when compared to the alternative: a life of self-injury, physical restraint, and inability to play with peers or go out in public."[103] (emphasis added)

---

[103] Mazur, James E. *Learning and Behavior, 6th Edition.* New Jersey. Pearson Prentice Hall, 2006.

PART 3 – Answering the case against punishment

> The reward-only ideologues deny effective treatment to many, many people, and thereby condemn them to lives of misery.

So again, we see that the punishment worked extremely quickly and effectively, and that it had only positive side-effects, whereas the positive reinforcement strategies (such as rewarding alternative behaviours) failed. A great outcome all-round for Sharon *which the reward-only ideologues would have denied her, and which they continue to deny many others like her*.

Speeding fines

Speeding fines are a great example of so many aspects of the use of punishment in dog and human psychology. There must be consequences for breaking the rules. If we tried the *"reward good behaviour and ignore bad behaviour"* insanity of so many puppy schools and "positive" trainers (or human psychologists), imagine what the roads would look like. Fines very effectively improve behaviour on the roads, as do other consequences in so many areas of life.

> There must be consequences for breaking the rules.

What if we removed fines or other consequences for drunk driving? And perhaps rewarded an alternative behaviour? After all, that's the strategy of the "positive" ideologues (as failed with Sharon). So, maybe the police, instead of penalising drunk drivers, pull over random people and give them a reward if they *aren't* drunk! "Oh, great, you're sober, here's $100." But there are no penalties if you *are* drunk, just, "oh, you're drunk, you don't get your $100, off you go."

Does anyone seriously think that this would stop people driving drunk? No, of course not. (And for those of you who say, "well, it doesn't stop it now, people still drive drunk," imagine how much worse it would be if all the penalties were removed.) "Positive" strategies often *just don't work*.

There must be consequences for breaking the rules, and for undesirable behaviour.

Differences between people and dogs

Before I leave this subject of whether punishment is effective with people (it is), I want to touch on two major differences between training dogs and people, that are relevant to us as dog (or animal) trainers.

Firstly, dogs don't have language the way we do. This is despite Dr. Dunbar's insistence on teaching dogs ESL (English as a Second Language), or Zak George's

CHAPTER 5 – Is punishment effective when training people?

advice to use "proper grammar and teach it your language like you would a young child".

And secondly, dogs don't understand abstract concepts (despite Zak George's ill-informed statement that dogs are *"predisposed to learn very advanced concepts from human beings"*).

Because dogs don't have these type of language skills, we can't *explain* things to them. And this is a large part of why timing is so important in dog training. Timing *can* be important in training humans also, particularly in teaching complex motor skills like when I'm training competitive ballroom dancers, or in certain aspects when I'm teaching people how to train dogs. But at other times it's not very important at all, for example in speeding fines. We can receive a fine in the mail three weeks after the fact, understand the connection, and it will influence our behaviour. Or with kids we can say, "I just found out you did this three days ago, you're grounded", and this will influence their behaviour.

With dogs, the reward or punishment has to be basically instant, during the act (there are situations in which we *can* push the limits on this, provided we understand how to go about it, but that is more advanced).

The second difference between dogs and humans, that dogs don't understand *abstract* concepts, means that the consequences have to be pretty simple and direct, just like the way dogs work with each other. The mother bites the pup. There is nothing abstract about the punishment. She doesn't get in a huff with the pup, she doesn't stop talking to it for a day, she doesn't burst out sobbing and call it a jerk and walk off in a huff like Dunbar, she just deals with it, it's done, move on.

> The mother bites the pup. There is nothing abstract about the punishment.
>
> She doesn't get in a huff with the pup, she doesn't stop talking to it for a day, she doesn't burst out sobbing and call it a jerk and walk off in a huff, she just deals with it, it's done, move on.

The "positive" dog trainers who like to recommend things like "time-outs" are just plain wrong (like Dunbar turning his back on his dog and laughing). With dogs in the *real world* (not rats in the laboratory) these things will hardly ever help; firstly, because hardly any trainers understand how to use a time-out properly; and secondly, even when used properly they are a very slow, laborious way of training, if they work at all. The way the vast majority of trainers attempt to use "time-outs" means that the dog doesn't understand that this is connected with something it shouldn't have done. Say you put the dog outside as a "time-out" punishment, or into the dog's play pen. One problem is that you also put the dog outside, or into its play pen, at random other times, *for no reason at all*. So why would the dog *this time* think that it is in relation to some behaviour it was

PART 3 – Answering the case against punishment

engaged in? Because you explain it calmly in English using proper grammar like Zak George's "no, don't do that", or start sobbing and call it a "miserable worm" or a "jerk" or have a conversation with it like Dunbar?

No, in dogs, time-outs are generally only useful as a *management* tool, not a training one. They *manage* the behaviour, in that it cannot occur while the dog is locked away, but they generally do not *teach the dog anything*.

> Time-outs *manage* the behaviour, in that it cannot occur while the dog is locked outside, etc., but they do not *teach the dog anything*.

And, even if they did work, as they can much more easily for humans, it is because of the aversive nature of the time-out. It is a punishment. As Dr. Michael Perone rightly states:

"If time-out is not aversive, it could not possibly function as a punisher."[104]

Just ask any prisoner who is put into solitary confinement if it's aversive. I'm sure most would tell you they would rather a smack.

The only way time-outs will usually appear to work with dogs in the *real world*, is if the manner of putting the dog *into* the time-out was *punitive*. For example, imagine your dog is ripping up your curtains, and you grab it by the scruff of the neck *while it is in the act*, shake it while growling at it "bad dog" in no uncertain terms, drag it outside and close the door. The effectiveness would have nothing to do with being locked outside, it would be entirely due to the punishment that occurred while the dog was *in the act*. It would be just as effective if you left the dog *inside* after telling it off in this manner, as if you put it out. In fact, then the dog would have a chance to try it again, and for the consequence to be repeated, and so it would see the pattern and make the connection (remember the hat trick rule). I'm not necessarily saying this is the best way to approach it—but it would work. My point is that if someone claims that time-outs work for them, there is *something else at play*.

So those are the two major differences between people and dogs that are of importance to us—dogs don't have language, so *timing* is far more important. And they don't understand abstract concepts, and therefore the consequences we apply—either good or bad—need to be simple and direct.

---

[104] Perone, Dr. Michael. *Negative Effects of Positive Reinforcement.* The Behavior Analyst 26, no. 1 (Spring 2003): 1-14.

Conclusion – Is punishment effective when training people?

So, despite the fact that supposedly "qualified" counsellors and psychologists tell as that punishment *never* works or *just makes things worse*, or that there is *always* fallout with *any form* of punishment (and that they have "science" on their side), we have clearly seen that these are all completely false. Just as we found for dogs, punishment correctly applied is also highly effective for people, including students or children. Anyone with any common sense has known this all along. It is only the brainwashed or ideological types who fail to grasp these simple facts.

As Mazur said:

> "The question of whether punishment can be an effective way of controlling behaviour *has been settled*. This chapter has presented numerous studies, from both inside and outside the laboratory, that have demonstrated that punishment can change behaviour and, in many cases, *change it permanently*."[105] (emphasis added)

Whether with people, or dogs, punishment works. So Myth #1, that punishment "almost never really works", has been soundly debunked in both the canine and human arenas.

Let's now look at the "positive" trainers claim that punishment has *terrible side effects* and therefore should never be used.

---

[105] Mazur, James E. *Learning and Behavior, 6th Edition.* New Jersey. Pearson Prentice Hall, 2006.

PART 3 – Answering the case against punishment

6

Does punishment cause undesirable side-effects?

Now that we've seen just how effective punishment is, whether with dogs or with people, and debunked the claims that "it almost never really works" or "just makes it worse", let's now look at the second argument used by "positive" trainers. They argue that correction and punishment may cause undesirable side effects, and therefore we shouldn't use it. In fact, many of them claim punishment will *always* cause problematic side-effects. Consider Susan Garrett, a well-known reward-only trainer:

> "…there is *always* fallout with *any form* of punishment–even something as benign as a time-out."[106] (emphasis added)

There are two questions we need to ask in relation to this. First, does punishment actually cause the claimed side-effects? And secondly, if these exist, should that disqualify punishment from use?

Let me elaborate on these two questions slightly further. There is no doubt that, when used *incorrectly*, punishment can cause some undesirable side-effects. It is also true that reward, when used incorrectly (which is often *when used as directed by these trainers*), can also cause some or all of these things (and more commonly does), as will be explained. Does this mean we should never use reward or

---

[106] Garrett, Susan. *Punishment: Pros and Cons.* Susan Garrett's Dog Training Blog, 2010, https://susangarrettdogagility.com/2010/04/punishment-pros-and-cons/

PART 3 – Answering the case against punishment

punishment because they can cause problems *when misused*? Of course not, that's absurd. And particularly when we realize the fact just mentioned that reward—even often when applied *as the positive trainers suggest*—causes far more of these problems than punishment ever does (see *Chapter Nine – More downsides and side-effects of "positive" training*).

> Avoidance of a proper use of punishment results in a huge number of dogs getting put to death every year for entirely curable (and easily preventable) behavioural problems.

And when we consider that **an *avoidance* of a proper use of punishment results in an untold number of dogs getting put to death every year** for entirely curable (and easily preventable) behavioural problems (and in higher failure rates and less well-trained guide dogs for example), it is clear that sensible people should not be too hasty to throw punishment out the window simply for some misguided utopian ideology.

The question really is, does punishment *inevitably* cause problematic side effects even when used properly? And the answer is a resounding *no*. In fact, punishment *never* causes negative side effects when used properly, and in fact has some very *desirable* side effects. We saw some of them in the example of Sharon in the last chapter. Not only did punishment very quickly, and permanently, cure her of punching herself in the face 200 times per hour, but:

> "There were also improvements in other behaviors: Sharon's crying spells (previously frequent) ceased, so she could now go with her parents to public places such as shopping malls and restaurants..."[107]

And, Sharon no longer needed to be kept constantly restrained (what the "positive" crowd refer to as "management") due to the ineffectiveness of the positive-reinforcement approaches. As Fontes and Shanan point out in their paper, *"Punishment and Its Putative Fallout: A Reappraisal"*:

> "...there appears to be a *lack of strong empirical support* for the notion that these potential problems with punishment are necessarily *ubiquitous, long-lasting, or specific to punishment.*"[108] (emphasis added)

They also comment:

> "The **lack of undesirable side effects** associated with the use of punishment has also been noted in the applied literature (e.g., Brantner &

---

[107] Mazur, James E. *Learning and Behavior, 6th Edition.* New Jersey. Pearson Prentice Hall, 2006.
[108] Fontes, Rafaela M., and Timothy A. Shahan. 2021. *Punishment and its putative fallout: A reappraisal.* Journal of Experimental Analysis of Behavior 115 (1): 185–203.

CHAPTER 6 – Does punishment cause undesirable side-effects?

Doherty, 1983; Harris, 1985; Johnston, 1972; van Oorsouw et al., 2008). Indeed, the use of punishment-based interventions typically has been *related to **increases in positive behavior*** (e.g., Bostow & Bailey, 1969; Firestone, 1976; van Oorsouw et al., 2008; Risley, 1968). For example, Matson and Taras (1989) reviewed ***382 applied studies*** employing different punishment procedures during interventions with individuals with developmental disabilities and concluded that the results reviewed *did not provide evidence supporting the occurrence of undesirable side effects*. Instead, the ***majority (93%) reported positive side effects** during punishment interventions*, such as increases in social behavior and responsiveness to the environment."[109] (emphasis added)

Did you get that? There is a *lack of evidence* for "undesirable side effects", and, in fact, conversely there are instead typically *"increases in positive behaviour"*—to the extent that 93% of 382 *applied* studies "reported *positive side effects* during punishment interventions". Wow! 93%! The *complete opposite* to Susan Garrett and the many other reward-only ideologue's claims that there is *always* fallout with any form of punishment and that it should never be used.

Let's delve into this in some more detail. The "positive training" advocates generally list five main areas in which unwanted side-effects are going to occur:

1. Aggression.
2. Fear.
3. Learned helplessness.
4. Damages the relationship.
5. Prevents a dog being "operant".

> **Did you get that?**
>
> There is a *lack of evidence* for "undesirable side effects associated with the use of punishment".
>
> In fact, conversely, there are instead *"increases in positive behaviour"* – to the extent that *93% of studies "reported positive side effects during punishment interventions"*.

The simple answer is that no, a proper use of punishment does *not* cause *any* of these things, certainly not in the long term. It is a bit like most things—they can cause problems *when misused.* An axe, when wielded by an axe-murderer, can cause problems, but is indispensable for chopping firewood. So can water—you can't live without it, but you can also drown in it. Reward can teach a dog to sit,

---

[109] Fontes, R. M., & Shahan, T. A. (2021). *Punishment and its putative fallout: A reappraisal.* Journal of the Experimental Analysis of Behavior, 115(1), 185–203.

PART 3 – Answering the case against punishment

and it can also teach your dog to attack you or others (and often does). Correction, punishment, and reward are all useful tools in the dog trainer's toolkit, but they can all cause problems *when misused*. In fact, punishment rarely causes any of those things *even if* occasionally misused.

It is obvious from the examples earlier that punishment does not necessarily cause problems. If a pup is racing through the bush not paying attention to where he is going, maybe playing with another dog, and crashes into a tree (punishment), he learns to pay more attention to where he is going. He doesn't become aggressive towards trees, nor towards other dogs, nor does he become frightened of trees, or frightened in general. He will get a fright when it happens, but the effect is not lasting. Nor does he suffer from learned helplessness. Why not? Because he knows the rules, and the rules don't change. He knows that crashing into immovable objects is not a good idea, so he pays more attention. Where is the inevitable "fallout"? There simply isn't any.

Karen Pryor attempts to draw a distinction between "aversives" and "punishment", by basically arguing that as long as we don't *mean* it as punishment, then everything is fine:

> "…while all punishment is aversive, not all aversives are punishment. Life is full of aversive events. It rains. You stub your toe. The computer locks up. The train leaves without you. These things happen to all of us, and to our pets. Even the best caregiver cannot spare an animal or a child from all life's little aversives. What we try to avoid is *deliberate use* of aversives *to train behavior.*"[110] (emphasis added)

So, if we accidently step on our pup's toes when it gets underfoot, that is fine, "it's just one of life's little aversives". But if we do it carefully and deliberately to *teach* the pup not to get underfoot, this is now a problem? Or, if the dog is punished *by the environment* there will be no problems, but if we do it deliberately as part of a carefully developed method, based on a great deal of knowledge and experience, all of sudden there magically appears a huge list of "fallout" that is going to occur? The "positive" crowd has a serious problem of logic there.

Emma Parsons, the *Karen Pryor Certified Trainer*, claims (without any evidence to back up her claim) that, if we use punishment, we risk:

> "…the *disastrous* side-effects that *commonly* occur, including hyper vigilance, irrational fear, heightened irritability, impulsive/explosive behavior, hyperactivity, aggression evoked with minimal provocation,

---

[110] Pryor, Karen. *Reaching the Animal Mind.* New York. Scribner. 2009.

CHAPTER 6 – Does punishment cause undesirable side-effects?

withdrawal and social avoidance, loss of sensitivity to pleasure and pain, and depressed mood."[111] (emphasis added)

That's quite a list. If there really is this terrible "fallout" from aversives (there isn't), then anyone with the slightest bit of rational intelligence can see that it doesn't matter at all whether we mean them or not, the effect would be the same. (In fact, as you will soon learn, there is actually *more* chance of aversives causing harm when we *don't* mean them, because they are more likely then to be *non-contingent*.) Regardless of whether the aversive is just one of "life's little aversives" as Pryor calls them, or one that we apply, either they both cause terrible fallout, or they don't.

Consider a bitch disciplining her pups. The pups don't end up frightened of her, although they are *initially*, nor do they become aggressive. They know that if they keep trying to get a feed after mum says feed time is over, they will get bitten. If they stop when she says so, everything is fine. The rules are simple, clear, and understood, and learnt in little more than one application. The pup will be fearful and confused the first time the bitch disciplines him, but after that—when it understands the cause and effect—the fear quickly disappears. *The pup is in control of the outcome via its behaviour.* The pup doesn't end up with "hyper vigilance, irrational fear, impulsive/explosive behaviour, depression, etc., etc., etc." [112].

> The pup is in control of the outcome via its behaviour.

So, in what circumstances *might* some of these claimed negative side-effects occur?

When might punishment cause problematic side effects?

There are certain circumstances where aggression, fear, or learned helplessness might occur on a more systemic level. However, provided the dog understands what the punishment is connected to, and therefore has full control over whether the punishment occurs, this will not happen. This basically comes down to timing. Like the pup that crashes into a tree, and learns to avoid trees. He understands

---

[111] Parsons, Emma. *Click to Calm – Healing the Aggressive Dog.* Waltham, MA. Sunshine Books. 2005.
[112] As above.

PART 3 – Answering the case against punishment

> Provided the dog understands what the punishment is connected to, and therefore has full control over whether the punishment occurs, problems will not occur.

cause and effect, and so doesn't end up with any side effects or "fallout", other than to learn to avoid crashing into trees. He doesn't spend the rest of his life beset by "hyper vigilance, irrational fear, heightened irritability, impulsive/explosive behavior, hyperactivity, aggression evoked with minimal provocation, withdrawal and social avoidance, loss of sensitivity to pleasure and pain, and depressed mood," as Emma Parsons and others would have us believe.

Probably the best example I can think of is electric fencing on farms. It shows both sides of the equation perfectly. Being on a farm, electric fencing is often part of life. I occasionally touch one accidentally, so do the livestock, so do the dogs. Do the cows run around terrified all day because they got punished? No, they just learn not to touch the fence. As do I! Do I get traumatised by the strong punishment of touching the fence? No. Do I end up full of anxiety and fear? No. I just know to be careful not to touch the fence. Where is the inevitable "fallout" claimed by Susan Garrett, Emma Parsons and nearly every other "positive" ideologue?

However, if a pup or dog touches an electric fence *without understanding what happened*, you can get problems. My father (who isn't a dog trainer) had this happen. He had a two-wire electric fence in which the bottom wire was high enough off the ground that the pup could walk under the fence without touching the wire. However, as the pup grew, it slowly got closer and closer to the wire, until one day out walking around the paddock its tail touched the wire and the pup got a shock much stronger than any e-collar. It ran for home. For the next month or so the pup was very fearful to go into that paddock at all. The problem was that the pup *didn't know what had happened*. For dogs, or livestock, or people that touch an electric fence, and *realize what caused it*, there are no such problems. We, and they, simply avoid touching the fence with no ill-effects.

I have an exercise paddock for my dogs, which has normal farm rabbit-proof fencing around it, with one electric wire on top. The dogs only ever try to climb out once, and they never do it again. They don't get traumatised, they are happy to go out and run around the paddock, they just don't climb the fence again (even if it's switched off). Remember Karen Pryor's own very effective use of an "invisible fence" e-collar for her dog?

> "…a trained dog can be effectively confined and will never receive an actual shock."[113]

---

[113] Pryor, Karen. *Don't Shoot the Dog!: The New Art of Teaching and Training*. Rev. ed. New York: Bantam Books, 1999.

CHAPTER 6 – Does punishment cause undesirable side-effects?

If you come home to find your couch destroyed, and you smack your dog, it will likely become frightened of you in general. It might cringe for no reason when you walk past. Why? Because to the dog your actions are *random*—it doesn't understand what the punishment is for (you have to catch the dog *in the act*). And so, it thinks that you just *randomly* abuse it. Conversely, however, if you punish your dog *in the act* of destroying the couch, these issues will be short-term at most. The dog soon connects the behaviour with the consequence, and it simply stops doing it.

It's no different than us with speeding fines. We know that if we don't speed, we won't get a fine. If we do, there is a good chance we will—we take our chances! We don't have to drive around all day stressed out and worried, so long as we obey the speed limit. *However*, if the police took to pulling us over and giving us *random* fines, "Hello Sir, I don't like your shirt (haircut, colour of your car, I'm just in a bad mood), here's a $500 fine", then we would definitely be stressed out in general driving around, because we know there is *nothing in our power* to avoid the fine.

So, provided the dog *understands the connection* between the consequence and the behaviour, with basically comes down to *timing* (while the dog is in the act), and a few repetitions so that the dog sees the pattern, none of these side effects ever occur. However, if you come home in a bad mood and kick the dog, you will certainly create problems. That is *abuse*, not training, and this is when punishment can cause problematic side-effects. It isn't complicated.

So let's look at these individual claims one at a time—that punishment inevitably causes aggression; fear; learned helplessness; damages the relationship; and stops a dog being "operant".

Aggression

It's often parroted by the ignorant "positive" ideologues that punishment will cause aggression, and/or make it worse. Therefore, they argue that you should never use any form of punishment for anything, or risk making your dog aggressive.

This, like most things we've already heard from the reward-only crowd—and are going to continue hearing throughout this book—is simply not true. Punishment, when used properly, is, in fact, a highly effective *cure* for aggression, whether that is predatory, possessive, food, dominance, or whatever other type of aggression, and even often

> Like most things you are going to hear from the reward-only crowd throughout this book, this is simply not true.

111

PART 3 – Answering the case against punishment

for fear-based aggression (depending on the case, and nuances of its use).

The main time punishment might, in a small percentage of cases, actually cause aggression, is when it is *non-contingent*. In other words, when it isn't punishment at all, but abuse. If you come home and beat your dog up after a bad day at the office, with some dogs you might elicit aggression. Even this is not particularly common with most breeds—some breeds and individual dogs are certainly more prone than others to react aggressively in these cases. It's no different to us. If we felt cornered and the need to defend ourselves, some of us would fight back. Some wouldn't. Conversely, when we are punished *fairly*, for example a speeding fine, sensible people don't feel the urge to fight back. Sensible people realize that consequences are necessary to keep our roads safe. However, if we feel the police or government is *unfairly* punishing or abusing people, then that can certainly change.

When using punishment as a training tool, so long as we are *fair* and it is *contingent* on the dog's behaviour, and therefore the dog has full power to either stop or avoid the punishment happening again in the future, it will not cause aggression.

Azrin states:

> "Aversive shocks are known to produce aggression when the shocks are *not dependant on behaviour* and to suppress behaviour when the shocks are arranged as a dependant punisher...these results show that *aggression is eliminated by direct punishment of the aggression...*"[114] (emphasis added)

So, basically, if you abuse your dog and punish it for no reason ("not dependant on behaviour"), you might, in a small percentage of cases, cause an aggressive response. But, if used correctly, punishment will *cure* aggression. And the *real world* proves Azrin 100% correct.

> 99% of aggression is maintained, or aggravated, by reward and not by punishment.

In all of my experience treating aggressive dogs, I would estimate that 99% of aggression is maintained, or aggravated, by *reward* and not by punishment. For example, a pup growls when someone comes close to its food bowl, and the person backs off and leaves the pup alone. They have *rewarded* the growling, and the pup instantly recognizes this. Perhaps the owner tries again, and gets a bit closer this time. The pup growls a bit louder, and the owner backs off

---

[114] Azrin, N. H., Ulrich, R., Wolfe, M., & Dulaney, S. (1969). *Punishment of shock-induced aggression*. Journal of Experimental Analysis of Behavior, 12(6), 1009–1015.

CHAPTER 6 – Does punishment cause undesirable side-effects?

again. The aggression is escalating, purely through being inadvertently rewarded and shaped. As time goes by, the pup may learn to generalise this behaviour, perhaps when he doesn't want to hop off the couch or give up a toy. This is the main mechanism by which aggression becomes an issue—through *reward*.

However, when treating *pre-existing* aggression problems (many times caused by reward), punishment may, and probably will in serious cases, cause an *initial short-term* escalation of an *already existing* aggression problem, before then fixing it. However, according to Zak George and scientist Meghan Herron, punishment won't work and is only going to make aggression worse:

> "The logic [of punishing aggression]: If your dog knows he will suffer as a result of aggressive behaviour, then he is less likely to behave this way. The trouble is that this isn't how it works at all. In fact, the exact opposite is true: *a survey* in Applied Animal Behaviour Science found that when people used confrontational methods to punish their dogs, the dogs exhibited even more aggressive behaviours."[115] (emphasis added)

George and Herron are wrong. In actual fact, this is *exactly* how it works. If the dog knows that "he will suffer as a result of aggressive behaviour", then he is indeed "less likely to behave this way".

> *"2 sessions and our 2 dogs are so much improved. The fighting has stopped & the humans in the house are much more relaxed and in control. Simple & effective. Thanks so much Tully!"*
>
> *- Fleur Hastings*

As we have already seen from Azrin, from actual, repeatable experiments, not a survey like Herron's:

> "Aversive shocks... suppress behaviour when the shocks are arranged as a *dependant* punisher... *aggression is eliminated by direct punishment of the aggression*..."[116] (emphasis added)

---

[115] George, Zak, and Dina Roth Port. *Zak George's Guide to a Well-Behaved Dog: Proven Solutions to the Most Common Training Problems for All Ages, Breeds, and Mixes*. New York: Ten Speed Press, 2019.

[116] Azrin, N. H., Ulrich, R., Wolfe, M., & Dulaney, S. (1969). *Punishment of shock-induced aggression*. Journal of Experimental Analysis of Behavior, 12(6), 1009–1015.

PART 3 – Answering the case against punishment

So let's see what actually happens in *the real world*. Take a highly food-aggressive, large adult male Labrador I was called upon to help with. If anyone approached him while eating, he would attack. I was the owner's last hope, and they were at that stage seriously considering their vet's advice to put the dog down, after trying various other training avenues without result.

I stepped close enough to elicit the Labrador's aggression, and immediately punished him as he lunged towards me to attack, with a strong smack on the nose with a rolled-up newspaper. This caused the dog to further escalate his aggression, as I fully expected it to. I simply escalated immediately also, and repeated the punishment strongly a number of times, until the dog relinquished the food and backed off. I then said, "good boy," handled his food bowl a bit, then backed off, whereupon he went back to eating.

I moved in again, and the dog growled, although much less confidently than before. I again punished the behaviour strongly, whereupon the dog immediately relinquished the food without escalation. The third time, the dog relinquished the food when I approached and showed no aggression whatsoever. I patted and praised him sincerely. Dog's life saved! All with a few smacks from a rolled-up newspaper. Your grandmother out on the farm could have told you that. But, according to Zak George and other "experts", "this isn't how it works at all."

(With some dogs and some owners, we might also muzzle the dog in a situation like this. This serves both to prevent injury and also to give confidence. If an owner lacks the confidence to follow through with such techniques, then they can indeed make the behaviour worse. We might also have the dog tied to something to prevent it from reaching the trainer. These precautions are more applicable to larger, stronger, more dangerous breeds, or to more timid (insufficiently dominant) owners. Another option is to use an e-collar which removes the physical confrontation side of the equation—there are both pros and cons to each approach, and in general the physical confrontation gets superior results with additional positive side-effects, in cases where it can be effectively and safely utilised.)

Elsewhere I give the example of a highly possessive-aggressive Cocker Spaniel, and the procedure and results where the same, as they always are. I could give you countless examples of this type of thing.

So, I guess, like Zak George, or Meghan Herron in her "survey" (I'll debunk that study fully in *Chapter Ten – The Scientific Basis—Fact and Fallacy*), you could technically claim that the punishment "made the aggression worse". But only *if* you were being deliberately deceptive, or were simply ignorant of how training works in these circumstances. A few seconds after the punishment causes an escalation in the pre-existing aggression, it stops it. So Dr. Herron with her "survey" is simply wrong.

CHAPTER 6 – Does punishment cause undesirable side-effects?

And, if people corrected such behaviour early in the piece, such problems would have been nipped in the bud long before they escalated to these levels. By dealing with the behaviour appropriately and effectively *early*, less punishment of lower intensity is therefore required. Instead, due to the influence of the reward-only ideologues, owners are afraid to use effective discipline *early*, and so the aggression is rewarded and escalates—the problem then becomes much worse, and greater punishment is required to remedy it. This is one of the problems I have with the popular LIMA (Least Intrusive Minimally Aversive) training framework—in practice it tends to result in punishment being used as a last resort (or not at all), when in many cases it is far better being used as a first resort (see *Chapter Thirteen – LIMA, LIFE and PROTECT*).

> Due to the influence of the reward-only ideologues, owners are afraid to use effective discipline and so the aggression is rewarded and escalates.

On leash aggression is a very common problem I am called upon to deal with all the time. This varies from mild reactivity all the way to extreme aggression. This is often after the owner and dog in question have been kicked out of the reward-based obedience clubs, or have tried the "positive" trainers without success. I have never failed in helping any owner cure any level of this in a week or so (and it would be much faster if I was doing the training myself).

Of course, the reward-only ideologues are jumping up and down now, and screaming about (amongst other things) how this is only going to *suppress* the behaviour, and the aggression will simply come out in some *unpredictable* and *worse* way at some later time. This is absolutely false (and they have zero evidence for their claim). What actually happens is that once we have rapidly removed the aggressive behaviour with good training, the dog can now be taken out *on a lot more walks around a lot more dogs* (that were previously avoided), and so is now getting much more desensitization and socialisation training, and is getting into good habits of being around other dogs without being aggressive. Every walk from then on is practice being around other dogs and *not* being aggressive, which progressively becomes a new habitual way of being—the dog's underlying feelings disappear over time and are largely replaced. The calmness and lack of aggression is intrinsically rewarded through the avoidance of the punishment, every time the dog uses self-control and doesn't react aggressively.

On the other hand, what does only get superficial results (at best) are the positive-reinforcement approaches. The aggressive behaviour itself has not been directly addressed (it has simply been avoided and side-stepped), and *these* are the cases where it will simply surface again at other times. The reality (as it usually is) is the opposite to what the positive trainers would have us believe.

PART 3 – Answering the case against punishment

And even if the positive methods worked, their impracticality nearly always rules out their use. Even Pryor notes this impracticality (and she is talking about highly skilled—in the *clicker* sense—trainers here, not your average pet owner):

> "It sometimes takes a long time to clicker-train a dog, step by step, to be calm and confident around strange dogs on the street instead of attacking them on sight; but it can be done."[117]

In an article titled "Overcoming Your Dog's Leash Reactivity", Alex Sessa (CPDT-KA, CDBC, VSPDT) likewise notes:

> "Leash reactivity is one of the most common behavioral challenges that pet owners face…"[118]

But then notes that in regard to the long, complicated, extremely difficult and entirely impractical positive reinforcement methods which she herself outlines:

> "This type of behavior modification can be tedious, and requires great attention to detail in order to be truly effective… As you go through this training process, remember that there will be successes, backslides, and that truly changing behavior is not fast or easy."[119]

Actually, truly changing behaviour can be both *fast* and *easy* (when you know what you're doing).

Emma Parsons, the *Karen Pryor Certified Trainer* and author of *"Click to Calm – Healing the Aggressive Dog"*, has written an entire book on how to attempt to "heal" the aggressive dog with positive methods (which, of course, includes the obligatory chapter on "Why punishment doesn't work"). An integral part of this process is having a *lot* of friends with a *lot* of dogs that you can practice on *regularly* for a *long* time with your aggressive dog:

> "Ben would not have been able to practice interacting with stable, nonreactive dogs if hadn't been for all of you who allowed him to do so with your dogs: Israel Meir, Geoff Stern and RoseAnne Mandell, Anya Wittenberg, Jean Berman, Rick Wood, Ellen Brinker, Margaret Stubbs, Barbara Beckedorff, Susan Conant, Roo Grubis, Louis Massa, Cathy Shea, Virginia Parsons, Joel Wolff, and Penny Schultz."[120]

---

[117] Pryor, Karen. *Reaching the Animal Mind.* New York. Scribner. 2009.
[118] Sessa, Alex. *Overcoming Your Dog's Leash Reactivity.* AggressiveDog.com, July 26, 2021.
[119] As above.
[120] Parons, Emma. *Click to Calm – Healing the Aggressive Dog.* Waltham, MA. Sunshine Books. 2005.

CHAPTER 6 – Does punishment cause undesirable side-effects?

How many friends do you have with "stable, non-reactive dogs" willing to put this amount of time and effort into helping you train your aggressive dog? My bet is nowhere near enough. This large number of willing and available assistants that are necessary reminds me of Kelly Dunbar's "Open Paw's" ridiculous recommendations that:

> "…puppies be *handled, groomed, and trained* by at least five unfamiliar people *each day*, that is, *150 different people* during the first month in a new home." [121] (emphasis added)

And Ian Dunbar's:

> "I always recommend that puppies are *handled and trained* by *at least a hundred different people* in the breeding kennel prior to eight weeks, and then *another hundred different people* during their first month at home." [122] (emphasis added)

Um, really? Are they serious? *Handled, groomed* and *trained* by 150 different people in the first month?! And 100 different people prior to eight weeks?! These people live in a different world to the one I do, that much is certain. And, in relation to fixing aggression, the Dunbar's note:

> Um, really? Are they serious? 150 different people in the first month?! And 100 different people prior to eight weeks?!

> "Aggression is not something we can train away by ourselves. It's a problem that needs a LOT of reinforcement. You're going to need to throw a "socialization party". Invite as many people as you can that already know the dog and aren't afraid of him. Have them all come over and wait outside." [123]

This "socialization party" evidently involves a lot of people (who aren't afraid of your big savage dog who wants to attack them on sight), because the next section is titled:

> "How to get 100 social interactions in a day" [124]

And then he gives this advice:

---

[121] Dunbar, Ian, Dr. *Barking Up The Right Tree - The science and practice of positive dog training*. Novato, CA: New World Library, 2023.
[122] As above.
[123] The Dunbars. *End Your Dog's Aggression. PLUS Easy Socialization*. Email to Tully Williams, May 3, 2021.
[124] As above.

PART 3 – Answering the case against punishment

> "Have your guests take turns exiting then coming back in over and over. As the dog gets better at this game have your guests give him more complex commands with even better rewards. Next, take the game out to the sidewalk. Your friends will walk around the block and spread twenty-five to fifty yards apart. Then you walk in the opposite direction, so you're meeting known people on the street. Again, have your dog sit and greet your friends…and give him even more treats. By the end of the day, your dog will love the idea of greeting guests at home and outside." [125]

If you're anything like me, you're sitting there reading all this with one eyebrow raised in stunned disbelief, and then shaking your head, rolling your eyes, and laughing at the absolute insanity of what the Dunbar's recommend you do with your new pup or aggressive dog. If you have access to hundreds of different people willing to "handle, groom and train" your pup like this, and the time to fit this into your daily schedule, then you probably have enough time, and enough friends with dogs (who also have nothing better to do with their time), to attempt Emma Parson's suggestions also.

So, even if these suggestions might work in some abstract, theoretical way, once again the level of skill, knowledge, time and practicality means that, for the vast majority of people, it simply isn't going to happen. In fact, I seriously doubt that there is a *single person on earth* who could realistically follow *any* of the advice we have just heard. Dunbar elsewhere notes:

> There isn't a *single person on earth* who could realistically follow any of the advice we have just heard.

> "When you look at my criteria for training, *it's got to be easy*. Or else the owners can't do it… *It's got to be quick*, or else they probably won't."

Unfortunately, his own training methods (and those of all "positive" trainers) are anything *but* quick or easy. In fact, very often they aren't even *effective*—they simply don't work *even if* they could be followed. On the other hand, with a punishment-based approach, none of that is even faintly necessary. And it actually is quick and easy, and it is truly effective. And it is all these things with only positive side-effects.

Another good example of the positive side effects of punishment (and its effectiveness) is curing predatory aggression via punishment. A client had a rehomed dog that would have killed their cat given the chance. With a negative reinforcement/punishment-based approach, the behaviour was quickly and easily cured in one session. Then, because the cat and the dog could now be allowed in the same space safely together, and the dog was now practicing being around the

---

[125] The Dunbars. *End Your Dog's Aggression. PLUS Easy Socialization.* Email to Tully Williams, May 3, 2021.

CHAPTER 6 – Does punishment cause undesirable side-effects?

cat without killing it, over the next few weeks the dog and cat became best friends. The owner sent me a video of the two of them rolling around on the floor playing and having a great time. However, according to the "positive" crowd, all we had done is *suppress* the behaviour, and had failed to deal with ("heal") the supposed underlying cause. And so, according to them, the aggression was simply going to come out in an unpredictable and worse way at some other time. No, it isn't. It's permanently and completely cured, with only positive side-effects.

I have a young Turkish Kangal livestock guardian dog here on the farm at the moment, named Nikita. She has a *very* strong predatory drive as many of these dogs do, and, if given the chance, would have killed the chooks or the sheep or the goats initially. But with some fairly simple punishment-based training, Nikki and the livestock all now live happily together. The behaviour hasn't just been "suppressed", and it won't come out in some worse way later on. It is permanently *cured*. And this has been achieved with only positive side-effects (she and the animals are now friends and happily co-exist).

I have often helped people with male chocolate labrador pups that are highly food aggressive at 10 or 12 weeks of age. I don't know if that is a common trait in male chocolate labs (I probably don't have enough numbers of chocolate male labradors to be statistically significant), but I have come across quite a few of them. All the pet owners I meet make the same mistake—when the pup growls at them when they go near it while eating, the owners back off. And so they have simply rewarded the behaviour, and it will get worse and worse. If, instead, the very first time the pup growled at them, they gave it a good solid smack on the nose/head area (just like its mother would bite it), that would be that—problem solved. When I explain that, they usually reply, "Oh, I didn't think we were supposed to do that nowadays." Unfortunately, according to the people who don't know what they're talking about, you're not—you're supposed to "ignore bad behaviour".

Meghan Herron also makes the following unsubstantiated claim:

> "*In almost all cases*, dogs are aggressive because they are afraid and feel threatened in some way."[126] (emphasis added)

This is more absolute rubbish, and there is zero evidence for such a claim. There are many reasons dogs are aggressive, and after thirty years of working with them I can tell you that fear is one of the less common.

Is predatory aggression (ie. killing sheep or wildlife, or the chooks, or the cat) based on being "afraid and feeling threatened"? Hardly. Is food aggression, when

---

[126] George, Z., and Roth Port, D. *Zak George's Guide to a Well-Behaved Dog: Proven Solutions to the Most Common Training Problems for All Ages, Breeds, and Mixes*. New York. Ten Speed Press, 2019.

one of your dogs finishes its food first and goes and attacks the other one to get its food, because it is "afraid and feels threatened"? Not at all. Is dominance aggression where your Ridgeback wants to kill most dogs it comes across, due to it feeling afraid and threatened? No. There really isn't some terrified, scared little puppy cringing and crying away inside your five-year-old Bull Mastiff, that, if we could just somehow "heal" the poor darling's underlying fear, then all his aggression would go away.

Figure 8 – If only we could "heal" the poor darling's underlying fear

Herron is wrong once again. But many, like Zak George and Victoria Stilwell, parrot this type of pseudo-science because it fits their ideology. Stilwell promotes the same myth when she says:

> "...suppose my dog lunges aggressively at another dog she sees in the distance because she is fearful."[127]

If your dog is lunging aggressively at a dog it sees in the distance, then fear is certainly not the overriding emotion. On the other hand, if your dog is hesitant to go towards the dog in the distance, or is actively pulling *away* from it, then fear is likely the cause. Or, if your dog is cornered and acts aggressively, and this is the only time it does so, then fear is probably the cause. But if your dog is actively pulling towards, and aggressively trying to get at, a dog it sees in the distance, then fear is definitely not a significant factor.

---

[127] Stilwell, V. *Train your dog positively*. New York. Ten Speed Press, 2013.

CHAPTER 6 – Does punishment cause undesirable side-effects?

So, does punishment cause aggression? No, it does not. Not unless it is used in a *non-contingent* manner as Azrin explained—in which case it isn't punishment at all, but abuse. And even then, this isn't particularly common. A proper use of punishment certainly won't cause aggression. It will, though, very effectively cure it.

(On the other hand, in *Chapter Nine – More downsides and side-effects of "positive" training—Aggression*, you will learn even more about how positive reinforcement approaches cause aggression.)

Fear and anxiety

It is often claimed that an inevitable side-effect of punishment is that it will make a dog (or child) fearful, or anxious. When we talk about fear (or anxiety), we need to distinguish between two types—the first is the short-term, short-lived variety, while the second is a more endemic, persistent fear. There certainly can be an element of short-term fear involved in some forms of punishment. That does not necessarily disqualify them from use. Like most things, we weigh up the pros and the cons. And when used properly, the pros often far outweigh the cons.

The first time a bitch punishes a pup for continuing to feed from her after she has growled at it to stop, it will certainly be fearful. And much more fearful than would occur with most punishments we would ever need to use in dog training. The pup might even yelp and scream and run and hide in the corner. By the time the next feed time comes around, the pup will be a bit hesitant about getting under mum for a feed. But as it feeds and nothing untoward happens, it will relax. When the bitch walks away this time, the pup may give chase as before. But when she growls again, the memory will come flooding back, and the pup will desist in a hurry, expecting to be bitten. When he doesn't get bitten, the pup's learning is almost complete. "Aha! If I stop when mum growls at me, she doesn't bite me!" Next time he won't even bother chasing her when feed time is over, and she won't need to growl at him. *And very soon all the fear is gone.*

The pup is back to its normal, confident self, having learnt that the bite *wasn't random*. Remember the "hat trick rule"? **The first time can be confusing. The second time a pattern is emerging. The third time is confirmation, and confidence returns.** Now the pup knows that if it simply *follows the rules* there won't be a repeat. *And* he has now learnt what a growl means from another dog—"stop what you are doing or there will be consequences", and this will stand him in good stead throughout life. And there is no problematic "fallout".

> Remember the "hat trick rule"? The first time can be confusing. The second time a pattern is emerging. The third time is further confirmation and confidence returns.

PART 3 – Answering the case against punishment

As we heard earlier, Zak George attempts to convince people into thinking that this is *not* how bitches actually train their pups in the real world:

> "… while it is true that mothers may resort to physical corrections like that, as far as I know it's a myth that this is a primary way that they communicate with their puppies. If a mother dog goes beyond those quick subtle corrections, for example if she's causing harm or actually biting her puppies, this is considered a major red flag amongst breeders and those who raise puppies. Typically it means there's some external issue like something medical, or possibly something in the mother dogs environment that's causing her to feel uncomfortable."[128]

If you buy into this misrepresentation of the facts of nature, and if you therefore prevent your bitch from training her pups properly, your pup *will* learn the same lesson at a later date from another dog. *Only it may end up much worse.*

In fact, if we took the pup away from its mother after the first instance of this maternal punishing (perhaps because we believe Zak George, that this is a sign of a medical issue or something abnormal on the bitch's part), then the pup's learning would be *interrupted* and therefore *incomplete*. And so, the fear would remain longer and could perhaps become a problem (depending largely on the pup's temperament—a robust pup would be less impacted; a timid pup more so). *The repetition of the process is necessary for the pup to see the pattern*—and once it does, it now knows the rules, and has learnt its own power over the rules. And so all its confidence comes back, with no possibility of long-term problems. **It is even more important for the timid pup that this learning process isn't interrupted, and is** *allowed to be repeated*, **than it is for the robust pup.**

> If you prevent any interactions where your pup gets told off by an older dog and bitten, you are setting your pup up for major fights later in life.

If someone—following George's logic—takes the pup away from the mother after she bites the pup the first time, then the pup (particularly a timid pup) may possibly have ongoing issues from this. And so, the confirmation bias kicks in: the reward-only ideologue now claims, "see, look how many problems punishment causes—the poor pup is traumatised." In fact, if they had let it remain with its mother, and let the process be repeated *as nature intended*, the pup would rapidly have overcome this initial fear, and would now have both learnt some dog etiquette, and also developed some resilience. It would have developed confidence in its own

---

[128] George, Zak, and Dina Roth Port. *Dog Training Revolution: The Complete Guide to Raising the Perfect Pet with Love.* New York: Ten Speed Press, 2016.

CHAPTER 6 – Does punishment cause undesirable side-effects?

ability to control its environment. The pup's confidence would have *increased*, not decreased, as a result of this process of punishment. Without this process of punishment occurring, this never has a chance to happen, as is seen in so many molly-coddled, positive reinforcement trained dogs.

Likewise, if you prevent any interactions where your pup gets told off by an older dog and bitten, you are setting your pup up for major fights later in life. It will assume (never having learnt otherwise) that it can do whatever it likes to another dog without consequences. And then, sooner or later as an adult, when it rushes up to another adult dog oblivious of dog etiquette, it will quite likely start a major fight. All because the well-meaning, but ignorant, breeder or owner, didn't allow the pup to go through this fear and training, and learn that it should stop when another dog growls at it.

> The pup's confidence will increase, not decrease, as a result of this process of punishment. Without this process of punishment occurring, this never has a chance to happen.

If the bitch punishes the pup, the pup is fearful initially, but the pup quickly gets over the fear but remembers the lesson for life. It has also learnt that being bitten, though scary, is not the end of the world. This also works to inoculate it against future fearful events, and to build resilience. With this, the real scientific literature agrees. Lindsay says:

> The pup learns that being bitten, though scary, is the not the end of the world. And this also works to *inoculate* it against future fearful events, and to build *resilience*.

> "In fact, it appears as though the fear originally elicited by the avoidance signal gradually undergoes extinction but not the cue's ability to control avoidance responding."[129]

And again:

> "…animals trained to a high proficiency (asymptote) on avoidance contingencies do not typically show signs of fear prior to the emission of the required avoidance response. Instead they are often *very happy and relaxed workers*. They appear to know what is expected of them and do it without hesitation and ostensibly without fear."[130] (emphasis added)

And Mazur:

---

[129] Lindsay, Steven R. *Handbook of Applied Dog Behavior and Training, Volume 1: Adaptation and Learning*. Ames, IA: Blackwell Science, 2000.
[130] As above.

PART 3 – Answering the case against punishment

> "Well-practiced subjects continue to make avoidance responses while showing *no measurable signs of fear*, and extinction of avoidance responses is very slow."[131] (emphasis added)

Karen Pryor describes her evidently very effectively punishment-trained dog, Gus (before she became a reward-only ideologue):

> "The obedience classes consisted of traditional training using choke-chain collars… Gus did learn the meaning of those commands. I almost never had to jerk his collar in punishment. Gus was a gorgeous dog, and *he worked briskly, with his head up, his eyes sparkling, and his stub of a tail wagging continuously*. Judges loved that."[132] (emphasis added)

Head up, eyes sparkling, tail wagging… but remember the *Karen Pryor Academy* trainer Emma Parson's claims about any form of punishment:

> "…the *disastrous* side-effects that *commonly* occur, including hyper vigilance, irrational fear, heightened irritability, impulsive/explosive behavior, hyperactivity, aggression evoked with minimal provocation, withdrawal and social avoidance, loss of sensitivity to pleasure and pain, and depressed mood."[133] (emphasis added)

I guess Gus (or any of the thousands of dogs I have helped people train or trained myself) must not have read Emma's list. As Lindsay noted:

> "…instead they [dogs trained with the use of punishment] are often *very happy and relaxed workers*. They appear to know what is expected of them and do it without hesitation and ostensibly without fear."[134]

> The pup or dog has full control over whether or not it gets bitten or corrected, depending on its own behaviour.

So, basically, if the dog understands the cause and effect (the rules), then it has no reason to be afraid. It knows that if it chases mum after feed time, it is highly likely to get bitten. If it doesn't, it won't. If it walks nicely on the lead, there will be no correction. *The pup or dog has full control over whether or not it gets bitten or corrected, depending on its own behaviour.*

---

[131] Mazur, James E. *Learning and Behavior, 6th Edition.* New Jersey. Pearson Prentice Hall, 2006.
[132] Pryor, Karen. *Reaching the Animal Mind.* New York. Scribner. 2009.
[133] Parsons, Emma. *Click to Calm – Healing the Aggressive Dog.* Waltham, MA. Sunshine Books. 2005.
[134] Lindsay, Steven R. *Handbook of Applied Dog Behavior and Training, Volume 1: Adaptation and Learning.* Ames, IA: Blackwell Science, 2000.

CHAPTER 6 – Does punishment cause undesirable side-effects?

And even positive reinforcement clicker training can be fearful to some dogs initially, as Karen Pryor notes in an article titled "Overcoming fear of the clicker":

> "Some dogs do find the clicker sound startling, and therefore alarming, at first. Once they understand a) that click means treats and b) that they can make you click, *the fear goes away; it's just a beginner problem.* However, it's hard for the dog to make that connection if it's hiding under the bed."[135] (emphasis added)

Does that mean no one should ever use clicker training? Of course not. So why wouldn't the same apply to punishment (like Pat Miller's cap gun) that may, in some situations, initially induce some short-term fear, which rapidly disappears (so long as we continue the training) when the dog knows a) that the punishment means don't bite, and b) that they can avoid the punishment if they don't bite? After all, it's *just a beginner problem*. And when punishment cures the behaviour quickly and effectively (like Pryor's use of the spray bottle to cure her dog of getting in the trash), then why not?

If using correction or punishment, so long as we are clear, and the dog knows the rules, and therefore knows how to avoid it in the future, no problems or undesirable side-effects will *ever* result. And the correction or punishment will be highly effective. And keep in mind that punishment is often *not* fearful—it can just be uncomfortable, or annoying. A speeding fine is not a fear inducing consequence. I often say that a fly can be annoying, and you want it to go away—it doesn't hurt, and it isn't scary, but annoying things can be used as punishment or negative reinforcement. An e-collar used on a low setting is simply annoying. So, punishment certainly doesn't have to cause fear at all, but sometimes fear is a necessary component.

For example, consider Pat Miller's use of the cap gun that cured Happy of chasing horses. Some short-term fear prevented any more chance of the dog being kicked and injured while chasing horses, and prevented injury to the horse. There are always pros and cons. We have to weigh up the various considerations.

As another example, consider one of my client's dogs. A large, aggressive bull mastiff that the owners had been unable to walk for *years*. It was too strong for them to hold, and it would try to attack every dog it saw. They had taken it to obedience classes (positive reinforcement) and been asked not to return (a common occurrence). They had tried various harnesses; they had tried various treat training methods they had watched on YouTube; and all to no avail (not surprisingly). So the dog simply no longer got walked, and the situation had been that way for years.

---

[135] Karen Pryor, *Overcoming Fear of the Clicker?* Clickertraining.com, accessed May 10, 2024, https://www.clickertraining.com/node/40.

125

PART 3 – Answering the case against punishment

They contacted me after someone recommended me—they had tried other trainers previously without results. And yet, with some quick, effective training, the dog was cured in a week and back out being walked every single day. Was that evil and abusive of the owners or I? No, it was the kindest thing that could be done—and I'm certain that if you could ask the dog it would 100% agree. Avoiding all punishment (as advised by the "positive" crowd) and allowing the dog to languish in the back yard for the rest of its life—*that* is the irresponsible thing to do.

> Avoiding all punishment (as advised by the reward-only crowd) and allowing the dog to languish in the back yard for the rest of its life – *that* is the evil, irresponsible thing to do.
>
> That is what "veterinary behaviourists" like those from AVSAB, and other "positive" trainers, would recommend – what they euphemistically call "management".

Yet, according to Zak George and many of his ilk, if you're looking for a trainer:

"It's vital that you find an expert who does not use "quick fix" tools like choke chains and the other ones I've mentioned throughout this book. Remember, these tools aren't very effective in the long term, and they can make the aggression even worse."[136]

As we have already learnt, this is rubbish—punishment is actually highly effective long-term (as we just heard from Karen Pryor with her choke-chain trained dog Gus); and, it can indeed be a "quick fix" and when done properly nearly always is, including for aggression.

Or, you could try the "positive reinforcement methods" like the ineffective methods Zak George outlines (see *Chapter Eight – Does Reward Work?*). But how do you apply those when you can't even take your dog out the front door? And even if you can, are you willing to put the *years* of effort (as you will see from the mouths of the "positive" trainers themselves as we continue) that are necessary into that, only to find that it *still doesn't work*?

Or, alternatively, you could apply a simple bit of good training, and the problem is solved completely in a week or two. Now, the dog that hasn't been walked *for years* is back *walking every day*. Just like Sharon, the down syndrome girl mentioned earlier, whose quality of life was *drastically* (and quickly) improved by some mild, short-term punishment.

The question you have to answer is whether applying some short-term discomfort is worth the resulting long-term greatly improved quality of life?

---

[136] George, Zak, and Dina Roth Port. *Dog Training Revolution: The Complete Guide to Raising the Perfect Pet with Love*. New York: Ten Speed Press, 2016.

CHAPTER 6 – Does punishment cause undesirable side-effects?

My advice? Avoid any trainers who *don't* use punishment like the plague. Despite Zak George's claim, there is no such thing as an "expert" who relies solely on positive-reinforcement, or who claims to be "force free" or "positive". For many of the most important things you will simply be wasting your time and your money. You will also be *failing your dog* by not using punishment when and where necessary, so that it can now enjoy years of *great freedom.*

> There is no such thing as an "expert" who relies solely on positive-reinforcement, or who claims to be "force free" or "positive".

Back to the subject of fear, consider also that *socialisation* is about fear. It is about *deliberately* exposing pups to things *they fear*, in order to desensitise them to those things. So we deliberately put them in situations that are frightening in order to teach them that there is nothing to be frightened about. Short term fear is used to alleviate long term, more endemic fear. And they also learn that fear itself is not that bad.

Contrast that to the crazy notions in modern child rearing, or even university education, systems. "Trigger warnings" are used to warn students that something they might find upsetting is about to be discussed. The problem is that protecting students from things they don't like will not help them. Just like in socialisation, they are much better off being *deliberately exposed and challenged* by these things, and learning to cope with them, as everyone did in the past. They will have to sooner or later in *the real world* anyway.

> "Avoiding trauma-related triggers is a coping strategy that might reduce short-term anxiety but *in the long run prevents people from returning to living a rich, meaningful existence*. Decades of scientific evidence demonstrate that the "gold standard" intervention for emotional disturbances is to gently, gradually expose people to what they are afraid of while modifying their assumptions about danger in the world, and building up their *sense of agency* and *tolerance of distress* so that they can pursue important goals no matter what they feel.
> "Knowing avoidance is a symptom of the problem, experts pointed to the *ironic effects* of trigger warnings. Several studies show that providing explicit alerts to students that upcoming class content might induce unpleasant reactions offers little to no benefit, and *often backfire (only making the anxious more anxious)*."[137] (emphasis added)

As Mary Tyler Moore said:

> "You can't be brave if you've only had wonderful things happen to you."

---

[137] Kashdan, T. (2024, January 29). *Rethinking Trigger Warnings*. Retrieved from https://toddkashdan.medium.com/rethinking-trigger-warnings-998d9ef9a89

PART 3 – Answering the case against punishment

Herodotus:

> "Adversity has the effect of drawing out strength and qualities of a man that would have lain dormant in its absence."

The same applies to dogs. Often the *lack of* discipline, i.e., the *lack of* any correction or punishment, can contribute to fear. Dogs, like humans, are often frightened of the unknown. A dog or pup with a timid nature, that has never been corrected, and its mother was never allowed to correct it, and which has been protected from the consequences of the *real world*, will often show a lot of fear at even the smallest corrections (like Cindy Benson's poor dog with a "stern look" or "sigh"). Many people assume that "he was attacked by another dog", or "he was abused by his last owner". But, in fact, it is often the case that the dog in question hasn't been exposed to correction or discipline enough, and so remains timid and fearful.

Preventing these occurrences is the *real* abuse.

In fact, with the sound use of correction and punishment, the timid dog will become greatly *less* afraid and timid. The worst thing you can do is to molly-coddle a timid dog, because this rewards their fear—their fear will increase, not decrease. For example, if we let them go and hide from things they are scared of (like with trigger warnings for people), this simply rewards their fear. Or, even worse, if we try to reassure them with, "It's okay, good dog," as we pat them when they are scared (and don't forget Zak George's advice to use proper grammar!). We are simply rewarding their fear.

In fact, punishment used correctly not only doesn't *cause* long-term fear, but it can actually be used to *cure* it and increase confidence rather than decrease it. I use it for that purpose regularly, and have turned timid wrecks into confident animals with absolute trust in me in a short space of time, both dogs and horses. Yes Zak, yet more "quick fixes".

I have a video on my *Real World Dog Training* website (*realworlddogtraining.com.au*), showing me breaking in a young horse named Madness. Horses are a great example when talking about fear, because they are naturally fearful, nervous animals (in fact, many of the techniques I use in working with fearful dogs come from my experience with training horses). In the video, there is a strong element of punishment and negative reinforcement (pressure/release) in training her to tie up, with the fearful noises and objects such as flapping bags and opening umbrellas that I present her with. When she gets

CHAPTER 6 – Does punishment cause undesirable side-effects?

scared and pulls back, the noose around her girth tightens and punishes her strongly. When she jumps forwards *towards the scary object*, the pressure releases, and so rewards her.

Figure 9 - The author on "Madness" (2003)

PART 3 – Answering the case against punishment

This will horrify the "positive" crowd of course, but it clearly refutes so many of their arguments. If you watch the video right through, you will clearly see that she quickly gets over this initial very short-lived fear, and those objects no longer remain so fearful to her. Also, you can see that she trusts me implicitly. She is not frightened of me in the slightest, and is calm and relaxed even when I stand on her back and crack a stock whip, or sit on her bareback and wave my hat in her face. Once again, the "positive" trainers will claim this only "suppresses" behaviour, and that it will come out in some worse way in the future. This is simply not true. When done properly (of course, like everything, it can be done poorly) the punishment/negative reinforcement training works brilliantly—Madness was then a horse I could tie up anywhere while out working on the farm, even to a barbed wire fence, and know that she would not pull away or get frightened by some object blowing in the wind. I had very effectively, and very quickly, cured her fear in that situation (and also, to a large degree, in other situations also).

It is actually the "positive" or "gentle" methods, that try to avoid any direct confrontations with the horse's behaviours by trying to handle them quietly and gently and feeding them treats and playing with them, that are actually the ticking time bomb, and often get inexperienced horse people injured. (And by the way, just like with dogs, most horse problems can be remedied far faster and more easily than is generally realized even by experienced horse people.)

So, firstly, this example refutes the claim that punishment doesn't work, or that it only "makes things worse". It also refutes the claim that punishment causes long term fear and anxiety. Short term, yes. Long term—exactly the opposite. In fact, done properly, it creates a very calm, relaxed, trusting horse. The environment is no longer such a scary place. The horse has learnt that it can *brave out* these scary objects, and soon learns not to fear them at all. So, a method based primarily on punishment *cures fear*.

> A method based largely on punishment *cures fear*.
>
> The "positive" trainers have no idea how this can even be possible. It is so far outside the realm of their limited experience and knowledge as to be incomprehensible to them.

The "positive" trainers have no idea how this can even be possible. It is so far outside the realm of their limited experience and knowledge as to be incomprehensible to them. For example, according to Zazie Todd, another reward-only ideologue and author of various pieces of junk pseudo-science:

"Some behaviour problems are due to fear or stress, but aversive methods do not resolve this (and may even make it worse)."[138]

---

[138] Todd, Zazie. 2018. *Why Don't More People Use Positive Reinforcement to Train Dogs?* Companion Animal Psychology. April 25, 2018. Accessed April 10, 2024. https://www.companionanimalpsychology.com/2018/04/why-dont-more-people-use-positive.html.

CHAPTER 6 – Does punishment cause undesirable side-effects?

Whether with horses or dogs, this is complete nonsense. Here is a testimonial:

> "Fantastic and fast dog training. Our Ridgeback Sam had a phobia of bridges (refused to cross the Mitchell St train bridge, or any others), and was hard to get into the car. Cured in 45 Minutes. Highly Recommended."
>
> - Andrew Wilkinson

Same thing. An aversive punishment/negative reinforcement technique cured the serious, long-standing phobia basically instantly, and permanently. It was actually much faster than the forty-five minutes mentioned—most of the time was involved with talking to the client. The training itself was less than five minutes. Zazie Todd and those like her are simply ignorant and inexperienced, but they produce "science" and/or make claims and write books supposedly proving the opposite to the actual reality.

So, in summary—correctly applied punishment does *not* result in long-term, endemic fear or anxiety. In fact, in general, the effect is the exact opposite. Previously very stressed and anxious dogs rapidly become far calmer, more relaxed, and more confident. Yes, with some instances of punishment there may be some short-term fear, but provided the dog understands the cause of the punishment, and *therefore has complete control over whether it occurs again or not*, the dog can go about its business with complete confidence. Just like us with speeding fines.

A little bit of discipline, some short-term discomfort or even occasional fear, and a lifetime with a well-trained dog that can enjoy enormous freedom. And a dog that now has increased confidence that it can control the uncomfortable things in life. The same applies to kids—I'm glad my parents disciplined me when I needed it. I feel enormously sorry for the epidemic of anxious kids today whose lives are being ruined by the reward-only lunacy, both in the home and in the school.

> A little bit of discipline, some short-term discomfort or even occasional fear, and a lifetime with a well-trained dog that can enjoy enormous freedom.

Learned helplessness

Besides these two unfounded claims, that punishment is going to cause aggression and fear/anxiety, it is also sometimes claimed that if you use punishment you will cause "learned helplessness". Others, like George and Herron, or Stilwell, even

PART 3 – Answering the case against punishment

claim that if punishment works (nearly always after saying that it doesn't), it is *because* you have caused *learned helplessness*. Once again, they are wrong.

Zak George:

> "Sadly, many people have unrealistic expectations, and they expect a dog to, say, stop barking at a person walking in front of the house or to stop pulling on a leash *in just a few training sessions*. When the dogs don't, these people resort to *quick fixes* such as choke, prong or electric collars... People are left to believe that a dog can be trained in thirty days... *they don't really work. Too often, learned helplessness... is interpreted as training success.*"[139] (emphasis added)

Actually, with a good trainer dogs *can* be trained in a just a few training sessions (often in just one), and these methods do really work, as we've already discussed in some detail. Only poor trainers take months or years to teach something that should take minutes, hours, or maybe a week or two. Consider the following from one of my clients:

> "We cannot express enough the gratitude we feel for Tully. Our two dogs were noisy, anxious and disruptive. Essentially real ankle biters. ***After only a few visits*** they are now quiet, calm and respectful. Our household is now more peaceful and relaxed. We no longer fear the sound of the doorbell because of the barking and aggression that occurred. We also no longer fear letting them outside as the constant barking caused problems with the neighbours. Now they are quiet. It was just ***such an incredible change in such a short time***, and we are just so blown away by the difference in our dogs and the mood of the household. Thank you so much Tully." (emphasis added) - Wendy Richards

So, the methods work. And yes, they can definitely be a *very* "quick fix". The anxiety is replaced by calmness. The barking and aggression are cured. The question here is, is this success a result of "learned helplessness" as George and Herron claim? Or did the training cause "learned helplessness" as an unwanted side-effect? The answer to both is *no*.

Firstly, we need to understand exactly what "learned helplessness" is. Martin Seligman coined the term after some experiments he conducted back in the 1960's.

---

[139] George, Zak, and Dina Roth Port. *Zak George's Guide to a Well-Behaved Dog: Proven Solutions to the Most Common Training Problems for All Ages, Breeds, and Mixes*. New York: Ten Speed Press, 2019.

CHAPTER 6 – Does punishment cause undesirable side-effects?

Some dogs were given random electric shocks that the dog could end either by jumping over a barrier, or by pressing a lever. The dogs quickly learned to escape the shock by jumping over the barrier or by pressing the lever (contrary to claims that punishment or aversives can never teach behaviours – *see Chapter Seven – Can punishment create behaviours?*). None of these dogs developed "learned helplessness".

The second group of dogs were given random electric shocks, but the shocks continued *regardless of anything the dogs did* to try to make them stop. Hardly surprisingly, the dogs in the second group eventually gave up trying. This is what is meant by *learned helplessness*—when dogs have no control over the random and continuous punishment—i.e. nothing they can do makes any difference—and so they give up trying. **That is not training—it is abuse**.

So, let's look at a few examples of punishment in training, and you can decide for yourself which category they fall into. Take Karen Pryor's method of curing her dog of getting into the rubbish bin. If your dog tries to steal food out of the rubbish bin, and you spray it with a water bottle, does the dog have control? Definitely. It simply learns not to steal rubbish out of the bin, and so it doesn't get sprayed again. Just like the dogs that learnt to jump the barrier or press the lever in Seligman's study. No learned helplessness here.

Or, if we apply a punishment when a dog pulls on the lead (for example a correction on a collar). Done properly, the dog will simply learn not to pull on the lead. It is fast and effective. The punishment is *contingent on the dog's behaviour*, and the *dog has full power over it*. Don't pull and you won't get corrected. Because the punishment is *contingent*, learned helplessness will never occur—it isn't even in the picture.

So when Zak George claims that sound training methods—such as those I use all the time—are actually only working because they are causing learned helplessness, as we read before:

> "*Too often, learned helplessness… is interpreted as training success.*"[140] (emphasis added)

He is completely wrong. He also quotes Dr. John Ciribassi (past president of the *American Veterinary Society of Animal Behavior* of which Dunbar is a member and was involved in founding) from an interview conducted by George's co-author Dina Roth Port:

---

[140] George, Zak, and Dina Roth Port. *Zak George's Guide to a Well-Behaved Dog: Proven Solutions to the Most Common Training Problems for All Ages, Breeds, and Mixes*. New York: Ten Speed Press, 2019.

PART 3 – Answering the case against punishment

> "Learned helplessness occurs when an animal cannot escape from an aversive situation, and it can no longer effectively fight back to alleviate a punishing circumstance… Animals are not learning at these times. Instead, they emotionally close down. Placing any animal in this situation is inhumane—I compare it to a person in an abusive situation who has no options to escape the abuse."[141]

When Ciribassi claims the dog "can no longer effectively *fight back* to alleviate a punishing circumstance," he is simply using emotive language. Learned helplessness has nothing to do with "fighting back"—it is simply about whether the dog can learn to avoid the punishment or make it stop, compared to nothing it does making any difference. In the experiments of Seligman, jumping over a low barrier or pressing a lever are hardly "fighting back"—the dogs just learnt a simple behaviour.

With us, we learn not to speed to avoid a speeding fine. We aren't "fighting back". We don't develop "learned helplessness" due to this punishment—we just learn to obey the speed limit.

George also says the following (after earlier quoting Herron's survey and claiming that punishment *doesn't work* and only *makes aggression worse*):

> "Yes, some people who use these methods might argue that their dogs do *decrease their aggressive behaviours*. So what about those cases? *That goes back to learned helplessness*… "Sometimes people can scare their dogs enough that the animals *achieve a state of learned helplessness*—they just sit and take it," Dr. Herron said."[142] (emphasis added)

No, Zak George and Dr. Herron, they don't just "sit and take it". What the dogs do is stop the aggressive behaviour in order to avoid the contingent punishment occurring again in the future. There is nothing whatsoever to do with learned helplessness here either. The punishment is *contingent on behaviour*, and the dog can both stop and then avoid the consequence in the future by altering its behaviour. Just like the dogs could jump the barrier or press the lever, and never developed learned helplessness.

As we heard from researcher Azrin:

---

[141] George, Zak, and Dina Roth Port. *Zak George's Guide to a Well-Behaved Dog: Proven Solutions to the Most Common Training Problems for All Ages, Breeds, and Mixes*. New York: Ten Speed Press, 2019.
[142] As above.

CHAPTER 6 – Does punishment cause undesirable side-effects?

> "Aversive shocks are known to produce aggression when the shocks are *not dependant on behaviour* and to suppress behaviour when the shocks are arranged as a *dependant* punisher…these results show that *aggression is eliminated by direct punishment of the aggression*…"[143] (emphasis added)

He is exactly right. And he and his colleagues actually conducted real experiments throughout their careers, not the pointless surveys of the general dog owning public as Dr. Herron's so often are, and which Zak George likes to frequently cite (see *Part 5 – The scientific basis—fact and fallacy*).

So the claims of Zak George and Dr. Meghan Herron, and various others, are *completely false*. In the experiments, the dogs that could end the punishment by eliciting a certain behaviour *never developed learned helplessness*. Learned helplessness is *never* caused by a proper application of punishment. It is only caused by non-contingent abuse that continues regardless of anything the dog does. It is simply a red herring used to further the reward-only ideology.

> Learned helpless is *never* caused by a proper application of punishment.
>
> It is only caused by non-contingent abuse that continues regardless of anything the dog does.
>
> It is simply a red herring.

I won't go into much detail here, but in *Chapter Nine – More downsides and side-effects of "positive" training*, I will delve into how *positive reinforcement* can, and does, cause learned helplessness. Even "positive" trainer Susan Garrett recognizes this fact:

> "I have seen many "shut down" dogs with handers that would never so much as raise their voices to their dogs."[144]

And in a study titled: *"Learned helplessness, depression, and positive reinforcement"*, it was found that:

> "…helplessness could be induced using positive reinforcement"[145]

So, we've seen that contingent punishment (good training) does not cause aggression, or fear, or learned helplessness, and in fact can be used to cure all three.

---

[143] Azrin, N. H., Ulrich, R., Wolfe, M., & Dulaney, S. (1969). *Punishment of shock-induced aggression*. Journal of Experimental Analysis of Behavior, 12(6), 1009–1015.
[144] Garrett, Susan. *Punishment: Pros and Cons.* Susan Garrett's Dog Training Blog, 2010, https://susangarrettdogagility.com/2010/04/punishment-pros-and-cons/
[145] O'rourke, T., Tryon, W., & Raps, C. (1980). *Learned helplessness, depression, and positive reinforcement.* Cognitive Therapy and Research, 4, 201-209.

PART 3 – Answering the case against punishment

On being "operant", and spoilt brats

There is another argument sometimes levelled at the use of punishment, a bit more abstract this time. It suggests that punishment prevents a dog from being "operant". For example, Jane Killion (author of "*Pigs Might Fly! Training success with impossible dogs*") states that:

> "The effect of punishment is to make your dog unwilling to try behaviours, which is the exact opposite of being operant."[146]

She is wrong (and provides no evidence for her claim). But, firstly, what exactly is "operant"? There are a couple of different ways the term is used, but for our purposes it is simply a fancy term to say that a dog engages in a lot of different behaviours in a given space of time. It is sometimes referred to as the dog's "operant level"—basically its baseline activity level.

It can simply manifest as a dog that runs around like a spoilt brat doing whatever it likes. Your dog doing laps of your lounge room at full speed, jumping off the couch, knocking over the rubbish bin, stealing your shoes, and generally causing mayhem, could be described as being highly "operant". Your lazy greyhound that spends a lot of time sleeping and lazing around is not particularly "operant". It has a low "operant level".

This claim is one of those things that is perhaps partly, but not really, true. In fact, I see this as a *very beneficial* effect of using punishment. The dog generally calms down and stops behaving like a crazy, spoilt brat. We've all seen kids running around completely out of control, making terrible nuisances of themselves. Is that really what you want your dog to be like? It is neither good for them, nor for anyone around them.

Imagine we have a dog that is crazy in the house as I just described. Then we introduce some rules, and *consequences* for breaking the rules. Effectively "if you bite me as you run past, you will get a smack" (as the dog bites you). All of a sudden, if that smack is firm enough, the dog not only stops biting you, but also calms down. (If the smack isn't firm enough, the dog might simply think that it's a fun game, and so it would reward the behaviour and make it worse.) Why does the dog calm down? Have we destroyed it from ever trying anything new again as Killion suggests? No, of course not.

---

[146] Killion, Jane. *When Pigs Fly!: Training Success with Impossible Dogs*. Wenatchee, WA: Dogwise Publishing, 2007.

CHAPTER 6 – Does punishment cause undesirable side-effects?

What we have done is caused the dog to *think before it acts*. It now has to *weigh up* the consequences of its actions, and it can't do that if it is screaming around half-crazed doing whatever pops into its head.

So it *must* slow down and *think* before it acts. This is often a *dramatic* change in spoilt, reward-only dogs. When people have a three-year-old dog which they describe as "he's still just a big pup", what they really mean is he's just a big, spoilt treat-brat, badly in need of some clear boundaries and discipline.

> What we *have done* is caused the dog to *think before it acts*.
>
> It now has to *weigh up* the consequences of its actions, and it can't do that if it is screaming around half-crazed doing whatever pops into its head.

There are all sorts of punishments in society. Does this stop you or I from trying things? No. We just consider things first, and as long as we aren't going to be breaking laws that will end up getting us in trouble, we are happy to try all sorts of things. We don't, however, go around doing whatever pops into our head. We don't drive like lunatics, take what we want from the supermarket without paying, punch anyone we don't like, etc. There are rules, and there are consequences for breaking those rules. That's a good thing, not a bad thing.

Consider this dog who was highly "operant" at the time I was called in:

> "We adopted a very boisterous Staghound cross, and as first-time dog owners with two indoor cats we were completely out of our depth. After just a few sessions with Tully our 'girl' is toilet trained, no longer chases the cats, walks on a lead, responds to commands and is generally a well-disciplined dog. We still can't quite get over the change his training has made to her and us! Tully has a relaxed approach and gives very clear and easy to understand instruction. If you follow his advice, you can't go wrong. Our home has gone from a place of chaos to peace and calm again. We highly recommend Tully and the service he provides." – Helen Kitching

A simple bit of discipline, some rules and consequences for breaking the rules, and dogs calm down and start to think before they act. As I said, that's a good thing, not a bad thing. So this is a beneficial side effect of punishment, whether with dogs, or kids. And it will not prevent your dog from "trying things".

(Nor will it prevent your dog from being "operant" in the even more abstract sense of the "operant training game" type training that some clicker trainers like when teaching tricks. A dog trained to be obedient using punishment will still be operant in that sense also.)

My dog won't like me

In addition to these false claims that punishment is going to cause aggression, fear, anxiety, learned helplessness, and make your dog "unwilling to try new things", people are also often worried that, if they discipline their dogs, then their dogs won't like them. The "positive trainers" parrot these types of untruths, claiming that discipline will "ruin the relationship" or "destroy the bond", and so on. After personally owning and training many hundreds of my own dogs, and helping clients with thousands more, I can guarantee that, in fact, *the opposite is the case.*

> After personally owning and training many hundreds of dogs myself, and helping clients with thousands, I can guarantee that, in fact, *the opposite is the case.*

I often relate dog training to the student-teacher relationship when at school. Let's consider three broad styles of teachers. There are the overly permissive teachers, whose classes are often unruly and undisciplined. The students learn little, but have a good time doing what they like (unless they actually want to learn). Such teachers are sometimes liked by the students, but usually not, and certainly not respected. (Unfortunately, more and more teachers and classes fall into this category, because nowadays teachers have been indoctrinated—particularly the younger, newer teachers—by this sort of "reward-only" psychology nonsense.) These are probably the teachers who try to get their students to like them by not resorting to "punishment", but instead try to "reward good behaviour and ignore bad behaviour" with gold stars, certificates of achievement (when the students really haven't achieved anything—and they know it), and the like.

At the opposite extreme are the angry, aggressive teachers whose classes are generally silent and unhappy. The students may work hard, but these teachers are generally disliked. They may be feared, but again, they are not really respected.

Then there are those teachers with a more balanced approach. I remember an extremely popular maths teacher at high school, Mr. Jansen. He was very strict, and his classes were always very well behaved. If someone stepped out of line, look out, he would leave them in no doubt as to the error of their ways. But when everyone was well-behaved (which was most of the time in his classes), he was very friendly, encouraging, and great fun. He was certainly my favourite teacher, and also one of the most popular teachers in the school. He was liked, and he was respected.

CHAPTER 6 – Does punishment cause undesirable side-effects?

I suggest that a good dog trainer should be exactly the same. As I often say, you can—and *should*—be very firm with your dogs when required (and it won't be required much if you do it properly), as long as you are *fair*. Your dogs will love (and respect) you for it.

Dogs who are just doled out treats all the time may like being around you (when you have treats), but not view you as much more than a treat giver. Many pet owners have probably noticed that when you train with treats, anytime you go to pat your dog it is actually trying to avoid the pat and simply see if there's a treat in your hand. **Its motivation is no longer to please you like it is with a dog that respects you, but simply to get a treat.**

> You can—and *should*—be very firm with your dogs when required (it won't be required much if you do it properly), so long as you are *fair*.
>
> Your dogs will love (and respect) you for it.

As Carol Ray, a former SeaWorld trainer of Tilikum the killer whale (who killed two of his positive reinforcement trainers over a number of years), says:

> "I spewed out the party line during shows—I'm totally mortified now. There's something like, "look at Namu, Namu's not doing that because she has to, Namu's doing this because she really wants to." Oh, my, gosh, like some of the things I'm embarrassed by, so embarrassed by. At the time I think I could have convinced myself that the relationships that we had were built on something stronger than the fact that I'm giving them fish. You know, I like to think that, but I don't know if that's the truth."

As Carol Ray eventually realized, the whales weren't performing because they "loved" their positive reinforcement trainers. They were performing because they were food deprived and that was the only way to get fed, and there was nothing better to do. They were forced to perform—and forced to train—in order to eat. **For those who think that "positive" training is non-coercive, think again.** (This is discussed in greater detail in *Chapter Nine - More downsides and side-effects of reward-based training*.)

> The killer whales weren't performing because they "loved" their positive reinforcement trainers.
>
> They were performing because they were food deprived and that was the only way to get fed, and there was nothing better to do.

If you are just mean to your dog, it will definitely not like you. Or if you punish it in a way that it cannot possibly understand (for example, hours after the fact), it will simply be scared of you. **But, if you are *fair*, you can be firm with your dog when required, and it will love and *respect* you for it** (the "positive" ideologues hate the "respect" word, like they hate the "dominance" word).

139

PART 3 – Answering the case against punishment

Very often, clients make the comment as I'm leaving, that, "The dog wants to come home with you!" And this is after I have been using some correction or punishment-based approach or another (in addition to reward, though usually not treats) to rapidly cure various issues. And the less clear boundaries and discipline the dog has had, the more marked is this response. Here are two examples:

Wild, destructive, out of control Doberman Pinscher

One was an aggressive, destructive Doberman Pinscher. The dog was quite a rogue by the time I was called in. The owners (a lovely older couple) couldn't put a lead on the dog out in its external dog yard—it would get aggressive and attack them when they tried (which I showed them how to overcome in a few minutes using a negative reinforcement approach that involved basically no physical contact). They could, however, get the dog into the house by opening the house door, and then opening the dog yard door, and standing back they would let the dog run up into the house on its own.

Once in the house, for some reason the dog was happy for them to attach a lead. And then they would tie the lead to a pile of barbell weights next to a dog bed (what AVSAB would approvingly call "management"!). Otherwise, the dog would destroy the house. One of the things they wanted to achieve was the dog staying on its dog bed without destroying the house, and/or for it to just behave calmly in the house.

After explaining to the owners their various options, I told the Doberman (which was tied up near its bed, but wasn't on it at the time) "On your bed". Of course, he didn't respond, because he had never been taught the command in the first place. However, one of the advantages of a punishment-based approach when indicated is that not only can we use it teach the *meaning* of a command very rapidly, but we also teach *obedience* to the command at the same time. This is in contrast to another completely misinformed opinion of Dr. Dunbar, who states that:

> One of the advantages of a punishment-based approach is that not only can we use it teach the *meaning* of a command very rapidly, but we also teach *obedience* to the command at the same time.

"At best, consistent and well-timed aversive punishment only communicates, "stop what you are doing," and at the worst, "No," "Stop everything," or "Freeze". A huge reason that aversive stimuli are largely ineffective is that they lack instruction and do

CHAPTER 6 – Does punishment cause undesirable side-effects?

not inform animals *which* specific behavior to stop or *what* to do instead."[147]

So according to Dunbar, "*at best*" punishment only communicates "stop what you are doing", and it doesn't "inform animals what to do instead." This statement shows once again a glaring lack of knowledge of the principles of learning. You are about to hear an example of punishment that doesn't just function to stop the dog doing something, but in fact teaches it *exactly* what to do instead (which is often the case). (Even Martin Seligman's experiments with learned helplessness demonstrated punishment effectively teaching dogs to either press a lever or jump over a barrier, again disproving the reward-only ideologues assertions.)

After the Doberman ignored my command, "On your bed", I got up out of the chair I was sitting in and walked towards him, continually repeating "on your bed". I then physically moved directly into the dog's space to slowly and calmly pressure him (chase him) onto the bed. If he tried to duck around me, I stepped to physically block him. Being tied up he couldn't get away. I kept it up until he jumped onto the bed (negative reinforcement). I then rewarded him with a sincere "good boy" and by backing off a bit. I then instructed him to "stay there".

I headed back to my seat, whereupon, of course, he jumped off the bed. I turned around and repeated the procedure. Then, the next time he hopped off the bed, I started to get firmer. Now I growled at him "on your bed" in a firmer tone, and a few *light* taps on the nose with a rolled-up newspaper. The consequence for hopping off the bed after being told to stay there was increasing. The next time, he got a *firm* smack on the nose with the newspaper and he jumped straight back onto the bed (contrary to Dunbar asserting that punishment can never tell a dog what to do).

From that point on, he remained sitting on the bed without further instruction. In fact, he sat bolt upright as Doberman Pinschers commonly do, and never took his eyes off me. I sat and chatted with the owners for the next 10 or 15 minutes, as they took notes on the course of action they needed to take from that point on to consolidate this training. The dog hardly moved a muscle during this time— Pinschers make very good statues!

As an aside, Dr. Dunbar (and Pryor, Garrett, Stilwell, George, and countless others) have also made claims about how much faster treat based training is (despite, as we shall see, the sometimes *years of work* necessary as they themselves, and others, point out). Dunbar often cites the following example:

---

[147] Dunbar, Ian, Dr. *Barking Up The Right Tree - The science and practice of positive dog training.* Novato, CA: New World Library, 2023.

PART 3 – Answering the case against punishment

> "For example, when a sit-stay improves from 0.2 to 1.2 seconds after just three trials, that's a colossal 600% improvement. Just four more equivalent 600% improvements and the Jack Russell will be sit-staying for four minutes and 29.2 seconds. Not too shabby!"[148]

The astute observer will realize the folly of his claim (and that's ignoring the problems with his math, which I will indeed ignore as it's beside the point). If he takes it one step further, you will now have a sit stay for 26 minutes. Just three more such "600% improvements" and you will have a sit-stay of nearly *4 days*. Only one more similar improvement and it will be *23 days*. Wow, what a brilliant "stay"! Unless Dunbar really doesn't understand maths (or dog training), then he should know full well that the improvement will not be exponential—you may well get the stated 600% improvement when going from 0.2 to 1.2 seconds, however that rate of improvement will *rapidly* decline.

Dr. Dunbar often states that he wants figures and maths (even though he uses them in a fundamentally flawed manner as we see here). So, how about this Doberman Pinscher that went from having no "stay" or "on your bed" *whatsoever*, to basically understanding, "On your bed," and remaining on the bed for a good ten or fifteen minutes, after just a few minutes training? That's a truly *massive* improvement after a tiny amount of training—a "quick fix" as Zak George would derogatorily describe it and warns everyone about. As Jamie Dunbar says, "if the behaviour improves then the method is good". But, according to them, *punishment hardly ever works*.

To return to the story of the Dobermann, after the dog had held a ten-to-fifteen-minute perfect stay with no attention from me, I then stood up, went over to him, gave him a good pat and "good boy", and unclipped his lead. I then encouraged him to follow me off the bed with "come on", which he did (if you want a good "stay" then you always have to tell your dog when he is now permitted to move).

I sat back in my seat without saying another word to the dog. He wandered around calmly for a few minutes, unlike the "highly operant" tear-away destructive monster he usually was, and then he came and lay right at my feet, leaning against my leg, at which the owners simply looked on in stunned amazement. Another pat and "good boy", and he remained there for the rest of the session until I stood up to leave. At that point he would happily have followed me home. Was he traumatised? No. Was he terrified of me? No. Did he hate me? No. Had I "ruined the relationship"? No. On the contrary, *the exact opposite occurred*, exactly as expected. He bonded to me almost instantly. And this was not an isolated

---

[148] Dunbar, I. (n.d.). *Barking Up the Wrong Tree — For 110 Years?* Dog Star Daily. Retrieved from https://www.dogstardaily.com/blogs/dr-ian-dunbar/barking-wrong-tree-110-years

CHAPTER 6 – Does punishment cause undesirable side-effects?

occurrence. It happens regularly, to varying degrees—and the less rules and discipline the dog has had, the more marked is this effect.

You will NOT ruin the relationship by being firm with your dog (or children, or students) when required. In fact, so long as you are fair and balanced, it will have the opposite effect. Don't believe a word the "positive" ideologues say.

> You will NOT ruin the relationship by being firm with your dog (or children, or students) when required.
>
> In fact, so long as you are fair and balanced, it will have the opposite effect.

Possessive aggressive Cocker Spaniel

Second example: this time a highly possessive-aggressive Cocker Spaniel (a reasonably common problem in this breed). A young couple called me out because their dog would steal their socks, and they couldn't get them back. If they tried to approach their dog when it had the sock, it would growl and attack and bite them if they got too close.

After the usual initial discussion, fact finding, and explaining things to the owners, we found a sock and allowed the Cocker Spaniel to take possession. I then approached him, and sure enough he started growling. I got closer, and he started making lunges towards me showing his teeth and making his intentions very clear. I calmly told him "No," before moving in closer. As he lunged forwards, growling and snapping his teeth, he received a fairly solid whack on the nose/head area with grandma's training aid, the rolled-up newspaper. This caused him to escalate (which it often does initially), at which point he received a much firmer application or two of the rolled-up newspaper. At this point he decided discretion was the better part of valour, relinquished the sock, and moved away. I immediately told him "good boy", and moving away from the sock and from him I squatted down, and encouraged him to come to me. He was a bit unsure (remember the "hat trick rule"), but he warily approached me, whereupon he received a good pat, his tail wagging like mad (remember the rebound effect mentioned earlier?).

At this point, I went and sat back in my seat. The Cocker looked at the sock (which was still in the same place) a couple of times, looked at me, then wandered back over to the sock. He picked the sock up in his mouth while keeping an eye on me, although not so keenly as before. I said "No," quietly, half rising out of my chair. He immediately dropped it, at which point I enthusiastically praised him (though not with the high-pitched, baby-voice hysterics many "positive" trainers seem to relish in) and encouraged him over. He came straight over and received another good pat.

PART 3 – Answering the case against punishment

As I chatted with the owners further, once again he just sat by my side, receiving the occasional pat. Then, he wandered back over to the sock once again. At this point I was ready to repeat the training, however I waited to see exactly what he would do. He picked up the sock, turned around and brought it over to me, and dropped it at my feet! He then lay down beside me, ignoring the sock, at which point he again got a pat and praise. The owners only needed to repeat this training (he had to learn the same rules applied with them), and the problem was solved permanently basically instantly. Yet another "quick fix" (effective training) for Zak George that he warns dog owners to avoid.

> The disinformation amongst the "positive" training ideologues is enormous and deplorable.

So much for the "positive" ideologues claims that punishment "never really works", "just makes it worse", "ruins the relationship", "causes aggression", "causes learned helplessness", and can't "show the dog what to do". The disinformation amongst the "positive" training ideologues is enormous and deplorable.

Troublesome dance students

Earlier, I related the incident with three misbehaving girls in a debutante dance class we were teaching, and how some simple, effective punishment, "you will sit the class out if you do that again; if you do it after that you will sit the next class out also; and if you do it a third time you are out of the deb entirely," fixed the behaviour instantly and permanently. There is another aspect to this that I left for now to relate, in relation to the question of whether you will "ruin the relationship" by using punishment, or if your dog (or child, or student) "won't like you".

As I talked about respecting and liking teachers who are fair but firm (like Mr. Jansen, very strict but also very encouraging, and definitely my favourite teacher), exactly the same happened here. Firstly, as a result of the discipline, the girls were a pleasure to have in the class from that point forwards. As my good friend Rod Cavanagh, author of *Australian Sheep Dogs – Training and Handling*, has said:

> "Discipline creates obedience which creates opportunities for praise."

Secondly, two years after teaching that deb group, we were teaching another deb group from the same school, just a couple of years down the line. When we turned up to the actual night of the debutante ball, as we walked into the venue, these same three girls were there to watch and support their friends. When they saw my wife and I walking in, they raced over from the other side of the stadium and gave us big hugs! This was by far the most effusive welcome we had ever received. Plenty of other students are friendly and welcoming, but this was another level entirely. So, the punishment certainly didn't do any harm to the "relationship"—once again, exactly the opposite. That's the most enthusiastic welcome we ever received, from the most problematic students we ever had. They just needed some clear boundaries, and to know that consequences existed should they cross them (just like the Dobermann Pinscher and the Cocker Spaniel). So, claims that you will "ruin the bond" or "destroy the relationship" are completely false. If you use correction and punishment properly, as part of a balanced approach, you will in fact do exactly the opposite.

Conclusion – Does punishment cause undesirable side effects?

So, we have now debunked Myth #2, that punishment causes all sorts of inevitable, undesirable, negative side-effects. Despite the "positive" trainers claims, we've seen that it *doesn't* cause aggression (in fact, it can be a highly effective cure for it); it *doesn't* cause learned helplessness (unless it is simply continuous abuse); it *doesn't* cause any long-term fear or anxiety (and can be a highly effective cure for it, and will result in a much calmer and less anxious dog); and it won't "ruin the relationship" or "destroy the bond" or any such thing—in fact, used properly, it will rapidly *improve* the bond.

As I mentioned earlier, the *real* science agrees with the *real world* evidence:

> "The **lack of undesirable side effects** associated with the use of punishment has also been noted in the applied literature (e.g., Brantner & Doherty, 1983; Harris, 1985; Johnston, 1972; van Oorsouw et al., 2008). Indeed, the use of punishment-based interventions typically has been **related to increases in positive behavior** (e.g., Bostow & Bailey, 1969; Firestone, 1976; van Oorsouw et al., 2008; Risley, 1968). For example, Matson and Taras (1989) reviewed **382 applied studies** employing different punishment procedures during interventions with individuals with developmental disabilities and concluded that the results reviewed *did not provide evidence supporting the occurrence of undesirable side effects*. Instead, **the majority (93%)** reported *positive side effects during punishment interventions*, such as increases in social behavior and responsiveness to the environment."[149] (emphasis added)

So, with that myth debunked, let's now consider one final criticism levelled at the use of punishment—that it doesn't "create" behaviours, it only stops them.

---

[149] Fontes, R. M., & Shahan, T. A. (2021). *Punishment and its putative fallout: A reappraisal.* Journal of the Experimental Analysis of Behavior, 115(1), 185–203.

7

Can punishment create behaviours?

So, aside from the reward-only ideologues claims that punishment doesn't work, and that it causes all sorts of terrible side effects like aggression, fear, learned helplessness, and "ruins the bond" (all of which are completely false, as we've discovered), another criticism often directed at punishment is that it can't *create* behaviours. That is, it can't teach the dog to *do something*. Karen Pryor claims:

> "...aversives stop behavior, they don't start it."[150]

We've already heard some examples from Pryor herself that disprove her own claim. She used the example of someone moving away from a cold draught to illustrate negative reinforcement—the cold draught is the aversive that causes someone to do something—to move away. She also used the example of "effecting change in a hard case" where, once again, an aversive (Pryor standing disconcertingly in the visitor's space) created the desired behaviour (putting her wet clothes in the laundry). So already her claim is disproven. The aversive of cold might also cause us to put a jumper on or start a fire; the aversive of a fly landing on us causes us to shoo it away; the aversive of hunger (used all the time by "positive" trainers like Pryor) causes us to seek out food. And perhaps Pryor would be surprised to learn that even Murray Sidman, who has had such an influence on her views, recognized that punishment can create behaviours:

---

[150] Pryor, Karen. *Reaching the Animal Mind: Clicker Training and What It Teaches Us About All Animals.* New York: Scribner, 2009.

> "Or we *construct new behavior* by teaching people how to prevent or escape the shocks we inflict on them" [151] (emphasis added)

In fact, to get even more advanced, Sidman points out that aversives can even become positive reinforcers, and be used to teach behaviour just like treats:

> "The shock had become so effective a positive reinforcer that we could use it to teach the animal something new"[152]

So, Pryor is wrong on every level when she claims "aversives don't start behaviour". However, she isn't alone because this is a widespread misconception. Dr. Gary M. Landsberg, BSc, DVM, MRCVS, DACVB, DECAWBM, and Dr. Sagi Denenberg, DVM, DACVB, Dip. ECAWBM (Behaviour), MACVSc (Behaviour), from the North Toronto Veterinary Behaviour Specialty Clinic, writing in the *Merck Veterinary Manual*, likewise claim:

> "Punishment cannot be used to achieve desirable behaviors, only to stop what is undesirable."[153]

We also heard something similar before from Dunbar:

> "At best, consistent and well-timed aversive punishment only communicates, "stop what you are doing," and at the worst, "No," "Stop everything," or "Freeze". A huge reason that aversive stimuli are largely ineffective is that they lack instruction and do not inform animals *which* specific behavior to stop or *what* to do instead." [154]

There is so much wrong with this statement of Dunbar's (like Pryor's, and Drs. Landsberg and Denenberg and the *Merck Veterinary Manual*) that we need to explore in detail exactly *why* these so-called "experts" are wrong. I already gave the example of the Doberman Pinscher, and teaching it "on your bed" and "stay" within minutes, and of Seligman's experiments where the dog learned to press a lever or jump a barrier in response to an aversive. Let me give you some more examples that further solidify the evidence (as if any more is needed).

---

[151] Sidman, Murray. *Coercion and Its Fallout*. 2000. United States. Author's Cooperative.
[152] As above.
[153] Landsberg, Gary M. *Treatment of Behavioral Problems.* In Merck Veterinary Manual. Last modified May 2014. Accessed May 27, 2024. https://www.merckvetmanual.com/behavior/behavioral-medicine-introduction/treatment-of-behavioral-problems#Operant-Conditioning:_v21352234.
[154] Dunbar, Ian, Dr. *Barking Up The Right Tree - The science and practice of positive dog training*. Novato, CA: New World Library, 2023.

PART 3 – Answering the case against punishment

If I am teaching a pup to "sit" using negative reinforcement (oh no, how terrible!), I tell it to "sit". When it doesn't, my method is to simply put an uncomfortable pressure (an aversive) with my hand downwards and rearwards on the pup's hindquarters, while squeezing with thumb and middle finger just in front of the pup's hip bones, while continually repeating the "Sit" command. This does two things: it creates a mildly uncomfortable, annoying pressure that the pup would rather not be there (an aversive punishment), and that it would like to go away (negative reinforcement). (Note also, that *continually repeating* the command during this process is indeed the best thing to do, despite various claims to the contrary—but more on that in future volumes.)

Because I am applying a downward and rearward pressure, and preventing any other options, it rapidly gives the pup the idea of what to do. It is basically the negative reinforcement alternative to Dunbar's favourite, "luring". I call it "suggesting". I do *not* push the pup all the way to the ground, which is vital—I simply push maybe halfway, and then hold. The pup will try to wriggle out, but that won't work. When he sits out from under my hand (which he very quickly will), I immediately release all that pressure, and praise and pat the pup (good boy!). (I certainly don't say "good *sit*" or any of that counter-productive pseudo-training, like Zak George and many others recommend.)

So, using an aversive, I teach the pup *what to do,* contrary to Pryor's claim that "aversives stop behavior, they don't start it"[155]. In fact, Pryor uses the same process herself when training a show dog to hold its tail in a certain way (while evidently not recognizing it as such, given that she claims "aversives don't start behaviour"):

> "Teach the dog to move her tail deliberately, at a standstill. You can use a target stick and touch the tail gently and try to have her move her tail away from the touch."[156]

So, she touches the tail with a stick (just like the uncomfortable pressure I apply on the pup's hips), so that the dog moves its tail away from the annoyance. This "starts the behaviour". Or:

> "One of my favorite shaping demos consists of teaching a dog to walk backwards. I usually start by leaning toward the standing dog, which

---

[155] Pryor, Karen. *Reaching the Animal Mind: Clicker Training and What It Teaches Us About All Animals.* New York: Scribner, 2009.
[156] Pryor, Karen. *On My Mind – Reflections of Animal Behavior and Learning.* Waltham, MA. Sunshine Books. 2014.

usually makes him lean away from me or take a step back. Click. I've prompted the behavior; he didn't do it on his own." [157]

The uncomfortable pressure (the aversive) is her leaning into the dog's space. The aversive creates the behaviour that can then be rewarded. So, in both of these examples of Pryor's, an uncomfortable aversive (a touch with a stick or moving into the dog's space) are used to create the behaviour, despite her claiming that "aversives stop behavior, they don't start it." There is no difference whatsoever, in principle, between this and someone applying an annoying low-level tingle on an e-collar (which can be set so low that most people can't even feel them) to help create a behaviour.

Also, consider the next steps when I'm teaching a pup to sit. I repeat the process just described a few times, and the pup sits more quickly and easily each time. Pretty soon, I just touch the pup's hindquarters and it sits, then it rapidly starts sitting just on the command.

So, the next stage is that we now introduce a stronger consequence for not obeying the sit command. Now, if we say "sit" and there is no response—perhaps because the pup is *distracted*—we give it a few light smacks (taps) on the rump. Having already learnt what pressure on the rump means, he soon understands that this means to sit. As soon as he does, "good boy," (never "good *sit*") and pat.

On the other hand, a pup trained with treats doesn't have any concept that a smack (or any other consequence) might mean he should sit—he hasn't been taught that. But our pup does. So, we are already a long way more advanced in gaining *obedience* to the command than the "positive" trainers. As all pet owners know, getting the pup to sit with treats isn't hard, getting it to listen *when distracted* is a whole different matter! However, in this negative reinforcement process we teach both the *meaning* of the command, and at the same time go a significant way towards gaining *obedience* to it.

So that smack on the rump clearly communicated to the pup that he should sit. This directly contradicts Dunbar's claim that punishment never communicates to the dog what it should do. The pup knows *exactly* what to do. And contrary to the claim that it only gets the dog to *stop* doing

> As all pet owners know, getting the pup to sit with treats isn't hard, getting it to listen when it doesn't want treats is a whole different matter!
>
> In the negative reinforcement process, however, we teach both the *meaning* of the command, and *obedience* to it, at the same time.

---

[157] Pryor, Karen. *On My Mind – Reflections of Animal Behavior and Learning*. Waltham, MA. Sunshine Books. 2014.

something, or "freeze", in actual fact it tells the dog it should have sat, and that it should now do so. And it will.

Let's take another simple example (there are a huge number I could choose from). You might not want your dog coming into the kitchen while you are cooking, for your dog's safety as well as your's. If it does, simply say "out" or some such command, maybe pointing out of the kitchen. If the dog doesn't respond, start moving towards it and chase it out of the kitchen. Go slowly and gently at first, while repeating the "out" command. Once it is out of the kitchen, you tell it "good dog" and go back in. The dog will no doubt follow you back in also. You chase it out again, "Out", repeating the command while doing so. Each time you should get a bit firmer, as the dog begins to understand.

With some more sensitive dogs, that would be that. Command taught, obedience gained, problem solved. With others, the consequence might need to get stronger. Next time the dog comes into the kitchen, a smack on the nose and chase it with a sterner "out". The punishment clearly communicates "go away". It doesn't say "stop" or "freeze", and it clearly shows the dog what to do instead.

We don't take the dog's collar and drag it out of the kitchen, or pick it up and carry it out—it must go out *itself*, which is why we *chase* it out. The punishment of being chased clearly teaches it what to do, as does the smack.

Or take a dog that isn't allowed on the couch. It gets on the couch, you tell it "off", and it ignores you. So, you give it a relatively gentle smack from *behind* it, and repeat this (slowly getting firmer) until the dog moves away from the annoying smacks (like Pryor's dog moved its tail away from the stick) and jumps off the couch. Don't stop until it does (with an aggressive dog—one that has learnt it can dominate you—it is especially important that you don't stop until it jumps off, otherwise you can simply reward and shape an increase in the aggression. If you aren't confident enough to apply this, there are other approaches you can take, however they are beyond the scope of this book). Once the dog jumps off the couch, "good boy". Again, this clearly communicates to the dog what you want it to do. This is all very simple, quick, and effective. It needs to be done hardly at all, and the problem is fixed. And, done as explained there are no unwanted side effects, only positives.

> When used simply as a "no" command to stop behaviours (which is a vital use of punishment), it tells the dog *very clearly* to stop *exactly what it is doing at the time*.

And to Dunbar's assertion that punishment does not inform the dog *which* behaviour to stop, this is even more ludicrous. It most certainly *does* inform the dog which behaviour to stop. When used simply as a "no" command to stop behaviours (which is a vital use of punishment), it tells the dog *very clearly* to stop *exactly what it is doing at the time*.

CHAPTER 7 – Can punishment create behaviours?

If the pup is trying to eat mum's food, mum bites him. He knows to stop trying to eat her food when she is eating. Or if our dog is jumping on us and we give it a smack, it knows to stop jumping on us. The dog puts his nose near the campfire, he gets punished, he knows to stop putting his nose in the campfire. We touch an electric fence, we know not to touch it again. Punishment, when delivered *during the act*, in fact tells the dog (or us) *exactly* what behaviour to stop doing—whatever behaviour is being engaged in at the time. It isn't complicated. (And this is despite Pryor claiming that "punishment does not have a predictable effect on the future."[158])

So punishment can not only communicate *exactly* what we want the dog to *stop* doing, but also what it *should do*, depending on how it is used. And that, as I mentioned earlier, is one of the reasons that the common "operant conditioning" definition of punishment, that it is something that *reduces* a behaviour, is so useless. If I give a dog a smack for not sitting when he's told, he will be *more* likely to sit next time, not less. So, punishment can *increase* the behaviour we are focused on—and according to Dunbar that makes it "a reward" and not punishment at all, because it *increases* the likelihood of the behaviour (see *Chapter Three – Some problems with operant conditioning*):

> "…anything (even if unpleasant) that increases behavioural frequency is a "reward"…"[159]

So, *everything* Dunbar said in the earlier quote, (and as I mentioned it isn't unique to him, he just happens to be one of the most influential "positive" trainers of the last 40 odd years), and the statements by Pryor and the veterinarian behaviourist "experts" in the *Merck Veterinary Manal*, are *false*. Punishment very clearly tells the dog *exactly* what to stop doing, and it can indeed create behaviours and inform the dog what we want it to do.

Conclusion – Answering the case against punishment

So, we have now answered our first two questions. Is punishment effective? Yes, it is highly effective when used correctly. It's also generally far *more effective*, far *quicker* and *easier*, and far more *permanent*, than "positive" approaches, for the things that most dog owners actually require.

---

[158] Pryor, Karen. *On My Mind – Reflections on Animal Behavior and Learning*. Waltham, MA. Sunshine Book. 2014

[159] Dunbar, Ian, Dr. *Barking Up The Right Tree - The science and practice of positive dog training*. Novato, CA: New World Library, 2023.

PART 3 – Answering the case against punishment

And does punishment cause undesirable side-effects? Does it cause aggression, or fear, or learned helplessness, or destroy the relationship? No, never, not when used correctly. In fact, it will generally do the exact opposite—we heard of the **93% of 382 applied studies** **that all found positive side effects to the use of punishment, not negative.**

We've also seen that, again despite the claims to the contrary, punishment can indeed be used to *create* behaviours (even though this is not a particularly important factor—no decent trainers rely solely on punishment, and so they will also use reward strategies when these are a better option).

> No trainers rely solely on punishment, and so they will also use reward strategies when these are a better option.

So, let's now turn the spotlight to the reward side of things, and see how *it* stacks up.

Part 4

INVESTIGATING REWARD

Chapter 8 – Does reward work?

Chapter 9 – More downsides and side effects of "positive" training

8

Does reward work?

Now that we've clearly seen that punishment does indeed work (despite the claims to the contrary), and in fact is *highly* effective as well as long lasting, quick, easy, and without unwanted side-effects or "fallout", let's now turn our attention to reward (positive reinforcement). Let's first consider whether or not "positive" training is effective, before moving on to look at its considerable downsides and side-effects. We need to consider this in three areas.

1. Is reward effective in teaching the *meaning* of commands?
2. Is reward effective in gaining *obedience* to those commands?
3. Is reward effective in *stopping* undesirable behaviours?

The answers are that yes, reward is sometimes effective in teaching the *meaning* of commands (but there are often far superior approaches); no, reward is not very effective at all in gaining *obedience* to those commands; and no, reward is not very effective at all in *stopping* undesirable behaviours.

We will also consider whether reward is effective in these areas in a theoretical sense, and whether that translates to being effective in a *practical* sense. In other words, sometimes reward might work in *theory*, either in the laboratory or in certain controlled or restricted circumstances (like captive dolphins), but for the vast majority of dog owners in the *real world* it simply won't work. As Ian Dunbar says:

PART 4 – Investigating reward

"That's what we need to do—change behavior *quickly*. When you look at my criteria for training, it's got to be *easy*. Or else the owners can't do it—the owner's not you, they don't have forty years' experience, they're not champion of the world—they're an owner… it's got to be *easy* or else they can't do it. It's got to be *quick*, or else they probably won't… And the sooner we as trainers get used to that, instead of holier-than-thouing them that you shouldn't have got a dog, the quicker we can help people out. And it's **got to be effective**."[160] (emphasis added)

He's dead right (although his implication that his methods do this is *dead wrong*). So we need to ask not only can reward training work in an abstract, theoretical sense, but also can it work in a *practical* sense, in *the real world,* for real dog owners? *Is it quick, is it easy, and is it effective?*

Teaching the meaning of commands

Nearly everyone knows that you can teach your dog to "sit" using treats (usually by "luring" the pup into position with the treat). It's usually fast and effective at teaching the meaning of this particular command.

But what if your pup isn't interested in treats? Well, Dunbar (and many others) simply advise you to *stop feeding your dog.* Sooner or later it hopefully will *become* interested. This is a strategy known as "nothing in life is free", or "closing the economy", or in more scientific terms—*deprivation* (there is more about this essential "positive" strategy in *Chapter Nine – More downsides and side-effects of "positive" training*).

In fact, using lure-based, positive-reinforcement training is the way I generally instruct pet clients (not sheepdog clients) to teach their pups the meaning of "sit" and "lie down" and "come here". It works in teaching the *meaning* of these commands. And in these things, it is reasonably comparable in speed and ease to non-reward-based methods.

However, in teaching many other commands, positive reinforcement is *far* less effective. For example, in teaching your dog "on your bed", or "go outside" (to leave the house, or the kitchen, or the bedroom), it is a very slow and much more difficult process. Possible? Yes. But why would you, when you can teach it so much more quickly and effectively by a different approach?

---

[160] Balabanov, Ivan. *Training Without Conflict Podcast Episode Four: Dr. Ian Dunbar.* YouTube, April 30, 2021. Accessed Feb 2024. https://www.youtube.com/watch?v=EF-INPvEhA8

CHAPTER 8 – Does reward work?

As we saw before, if you want to teach your dog to "get out" of the kitchen (or anywhere else), then simply tell it "Get out", and calmly and methodically *chase/pressure* it out (don't lead it out, don't call it out, *chase* it out) while repeating the command. Quick, simple, intuitive (natural), and effective. Why muck around teaching the dog with treats for weeks or months (and it will still not be reliable), when you can teach it in literally five minutes?

But yes, treat training can work in teaching the *meaning* of commands. And sometimes it can even be relatively quick and easy. For others, it's simply a waste of time. And in *the real world* time is important, as Dunbar mentioned:

> "…it's got to be easy or else they can't do it. It's got to be quick, or else they probably won't." [161]

Unfortunately, there's nothing quick or easy about Dunbar's own methods, despite his claims or those of other "positive" trainers. As we've seen, Zak George spends a lot of time criticizing "quick fix" methods, because he's at least more honest (at these times) in admitting how slow his methods are (when he isn't claiming how much faster they are!).

I said earlier that I generally use treat training for teaching "sit" and so on with pet clients. There are a few reasons for this:

1. It works at this stage of training.
2. Most people nowadays expect it (thanks to the "positive" marketing machine).
3. Kids enjoy it.

What I also teach, however, is that this is just the "first stage" of training—teaching the *meaning* of the commands. I then follow on by explaining that once their pup knows the *meaning* of the commands, we must move past the treats if we want *obedience*. That's the second stage of training. We move from saying, "if you want a treat, sit", to, "if you *don't* sit, there will be a consequence". That's how we gain obedience (see the next section—*Gaining obedience to commands*).

However, I hardly ever use these treat methods myself, on my own dogs. If these methods were actually quicker and easier and more effective, I would use them.

---

[161] Balabanov, Ivan. *Training Without Conflict Podcast Episode Four: Dr. Ian Dunbar.* YouTube, April 30, 2021. Accessed Feb 2024. https://www.youtube.com/watch?v=EF-INPvEhA8

PART 4 – Investigating reward

> I hardly ever use treat methods myself, on my own dogs.
>
> If these methods were actually quicker and easier and more effective, I would use them.

Instead, by using negative reinforcement methods—for example putting physical pressure on the dog's rump until it sits—we are *combining both stages*. The dog is learning not only the meaning of the command just as quickly (more quickly for many things), but is also learning that it isn't optional—it is learning that it *must* sit. Therefore, we are working towards gaining obedience simultaneously. This is something that teaching with treats simply doesn't do (regardless of whether you move on to "variable schedules of reinforcement" or swap treats for toys or "life rewards", or whatever other ineffective strategy the "positive" trainers suggest). This combination—of teaching the meaning and working on gaining obedience at the same time—is one reason why negative reinforcement is so much faster and more effective, and why I don't personally mess around with treats.

So, yes, positive reinforcement can be used to teach the meaning of commands. However, that doesn't mean it is always the fastest or easiest method—sometimes it is, but often it's painfully slow, complicated, and difficult.

Gaining obedience to commands

The real problem with reward-based training is when we actually want to gain *obedience* to our commands. And this is generally the most important part.

Pet owners know they can easily teach their pup to "sit" with treats, or "come here", etc., *when it wants to*. It's pretty simple. What they also know, however, is that this in no way translates into obedience *when distractions are present*, or when the pup simply doesn't want to. In other words, it works for these things *until you actually need them*.

> Reward can teach the *meaning* of commands or behaviours.
>
> Where it fails is in gaining *control* and *obedience* to those commands.

It works until your guests turn up, or until the pup gets out the front door and takes off down the street. It might work getting your pup to walk on a loose lead beautifully in the back yard, when you have treats and it's hungry. But when you get out on the street it all reverts to chaos. And in the *real world* (not in the world of trick training, or of captive dolphins, or of rats or pigeons in boxes, and so on), *this* is what counts.

This is the major area where *treat training fails*. And it often fails in both an absolute sense, and in a practical one.

Why is this so? The simple answer is that the dog is more interested in things other than the treats. What you simply have is *competing rewards*. The *environmental rewards are bigger* than your training reward. The dog is being *rewarded when it disobeys you*, because it gets to go and chase the cat or play with other dogs, or whatever the distraction happens to be.

Dunbar, as always, tries to make "positive" training sound super-simple, super-effective, and super-easy:

> "Indeed, once you have taught your puppy the positive consequences of cooperating, it will eagerly want to comply!"[162]

Sounds great! Except for the tiny little problem that when there are competing motivations it *simply isn't true*. Why? Because there are often bigger "positive consequences" for *not* cooperating, and so in various situations your pup will eagerly want to *not* comply. So, although you can teach the meaning of "come here" with treats in your house when there are no distractions, out in *the real world* when your dog sees a rabbit to chase it *couldn't care less* about your treat.

> Sounds great! Except for the tiny little problem that it *simply isn't true*.
>
> Out in *the real world* when your dog sees a rabbit to chase it *couldn't care less* about your treat.

Let's look at some specific training examples of the failure of positive reinforcement in its attempts at achieving obedience.

Recall

Earlier, I quoted Pat Miller from "The Power of Positive Dog Training":

> "The most notable example of my reluctant application of positive punishment involved a clients Pit Bull mix who was very dedicated to chasing horses... I **pulled out the cap gun and waited. Just as he ducked under the wire into the pasture, I fired.** Happy wheeled in his tracks and dashed for the porch... It was the first time we had succeeded in stopping him in mid-charge... I left the gun with the owners, and we all crossed our fingers. They only had to use it one more time. ***Twice was enough to convince Happy to leave horses alone forever.*** Mission

---

[162] Dunbar, Ian. *The Good Little Dog Book*. 3rd ed. Berkeley, CA: James & Kenneth Publishers, 2003.

accomplished…To this day, I ask myself if there wasn't another, more positive way to accomplish our goal… I wonder if we could have succeeded with Leslie Nelson's "Really Reliable Recall". I am convinced that there was a positive way to do it—I just couldn't find it at the time."[163]

So Pat Miller, a well-known author and influencer of the "positive" school (according to Wikipedia she has authored over 500 articles on dog training), couldn't teach the dog to recall or stop chasing horses, not without recourse to punishment. And yet, chasing horses, livestock, wildlife, cars, cats, or other dogs, are all very common problems to the average pet owner, and something I deal with successfully all the time. If someone of the experience and "knowledge" of Pat Miller couldn't do it, then what hope does the average pet owner have? So, positive training in these cases is a total failure.

Let's read Pat Miller's resume on the back cover of her book:

> "PAT MILLER has been a dog trainer for *over thirty years*. She is the founder of Peaceable Paws Dog & Puppy Training Centre and is on the Board of Directors of the *Association of Pet Dog Trainers*. She is a leading proponent of positive dog training techniques, and her columns on training are read by thousands in publications such as Whole Dog Journal."

Remember what Dunbar said (who is the founder of the *Association of Pet Dog Trainers* which Pat Miller was on the board of directors of):

> "When you look at my criteria for training, it's got to be easy. *Or else the owners can't do it—the owner's not you, they don't have forty years' experience, they're not champion of the world—they're an owner.*"[164] (emphasis added)

And yet, even with *over thirty years' experience*, Pat Miller, a Director of the *Association of Pet Dog Trainers* (the very association founded by Dr. Dunbar), couldn't teach a dog not to chase horses, or recall under that distraction, without the use of punishment. Yet Dunbar and others try to make out how easy it all is with "positive" methods. It most certainly is not. Yet, as we saw, a punishment-based approach (the cap gun) cured the behaviour basically instantly and permanently. Now *that* certainly meets Dunbar's requirements of *quick*, *easy* and *effective*.

---

[163] Miller, Pat. *The Power of Positive Dog Training*. 2nd ed. Hoboken, NJ: Howell Book House, 2008.
[164] Balabanov, Ivan. *Training Without Conflict Podcast Episode Four: Dr. Ian Dunbar.* YouTube, April 30, 2021. Accessed Feb 2024. https://www.youtube.com/watch?v=EF-INPvEhA8

CHAPTER 8 – Does reward work?

Jean Donaldson, in *"The Culture Clash"*, who is *highly* critical of any aversive approaches, and likes to use a lot of highly emotive language to attack it, also admits the shortcomings of positive training. She explains a long, complicated, difficult, and highly impractical process (including involving the help of other people), which includes kindergarten level exercises, elementary school exercises, high school exercises, and finally college level exercises. At the end of her penultimate "High School Level Recall" she claims:

> "Perfection of this exercise lends tremendous distraction-proofing to your recall." [165]

Fantastic! But wait, is it really? When Donaldson then moves on to her "College Level Recall", she suggests testing your dog by having a helper drag a furry toy behind them, and says:

> "Once he's spotted the furry toy and given chase, try your recall cue. There's a small chance he'll do it, but a better chance he won't." [166]

So much for "tremendous distraction-proofing"! Donaldson admits there's a "small chance" your recall command will work even after all the effort put in up to that point (a huge amount), and a "better chance" it won't! And even after the conclusion of the "College Level Recall" (the final step in her recall training process), Donaldson notes:

> "then [after all this work] there's *some chance* it [the recall] *might* kick in, in a *real situation*. Mastering this exercise *doesn't produce* a squirrel-proof recall on every dog. But it does *sometimes*." [167] (emphasis added)

Wow. So, when you really need it, your dog "might" come back, and all that hard, complicated, skilled, time-consuming training (that no-one in *the real world* with a family, and a job, and a life, will ever have time for) might "sometimes" work! Again, remember that this is coming from a so-called "expert" and highly respected name in the positive reinforcement training world. Ian Dunbar described her book as "quite simply the very best dog book I have ever read."[168]

If, after all that long, laborious training process that Donaldson outlines in the very best dog book Dunbar has ever read, and *might/sometimes* is the best she can hope for, what hope do general pet owners have? The answer is *none*. They won't get close to even that pitiful, *might/sometimes* result.

---

[165] Donaldson, Jean. *The Culture Clash*. 2nd ed. Berkeley, CA: James & Kenneth Publishers, 2005.
[166] As above.
[167] As above.
[168] As above.

PART 4 – Investigating reward

> If, after all that long, laborious training process she outlines in the very best dog book Dunbar has ever read, and *might/sometimes* is the best she can hope for, what hope do general pet owners have?
>
> The answer is *none*. They won't even *get close* to that *might/sometimes* result.

Pippa Mattinson in her book *"Total Recall – Perfect Response Training for Puppies and Adult Dogs"*, makes similar claims about how easy it is:

"…this book deals not only with creating *a superb recall the easy way* (starting with a new puppy), but also with helping those of you who have gone astray somewhere along the recall training path."[169] (emphasis added)

She then spends an entire 240-page book explaining this "easy" process! And it includes comments like the following:

"Is it worth the effort? While embarking on a training programme may seem a little daunting you will never regret the time you invest…"[170]

And:

"Is it all too much? Sometimes a dog owner will be **horrified at what is actually required** to proof their training effectively, wondering how they can **possibly be expected to go to all that trouble**."[171] (emphasis added)

So much for her claim of "creating a superb recall *the easy way*"!

Karen Pryor, for all her talk about how quick and easy and effective positive reinforcement training is, evidently does indeed also think that it's just *all too much* (like these owners mentioned by Mattinson who are *"horrified at what is actually required"* and *"wondering how they can possibly be expected to go to all that trouble"*):

"Dealing with squirrel chasing in the woods is just a shaping staircase; *if you want to do it, it can be done, but it involves a lot of steps.* ***For me, that's too much like work.*** My practical solution is a mix of training and management. The backyard is fenced… In the woods, my poodle, whose lust for squirrels is mitigated by his general timidity, can be off-leash… My 17-year-old border terrier, however, **stays on-leash** in the woods."[172]

---

[169] Mattinson, Pippa. *Total Recall: Perfect Response Training for Puppies and Adult Dogs.* Shrewsbury: Quiller Publishing, 2013 reprint.
[170] As above.
[171] As above.
[172] Pryor, Karen. *On my mind – Reflections on Animal Behavior and Learning.* Waltham, MA., USA. 2014.

So, even Karen Pryor, the "mother of modern dog training", while claiming it's "just a shaping staircase" (it isn't), finds it simply too hard and it's just *"too much like work"* to train a reliable recall. Her answer to this is simply to **never let her dog off-leash.** This is what AVSAB, and the positive training crowd in general, like to refer to as "management".

In contrast, if she had used sound methods she could have achieved a truly reliable recall with her dog in days. And done so without eliciting anywhere near as much fear as the dog she talks about that was so frightened of the sound of the clicker initially that it went and hid, terrified, under the bed. And, if properly trained when a pup, her terrier could have enjoyed *17 years of greatly increased off-lead freedom, which Pryor's ideology denied him*.

Yet, on the *Karen Pryor Academy* website, it makes this claim:

> "Karen Pryor Academy's approach is more effective [than punishment] in *every meaningful dimension*."[173] (emphasis added)

This obviously couldn't be further from the truth. (And don't forget also, that while claiming to be positive reinforcement trainers, *Karen Pryor Academy* trainers use aversive, force-based head collars in an attempt to keep their dogs under control on the lead; so once again, their "positive" training is not particularly effective at all otherwise they wouldn't need such tools. But more about lead training next.)

Pippa Mattinson (like Jean Donaldson or Zak George) in her book *"Total Recall – Perfect Response Training for Puppies and Adult Dogs"* also admits that in the end, her positive reinforcement recall training *simply won't work* with various dogs:

> "But with some dogs, training is not quite enough. If you have a dog with strong hunting instincts, you will need to supervise him to some extent throughout his life. Without that supervision he is very likely to revert to his old ways." [174]

So much for *"perfect* response training"! When she says that for some dogs "training is not quite enough", what she should say is that *positive-reinforcement training* is not enough. But remember her initial assertion about "creating a *superb recall the easy way*"? Mattinson also claims that punishment-based approaches don't work:

---

[173] Karen Pryor Academy for Animal Training & Behavior. *About Us*. Accessed May 27, 2024. https://karenpryoracademy.com/about/.

[174] Mattinson, Pippa. *Total Recall: Perfect Response Training for Puppies and Adult Dogs*. Shrewsbury: Quiller Publishing, 2013 reprint.

PART 4 – Investigating reward

> "The reality is that punishment of any description is rarely a useful or effective tool in the recall training process."[175]

Actually, it is highly useful and far more effective than any reward-based approach can ever be. I can help a client train a truly easy and reliable recall in a week. And do that regardless of breed or any other consideration, and all with a miniscule fraction of the huge commitment of time and effort, and with only positive side-effects (including greatly increased safety and freedom).

"Tully was fantastic. After just one session I was confident using the e-collar to train my dog to come to me on command and within a week he was a new dog!

I hope to return to get tips on how to stop them pulling when on the lead and trust that Tully would be able to help with that. Total professional and knows what he is talking about." - Patrick O'Keefe

---

Of course, Dunbar tries to make out the exact opposite, claiming that trainers who use punishment are deliberately using slow methods to retain clients:

> "Food lure/reward training is just so incredibly efficient and effective that most dogs are well-trained within minutes. By advocating more time-consuming and laborious techniques, the trainer is able to keep clients for longer and extract more money!"[176]

And:

> "In fact, food lures and rewards are so effective, their use should be mandatory."[177]

Really? Jean Donaldson obviously doesn't think it's fast. Zak George obviously doesn't. He keeps talking about the time and patience and relentless consistency that are required, and criticizing the correction-based "quick fixes" (while also likewise claiming to show us a "faster and easier way"!). Lelise Nelson doesn't think it is fast (as you will see shortly with her "Really Reliable Recall"). Mattinson and Pryor don't think it's fast—people are "horrified at what is actually

---

[175] Mattinson, Pippa. *Total Recall: Perfect Response Training for Puppies and Adult Dogs*. Shrewsbury: Quiller Publishing, 2013 reprint.
[176] Dunbar, Ian. *The Good Little Dog Book*. 3rd ed. Berkeley, CA: James & Kenneth Publishers, 2003.
[177] As above.

CHAPTER 8 – Does reward work?

required" and "wonder how they can possibly be expected to go to all that trouble", and it's just "too much like work" even for Pryor to train a decent recall.

So just how *incredibly fast* are these lure/reward methods? In relation to just how much work and time we are talking about, consider the following. In relation to teaching the recall using positive reinforcement, Jane Killion (author of "*Pigs might fly – training success with impossible dogs*"), a well-known reward-based trainer, asks her readers:

> "Are you beginning to get a feel for the *enormity of reinforcement* that is necessary to build a recall?"[178] (emphasis added)

And this not from a critic of the process, rather it is from someone who is promoting its use. And then a few pages later:

> "If you are going to take your dog to a park or trail and let him off leash, you may need *literally years* of substantial reinforcement history of coming when called.*"* [179] (emphasis added)

Years?! That's a long time to risk your pup or dog getting run over, or killing wildlife, or getting kicked by the horse, or having it take off down the street and chasing it for hours around town, or being unable to let it off the lead. And that's even *if* you believe this type of training will actually *work at all* even after these long *years* of training (it won't).

Just how long are you going to persevere with it before you give up? Most puppy school trainers and so-called "positive" trainers don't tell their clients that! This hardly fits Dunbar's requirements of "quick and easy so that owners can do it" criteria. Or his claim that "Food lure/reward training is just so incredibly efficient and effective that most dogs are well trained in minutes".

> Just how long are you going to persevere before you give up?
>
> Most puppy school trainers and reward-based trainers don't tell their clients that!

And, if we go back to Jamie Dunbar's assertion that if the behaviour improves then the method is good, what constitutes improvement? In the above scenario, the improvement is absolutely *miniscule* each training session, if *years* is how long it takes. And if it never gets you a reliable recall, can you really claim the method works?

---

[178] Killion, Jane. *When Pigs Fly!: Training Success with Impossible Dogs*. Wenatchee, WA: Dogwise Publishing, 2007.
[179] As above.

165

PART 4 – Investigating reward

On the other hand, with sound methods of training, a truly reliable recall is eminently possible in a week, if not less. Massive improvement almost instantly. (And as we learnt earlier, there are only positive side effects when done properly.)

When Jane Killion talks about "training success with impossible dogs", she is talking about dogs that are often impossible *for "positive" trainers*. They aren't impossible at all for good trainers, who actually understand *real world* training and motivation. They aren't even particularly difficult. I have virtually no more trouble training a highly aggressive Amstaff to walk nicely on a lead, or to recall, than a mildly reactive Cavoodle.

But Jane devotes an entire book on how to get some basic obedience out of these dogs, involving very long, complicated, difficult, and incredibly impractical methods, for things that some sound training would solve quickly and easily in a week.

So, in contrast to a sound application of correction or punishment or negative reinforcement—which truly is *quick* and *easy* and *effective*—reward isn't looking so good.

Jane Killion also states that:

> "In twenty-four years of having bull terriers, I have never owned or seen one that I would trust indefinitely off leash... Thirty foot leashes are cheap and easy to find..."[180]

And yet, her book is sub-titled "Training Success with Impossible Dogs". This certainly doesn't sound like "training success" to me—anyone can put their dog on a thirty-foot long-line with *zero* training. And even if these elaborate and highly time-consuming methods actually end up getting the desired results (even these "top" trainers admit they often don't), are they practical *in the real world*? Of course not. Unless you're a mad-keen, obsessed dog trainer who just wants to train your dog for an hour every day for the next two years, and even then still not be confident to let your dog off lead, how *can* they be practical? What percentage of general pet owners have *any* chance of making this work? Zero. So that's an ***awful lot of dogs that can never be given much freedom***, like Pryor's terrier, or that have the safety which a decent recall provides.

> What percentage of general pet owners have any chance of making this work?
>
> Zero.
>
> So that's an awful lot of dogs that can never be given much freedom, or that have the safety that a decent recall provides.

---

[180] Killion, Jane. *When Pigs Fly!: Training Success with Impossible Dogs.* Wenatchee, WA: Dogwise Publishing, 2007.

Elsewhere, Jean Donaldson states that:

> "When all else fails with a predation problem [chasing squirrels, etc.], one of the toughest things to fix, and the dog's life or quality of life hangs on the balance, I personally still might employ negative reinforcement to train a recall." [181]

So, there we have, once again, an admission of the failings of "positive" training, and of the effectiveness of negative reinforcement (also, note that dogs *are* predators—so predation is hardly an uncommon problem). But Dunbar says reward is "so effective it should be mandatory", and so fast that "most dogs are well-trained within minutes". Perhaps he just has an incredibly low standard in his definitions of "effective" and "well-trained".

Yet, for just about every dog owner, a good recall is one of the most important things in order to keep your dog safe, and to give it a good quality of life that involves lots of freedom.

And by the way, predation problems aren't particularly hard to fix—that is, unless you're a "positive" trainer. I deal with them on a regular basis. For example, a couple I helped who owned two Labradors. They had moved out of the city onto forty acres in the country, and assumed their dogs would be able to have a great time and lots of freedom. Instead, the dogs basically had to be confined to the house (AVSAB's recommended "management" strategy, or Pryor's) or they would disappear chasing rabbits, or kangaroos, or the neighbour's sheep. What *practical, effective* solutions that actually work for real people *in the real world*, would a "positive" trainer have been able to offer? Even the "top" "positive" trainers in the world have no answers. Karen Pryor made use of an invisible fence electronic dog containment system for her own dog (or just never let it off the lead), so she evidently had no positive reinforcement solutions.

Yet Zak George claims there is *never* a need to use any aversive methods (all the while teaching many completely ineffective reward-based methods for a whole variety of behaviours):

> "Punishments and aversion are actually *never required*. There is *no known instance in the animal kingdom* where using physical punishment is required to create new behavior or to stop unwanted behavior. And so, if one is making the claim that tools like choke chains, prong collars, and electric collars are necessary to deliver aversion to a dog, to discourage

---

[181] Donaldson, Jean. *The Culture Clash*. 2nd ed. Berkeley, CA: James & Kenneth Publishers, 2005.

future behavior, then they would be responsible for uncovering new science."[182]

Did you get that? "There is *no known instance in the animal kingdom* where using physical punishment is required to create new behavior or to stop unwanted behavior". Yet, in the animal kingdom, punishment is one of the main ways animals interact with their young, and with each other, and with the environment, all the time. In fact, instead of the "reward good behaviour, ignore bad behaviour" mantra of the positive ideologues, in the natural world the reality is the opposite—animals generally *punish bad behaviour and ignore good behaviour*. (For this very reason I was originally considering "Natural Dog Training" as the title for this book.) Consider this quote from scientists Fontes and Shanan:

> In the natural world animals *punish bad behaviour and ignore good behaviour.*

> "Punishment is a biological, behavior-regulation mechanism that is critical for learning to stop engaging in maladaptive behavior (e.g., Todorov, 2011; Vollmer, 2002). Regardless of whether or not one believes that punishment should ever be part of explicitly arranged contingencies, it will always be a part of natural ones."[183]

So, George is wrong from both a basic, *real world* experience perspective, and also from the scientific one. What about this science from Marschark & Baenninger, about the necessity for negative reinforcement and punishment:

> "While positive reinforcement can be used exclusively for the training of certain behaviors, it is suggested that in the context of instinctive motor patterns, negative reinforcement and punishment may be desirable and *necessary additions* to positive reinforcement techniques."[184] (emphasis added)

So it seems George doesn't know his science too well. And even Donaldson admitted that she would use negative reinforcement when her "positive" methods failed. And, George also says that off-leash training "takes months and months" (as we've seen above, more like *years*), and, in fact, might not work at all:

---

[182] George, Zak. *State of Emergency: The Dog Training Crisis is Here. The Case Against Aversive Tools.* YouTube, March 26, 2024. Accessed Feb 2024. https://www.youtube.com/watch?v=BWM_gHMQaUQ
[183] Fontes, Rafaela M., and Timothy A. Shahan. *Punishment and Its Putative Fallout: A Reappraisal.* Journal of the Experimental Analysis of Behavior, December 2020.
[184] Marschark, Eve & Baenninger, Ronald. (2002). *Modification of instinctive herding dog behavior using reinforcement and punishment.*

CHAPTER 8 – Does reward work?

> "…you must take a massive step backward in your training and *tightly control the environment*… A huge part of off-leash training is having *perfect* control of the environment… *Not all dogs* will excel at off-leash training, and progress might be *very slow* for some, especially those high-energy dogs… Also keep in mind that *some dogs will never be okay off leash*, and that's fine too. Only you will know when the time is right to let your dog off leash in certain situations, *if ever*."[185] (emphasis added)

So, according to Zak George, there is "no known instance where punishment or aversion are ever required to teach new behaviour or to stop unwanted behaviour", but "some dogs will never be okay off leash". Which is it? He says you *never* need punishment, and claims there is no science that says you do (we've just seen this is false), but that without it he admits you can't get many dogs to recall off leash. I think George just discovered some "new science"! (And not the junk pseudo-science he likes to quote—but we will deal with that later.)

So, once again we have the admission that "positive" methods may simply not work. In actual fact, this is very true for *most* dogs. And as I asked earlier, how many months and years of this training (and your life) are you going to put in, and waste, before you give up? Remember what George said?

> "Only you will know when the time is right to let your dog off leash in certain situations, if ever." [186]

I can tell you that I have *never failed* to teach a very large number of clients to teach their dogs a reliable recall *quickly and easily* (provided they were willing to follow simple instructions, and provided some reward-only ideologue or their vet didn't get in their ear and talk them out of it), usually in a single session. That's a lot of dogs that can now have a lot of freedom almost straight away, and greatly increased safety.

> That's a lot of dogs than can now have a lot of freedom almost straight away, and greatly increased safety.

So, here we are, comparing positive reinforcement methods that *might sometimes* work if you're lucky, but usually don't, even after a *huge* commitment of time and energy (years), compared to methods that do, always, every time, succeed. And which do so in a very short space of time.

As Jamie Dunbar says, "if the behaviour improves, the method is good". And as Ian Dunbar says, the methods *must be quick, easy and effective*. As we've clearly

---

[185] George, Zak, and Dina Roth Port. *Zak George's Guide to a Well-Behaved Dog: Proven Solutions to the Most Common Training Problems for All Ages, Breeds, and Mixes*. New York: Ten Speed Press, 2019.
[186] As above.

PART 4 – Investigating reward

seen, "positive" training fails dismally on these criteria. If I'd relied on their ineffective "methods", those Labradors would still be locked inside the house the majority of the time (AVSAB would approve). Instead, they have great freedom and a great life.

Why don't the reward-based methods work? As I mentioned earlier, it is simply due to what most pet owners realize—the *environmental rewards* are worth more to the dog than your treat (or toy, or whatever reward you are attempting to use).

You simply have competing rewards—and *the environment's rewards are bigger.* The dog is *rewarded massively* when it ignores your recall command and chases a rabbit, or goes to play with other dogs. The only way you can outweigh that is with a *consequence* the dog doesn't like—a correction or punishment. Just like us with speeding fines—there *must be consequences* for breaking the rules.

> The only way you can outweigh the environmental rewards is with a *consequence* the dog doesn't like – a correction or punishment.
>
> Just like us with speeding fines – there *must be consequences* for breaking the rules.

And that is why Zak George tells you that you must have "*perfect* control of the environment" to prevent it rewarding your dog. But, if you want to train a dog to be reliable off lead, by definition sooner or later you actually have to let it off the lead. Or, if you fail to control its environment perfectly and it gets off the lead, or out the door, your training is ruined. You've lost that "*perfect* control of the environment" and so, once again, the environment *immediately and massively rewards* your dog for ignoring you. As even Sidman notes:

"Nowhere outside the laboratory is such control possible." [187]

It is possible, with some dogs, to get a *passable* recall without *obvious* recourse to punishment, or at least not to *physical* punishment; just like it is possible to get a dog that walks on the lead acceptably without much training at all. You might have a timid little dog that is too afraid to venture far, doesn't like other dogs, doesn't want to chase things, doesn't like other people, and is highly food motivated. In that case, you may well be able to get away with only training with treats (or with not much training at all). But that's not 99% of dogs. Consider Karen Pryor's two dogs:

---

[187] Sidman, Murray. *Coercion and Its Fallout*. 2000. United States. Author's Cooperative.

CHAPTER 8 – Does reward work?

> "In the woods, my poodle, whose lust for squirrels is mitigated by his general timidity, can be off-leash… My 17-year-old border terrier, however, stays on-leash in the woods." [188]

So a positive reinforcement trainer of Pryor's knowledge and experience couldn't train her terrier to come back when called, or to stop chasing squirrels. But she can trust her timid poodle off-leash, of whom she says:

> "He's 10 years old now, and still panics at every sudden sound or strange sight." [189]

If Pryor ever raised her voice at him (which she at times admits she isn't against, and that it's "legal" for a "positive trainer" to do so, with the strange reasoning that this doesn't constitute punishment but is rather "a social act" or "communication" [190]!), then she has used punishment. And to her timid poodle Misha, that would have been *strong* punishment, and therefore effective. Consequently, Misha can be trusted off leash. To Pryor's terrier, on the other hand, any reprimand ("social act", "communication"!) would be water off a duck's back, and therefore ineffective.

So, basically any training, or not much real training at all, or the punishment of Pryor's stern voice, were sufficient for one her dogs. On the other hand, all of her training tools and knowledge and lifetime of experience were insufficient to teach a reliable recall for the other. With Misha, *any general pet owner could have achieved the same*, with no real knowledge of training at all. If only Pryor had called upon a good trainer for help, her terrier could have had a lifetime of off-leash freedom with a week's simple training.

And, as I alluded to there, often when people think (or claim) they have trained without recourse to punishment, they are simply wrong. They may have never resorted to *physical* punishment, but for a timid dog simply reprimanding them vocally can be a big enough punishment that it works effectively. Or a little tug on the lead. To most other dogs these things are simply water off a duck's back, but to *these* dogs they can be significant and effective punishments.

So, unless you can carry around a reward the dog wants more than chasing the rabbit, you have no hope if you rely on positive reinforcement. You can try the long, arduous, convoluted methods that are sometime touted to try to circumvent these issues,

> All I can say is, good luck!
>
> Unless you have countless months and years to waste, don't bother.

---

[188] Pryor, Karen. *On my mind – Reflections on Animal Behavior and Learning.* Waltham, MA., USA. 2014.
[189] Pryor, Karen. *On my mind – Reflections on Animal Behavior and Learning.* 2014.
[190] As above.

such as appealing to the "Premack principle" or "life rewards", and/or "literally years of substantial reinforcement history"[191] (a myth), and hope for a *might/sometimes* recall. All I can say is, good luck. Unless you have countless months and years to waste, don't bother.

The "Really Reliable Recall"

Earlier, we heard about Pat Miller, with her thirty plus years' experience (in addition to being on the board of directors of Dunbar's *Association of Pet Dog Trainers*), failing to teach Happy to stop chasing horses without recourse to punishment, or to come back when called with that distraction. She mentioned that perhaps the so-called *"Really Reliable Recall"* of Leslie Nelson might have done the trick.

Leslie Nelson is another "positive" trainer attempting to find a way to get a good recall. Her *"Really Reliable Recall"* is a great example of a solution that indicates the problem. If treat training really was easy (like the Dunbar's and all the rest try to make out), then what on earth would you need a separate emergency recall for? Nelson's approach is to create not an *everyday* reliable recall, but an *emergency* recall. It is to be saved for, and used only in, an emergency. The obvious question is, "Why?" Why not just train a good reliable recall for everyday use? Why this need for a special "emergency" recall?

> The "Really Reliable Recall" is a great example of a solution that indicates the problem.

The answer is that, obviously, her everyday recall was simply not reliable. Dunbar has a similar strategy (while telling us how quick and easy reward-only training is)—he actually has an "everyday name" for the dog, when you aren't too fussed if it obeys, and then a "formal name" when you actually want it to listen to you:

> "Achieving reliable compliance when we need it involves giving our dog at least two names…"[192]

And likewise, he has "everyday commands", and then "formal commands". Again, why this need for special commands and special names if "positive" training is so quick and easy and effective that it "should be mandatory"?

---

[191] Killion, Jane. *When Pigs Fly!: Training Success with Impossible Dogs*. Wenatchee, WA: Dogwise Publishing, 2007.
[192] Dunbar, Ian, Dr. *Barking Up The Right Tree - The science and practice of positive dog training*. Novato, CA: New World Library, 2023.

CHAPTER 8 – Does reward work?

The premise is that you choose a *different* command to your day-to-day recall, an *emergency* command, and then after a great deal of training *spanning perhaps a year or more,* hopefully if you ever need it, it might work.

Because you hardly ever use it, there is less chance for the environment to reward your dog for ignoring it.

And so, it might just work *once* when you have an emergency. But these trainers know that if you *keep* using this recall command day to day, its effectiveness will quickly fade. That's why it would have failed for Pat Miller's client's dog Happy. Even *if* they put the massive amount of time into it as Nelson outlines (see below), at best it would only work once or twice before it degraded.

And during that year of training, what would they do with Happy? According to the *American Veterinary Society of Animal Behaviour (AVSAB)*, you just need to "avoid situations that lead to the unwanted behaviour" and utilise "environmental management"[193]. In other words, either keep Happy locked up, or get rid of the horse. That pretty much sums up the main gist of the AVSAB's "training" solutions (just like keeping Sharon constantly restrained in a strait jacket to stop her punching herself in the face—until punishment quickly and permanently cured the problem). AVSAB too realizes the impotence of "positive" training, even while claiming that it "has been shown to be more effective than aversive methods."[194] (One of the main studies used to back up these fallacious claims will be debunked in *Chapter Ten – The scientific basis—fact or fallacy?*)

Nelson explains the time involved in teaching her *"Really Reliable Recall"*. She recommends giving your dog training sessions:

> "Three times a day, for the next two weeks... After the first two weeks, practice once a day or even 4 or 5 times a week for a year. After that, practice once or twice a week for life and you and your dog will have a recall to be proud of."[195]

Wow. Once again we see the enormity of the amount of work required. 4 or 5 times a week *for a year*, then once or twice a week *for life*. For a recall that you only use in an emergency. Brilliant. And yet, on the back cover of her DVD it claims "Works Immediately". And Nelson also claims it's "really pretty easy to teach"!

Let's look at Leslie Nelson's qualifications:

---

[193] American Veterinary Society of Animal Behavior. *Position Statement on Humane Dog Training.* 2021. https://avsab.org/wp-content/uploads/2021/08/AVSAB-Humane-Dog-Training-Position-Statement-2021.pdf.
[194] As above.
[195] Nelson, Leslie. *Really Reliable Recall Booklet.* Dogwise. Kindle Edition.

PART 4 – Investigating reward

> "LESLIE NELSON is the owner and director of Tails-U-Win! Canine Centre, in Manchester, Connecticut. The Centre has a staff of over 30 and conducts 50 plus canine classes a month.
>
> Leslie has a degree in education and has been teaching pet obedience classes since 1973. She has bred and shown numerous breed champions and performance dogs and continues to compete with her Afghan Hounds, Whippets and Standard Poodles.
>
> Leslie instructs both family dog and competition classes; conducts private lessons and behavioural consults. She is a sought after speaker and was featured in the training videos: *The Power of Positive Training* and *Proof Positive*. Her booklets, *Management Magic*, *The Really Reliable Recall* and her *Instructor's Manual* are used in training classes across the country."[196]

So, to the question of *practical* effectiveness of "positive" training, here again we have a well-known "positive" trainer, who ran a large canine centre in the USA with over 30 staff and 50 classes per month. And yet, her general recall or that of her clients was not good enough that she felt the need for a *second* recall command, to be saved just for emergencies. If this is the best Nelson can hope for, what hope does the general pet owner have?

Let's now move on, having discovered just how ineffective "positive" training is for recall, and consider whether "positive" trainers have any useful strategies for that other most basic of dog training tasks—walking nicely on the lead.

Walking on the lead

When it comes to walking on the lead, Jean Donaldson (remember—author of the best dog book Dunbar has ever read), offers a number of "solutions".

Her first is the "stop/go" type method. The theory is that if every time the dog pulls, you stop, and don't move again until the dog takes the tension off the lead, then you can teach it to walk on a loose lead. This is one of those methods that *might/sometimes* work, and Donaldson admits as much.

Sometimes, with some dogs, we don't need to do any particular lead training much at all—some dogs just naturally walk pretty well on the lead without *any* real training. So, basically *no* training sometimes works also, and those are the dogs for which the stop/go method will appear to work. Donaldson also admits that the

---

[196] Nelson, Leslie. *Really Reliable Recall Booklet*. Dogwise. Kindle Edition.

CHAPTER 8 – Does reward work?

"stop/go" method is only likely to work if you start from scratch with a fresh pup—not one with existing problems.

That's the only "positive" method she offers. The next solution she suggests is some form of harness, such as a head halter (which is like a horse halter for a dog, only the type she recommends tightens uncomfortably around the dog's nose). These work on two fronts—leverage, tending to turn (force) the dog's head back to face you when it pulls, and to some extent turn its body also, so the dog cannot pull as strongly; and secondly, punishment—the nose band tightens around the dog's nose. If you watch videos of Donaldson using one you will see her physically lift and hold the front of the dog off the ground with this device—in other words, *holding the dog's front feet off the ground via a noose around its nose*. That is pure and direct punishment (and force). If trainers are using these tools, can they really claim to be "force free" or "positive"? Hardly. *The Karen Pryor Academy* also uses these same or similar devices, as does Victoria Stilwell, etc. So much for "positive" and "force free"! If their reward-based methods really are so quick and easy and effective, why are they all using these force-based control and punishment devices?

> If reward-based methods are so quick and easy and effective, why are they all using these force-based control and punishment devices?

Then Donaldson moves on to discuss choke chains, which she calls (in her emotive way) "strangle collars". Most obedience trainers call them "correction collars" or "check chains", because that's the way they're generally used to teach dogs to walk on the lead—there is a sharp jerk on the collar when the dog breaks the rules—it is not choking the dog; if it is, you are *not* using it correctly. It is simply a consequence for breaking the rules, like a smack or a speeding fine, or the mother that bites her pups when they get annoying.

Donaldson is right to an extent—if *used poorly* check chains are next to useless. But if used correctly they are highly effective. I don't use them a lot with clients nowadays (e-collars are generally a much superior option—requiring less skill once the method is understood, no strength, no physical jerk, and are far more precise), but I have on countless occasions in the past taught even the biggest, most highly aggressive and reactive dogs to walk quietly and calmly past other dogs in little more than five or ten minutes. Other experienced, skilled trainers could tell you the same.

Even Karen Pryor (who Donaldson calls the "mother of modern dog training") notes the effectiveness of choker chain training when done properly (and there are better methods than the one she used). She describes her dog Gus learning to walk nicely on lead at obedience classes (previous to Pryor becoming a reward-only extremist):

PART 4 – Investigating reward

> "The obedience classes consisted of traditional training using choke-chain collars... Gus did learn the meaning of those commands. I almost never had to jerk his collar in punishment. Gus was a gorgeous dog, and he worked briskly, with his head up, his eyes sparkling, and his stub of a tail wagging continuously. Judges loved that."[197]

So, highly effective, and without any sort of negative "fallout". Gus working "briskly, with his head up, his eyes sparkling, and his stub of a tail wagging continuously" doesn't sound anything like *Karen Pryor Certified Trainer* Emma Parson's list of:

> "hyper vigilance, irrational fear, heightened irritability, impulsive/explosive behavior, hyperactivity, aggression evoked with minimal provocation, withdrawal and social avoidance, loss of sensitivity to pleasure and pain, and depressed mood"[198]

Two instances in particular have stuck in my mind. Over thirty years ago, when I was sixteen or seventeen, I taught my first one-on-one private dog lesson. The owners had two large adult Rottweilers who they were having a lot of trouble walking on the lead. They were big, strong, and reactive. But in ten minutes or so I had them both walking beautifully on the lead, and proceeded to teach the owners the technique. Even back then I was getting far better results (as were most obedience trainers) than any modern "positive" trainers are nowadays, with all their supposed "advances" in their supposedly "scientific" methods.

> 30 years ago obedience trainers were getting far better results than any modern "positive" trainers today, with their supposed "advances" and their supposedly "scientific" methods.

Many years and many dogs later, I turned up to a house in a fairly rough part of town. A big, heavily tattooed, strong and imposing character opened the door with his large Pit Bull cross. The owner wasn't interested in small talk, so I proceeded to put my lead and choke chain (correction collar) on the dog. The owner, who could no longer walk his dog due to its size, strength, and dog aggression, looked at all 60kgs of lightweight me and informed me confidently that *"you'll* never walk *this* dog!"

I then explained in detail the long and complicated process of how we were going to *heal his poor darling's devastating underlying fear* over the next year or two with the help of hundreds of his friends... oh, woops, no, that wasn't what I did at all! Actually, I just gave a bit of a wry smile and away I went down the street. In

---

[197] Pryor, Karen. *Reaching the Animal Mind.* New York. Scribner. 2009.
[198] Parsons, Emma. *Click to Calm – Healing the Aggressive Dog.* Waltham, MA. Sunshine Books. 2005.

five minutes the dog was walking better than any modern, treat-trained obedience champion, calmly on a completely loose leash regardless of other dogs. The owner was suitably impressed. The results are always the same, when used properly.

So, after "strangle collars", Donaldson moves on to "pinch collars" (sometimes called "prong" collars), and states that these are a better option. I suspect she might be correct in some cases. They look vicious; however I'm sure they would be a great tool in certain cases. However, in my state in Australia these are illegal (which is another case of government agencies banning tools—on the advice of ignorant "animal welfare" activists—that are highly useful in improving dog's quality of life, with the result that more dogs end up stuck in back yards and never walked, or end up being put to death). Prong collars simply pinch the dog's skin when it pulls—removing in some cases the need for the strong leash jerks that are required with a check chain.

And that is the sum total of Donaldson's solutions for dogs with leash problems— the "stop/go" process, a force-and-punishment based head halter, the choke chain or the pinch collar. And yet, this is probably the biggest problem for the biggest number of pet owners. So the "very best dog book" Ian Dunbar had ever read has no real "positive" solutions in either an absolute, or in a practical, sense.

Victoria Stilwell also offers no effective solutions. She likewise starts off with the stop/go method, before suggesting that "if you find the preceeding method *too slow*" then try "the reverse direction technique", which once again is completely ineffective. She then suggests bringing out the treats, and rewarding the dog when it is close to you, which will also fail with any determined puller (competing rewards). And that's the sum total of her suggestions.

Zak George is no better. He also starts off with the stop/go method, and says:

> "Eventually, after a lot of repetition, this lesson starts to sink in."[199]

I can tell you that no, it hardly ever does. And certainly not unless you (and the kids, and the husband) are 100% consistent. Even then it won't work, unless the dog *didn't really require training anyway*, which does sometimes happen.

George then outlines another method of basically teaching the dog to "look at you" using treats, first inside the house, and then *very* gradually outside. (Ask yourself why we would need another method *if the first one worked,* and is so "mind-bogglingly brilliant" as Dunbar describes the method?) George admits that the process is a long, tedious one:

---

[199] George, Zak, and Dina Roth Port. *Zak George's Guide to a Well-Behaved Dog: Proven Solutions to the Most Common Training Problems for All Ages, Breeds, and Mixes*. New York: Ten Speed Press, 2019.

PART 4 – Investigating reward

> "Leash training usually takes extensive follow-through for a solid year at least or you risk regression!"[200]

And admits that his own dog, was, in his words:

> "…a late bloomer when it came to the more advanced behaviors such as leash walking and listening at a distance."[201]

> *Teaching a dog to walk on a lead or come when called are very basic elements for good trainers, though they often pose insurmountable challenges to "positive" trainers.*

Personally, I don't consider leash walking an "advanced behaviour". It's one of the most basic, most needed behaviours for the vast majority of pet owners. (Indeed, compared to sheepdog training, all pet behaviours are pretty basic. Teaching a dog to walk on a lead or come when called are very basic elements for good trainers, though they often pose insurmountable challenges to "positive" trainers.)

George then goes on to explain that, before attempting to train your dog on the leash (if the stop/go method fails):

> "First, your dog must have an outlet for all of her energy. The key is to make sure she gets enough exercise during the day so that walks are much more manageable… the absolute best time to do a leash training session is instantly after a workout that tires your dog—the kind where *she's so wiped out that she couldn't run if she wanted to*." [202] (emphasis added)

For many people, taking the dog for a walk *is* the exercise! However, if you can't *completely* tire your dog out *before* taking it for a walk, he suggests:

> "…take her for a long enough walk that she expends enough energy to then focus on learning proper leash walking… you really just have to wait for her to tire a bit during the walk." [203]

All I can say, once again, is good luck with that. According to George's advice you just have to put up with all the pulling until the dog finally tires out, and then you can give it some training to walk nicely on the lead. If you have a big, high-energy

---

[200] George, Zak, and Roth Part, Dina. *Zak George's Guide to a Well-Behaved Dog: Proven Solutions to the Most Common Training Problems for All Ages, Breeds, and Mixes*. New York: Ten Speed Press, 2019.
[201] As above.
[202] As above.
[203] As above.

CHAPTER 8 – Does reward work?

dog, *you* are going to be worn out long before the dog is. And then, when you're both exhausted, *now* you start training.

And how are you going to do that if you're elderly, or out of shape, or injured, etc.? Or if your dog is so big and so strong, or so aggressive, that you simply can't hold it?

Figure 10 - Tire your dog out completely and THEN start training (every walk)!

Another glaring problem with this advice is that the behaviour of pulling on the lead is self-reinforcing, and you are simply allowing the dog to practice the bad behaviour. George admits this problem in general terms in other places:

> "…the more your dog engages in an unwanted behaviour, the more ingrained that behaviour becomes."[204]

---

[204] George, Zak, and Dina Roth Port. *Zak George's Guide to a Well-Behaved Dog: Proven Solutions to the Most Common Training Problems for All Ages, Breeds, and Mixes.* New York: Ten Speed Press, 2019.

PART 4 – Investigating reward

Dunbar describes it as:

> "...the colossal reinforcing properties of being allowed to pull on leash"[205]

Or Emma Parsons who correctly notes (while advocating the same ineffective stop/go technique):

> "Never allow your dog to pull you. If he succeeds **even once in a while**, you are inadvertently employing what trainers call "a variable ratio schedule of reinforcement," a method of occasional rewards that can make a behaviour stronger and much more difficult to eradicate."[206] (emphasis added)

She's right about that. Yet George suggests walking your dog *while allowing it to pull* and engage in all its undesirable behaviours, for long enough that it's completely tired out, *before* starting every leash training session. And doing that *every walk for a very long time*. That's an awful lot of "engaging in an unwanted behaviour", and causing it to become "more ingrained", and of "colossal reinforcing" of a bad behaviour.

And don't forget that Dr. John Ciribassi, past president of the *American Veterinary Society of Animal Behaviour*, fully endorses George's methods. George's methods for walking on the lead, like for many other things, *simply will not work*, even *if* owners follow his advice. And most owners simply will not be able to follow even a fraction of his slow, tedious method *in the real world*.

I've helped people who have had both shoulders operated on from walking their dog; people who've been dragged face down on the concrete with broken nose and smashed teeth while their dog tries to attack another dog; people who've been dragged out into lakes while their dog tries to get at the ducks; and so on and so on. How, exactly, is this advice going to help them? And how is it even going to help the general pet owner? Who has the time to do all this year or more's worth of training with work, kids, other hobbies, and a life? Even if it worked (which it doesn't)? Nobody.

> Who has the time to do all this year or more's worth of training with work, kids, other hobbies, and a life?
>
> Even if it worked (which it doesn't)?
>
> Nobody.

---

[205] Dunbar, Ian. *The Good Little Dog Book*. 3rd ed. Berkeley, CA: James & Kenneth Publishers, 2003.
[206] Parsons, Emma. *Click to Calm – Healing the Aggressive Dog*. USA. Sunshine Books, 2005.

CHAPTER 8 – Does reward work?

Dunbar advocates various tactics, starting of course with the singularly ineffective "stop/go" method (which he describes as "wait and reward") which he claims is:

> "…mind-bogglingly brilliant for teaching dogs to walk calmly on-leash"[207]

Another tactic he recommends (again—why would anything more be needed if "wait and reward" was so "mind bogglingly brilliant"?!) is pre-walking your planned route before you take the dog for its walk. He suggests you hide treats in various places, like up in trees, on fences, etc., so that you can grab them when you want to reward your dog. His reasoning is that this helps your dog to learn to be obedient even when it thinks you don't have treats (a problem every pet owner who tries treat training has encountered). The simple fact is that this still won't get you a dog that walks nicely on the lead. And who has time to go for *two* walks?! Once without the dog while you plant treats along the route, and once with? As Dunbar says (even while advocating methods like this that are tediously slow and still don't work):

> "If dog training techniques are excessively time-consuming, requiring the patience of Job, many people might not devote sufficient time."[208]

He's right. They won't. And so, the dog simply won't get walked. Or maybe it will get rehomed or put down. As George points out elsewhere:

> "What's tragic is that so many of these issues are part of the reason our shelters are overflowing with unwanted dogs. **Millions of people get rid of their pets, often because they just can't deal with certain disruptive behaviors**… a study in the Journal of Applied Animal Welfare Science found that **65 percent of people who relinquished their dogs reported some behavioural issues as a reason**. Sadly, **hundreds of thousands of such shelter dogs are euthanized each year.**" (emphasis added)

He's also right. However, it is his methods, and Donaldson's, and Dunbar's, and Pryor's, and those like them (and the *American Veterinary Society of Animal Behavior's* position statement) that are *the major cause* of this problem. The "positive" ideologues have the blood of hundreds of thousands, and more like *millions*, of dogs on their hands.

> The "positive" ideologues have the blood of hundreds of thousands, and more like *millions*, of dogs on their hands.

So, reward doesn't work well for teaching your dog to walk on the lead either, just like it failed on the recall. Why? Once again because the environmental

---

[207] Dunbar, Ian, Dr. *Barking Up The Right Tree - The science and practice of positive dog training*. Novato, CA: New World Library, 2023.
[208] As above.

rewards and the intrinsic rewards the dog gets out of the behaviour ("the colossal reinforcing properties of being allowed to pull on the lead") simply outweigh any reward you are using.

Except for a small number of dogs that probably didn't need much training anyway, for most dogs the "positive" strategies fail. And even if there is a possibility that they *might* work, the huge commitment in time, consistency, and absolute control of the dog's environment and serious restrictions on its freedom that we've heard about, still renders such methods practically useless.

Competing rewards—and Premack

As we've seen, the single biggest reason treat training almost always fails in training a *reliable* dog *under distraction*, or of curing bad habits, can mostly be outlined in two words: *competing rewards*.

Although a reward *increases the likelihood* (one definition of "positive reinforcement") of a particular behaviour, the environment is also rewarding the dog for *not obeying* our commands, or for engaging in behaviours we don't want, and so increasing the likelihood of *those* occurring instead. Only it's using far bigger rewards. Jean Donaldson points this out:

> "The environment is training the dog all the time, expertly."[209]

So true. This is why she had no useful positive reinforcement methods for training a dog to walk on the lead, and the best she hoped for with recall training in the presence of real distractions after an enormous amount of work was a *might/sometimes* recall. It isn't rocket science. It's simply a case of *balance*. Which is going to win out? *Our* positive reinforcement training, or the "expert" environment doing *its* positive reinforcement training with *far bigger rewards*?

> It isn't rocket science.
>
> It's simply a case of *balance*. Which is going to win out?
>
> *Our* positive reinforcement training, or the environment doing *its* positive reinforcement training with bigger rewards?

Most dog's motivation to play with other dogs or to chase wildlife, etc., is extremely high (they are pack animals and predators, remember) and therefore the reward of doing so is *extremely high*. Therefore, the rewards dished out by these environmental distractions *are going to trump our treats every time*. And not by a little bit either, but rather by a

---

[209] Donaldson, Jean. *The Culture Clash*. 2nd ed. Berkeley, CA: James & Kenneth Publishers, 2005.

huge margin. So your dog is *jackpot* rewarded for ignoring you, and for doing the things you don't want.

Therefore, the secret to training with rewards, according to Pat Miller, is this:

> "All you have to do is figure out how to *prevent your dog from being rewarded for the behaviours you don't want* and rewarded consistently and generously for the behaviours you do want." [210] (emphasis added)

Ah, so simple! All we have to do is *prevent* the dog from being rewarded for the behaviours we don't want. Unfortunately, this is nowhere near as simple as Miller makes it sound. The problem is, that in order to achieve long-term training success using positive reinforcement, you have to keep control of *everything* the dog wants, *all the time, everywhere.* Jean Donaldson also likewise says:

> "All you need to do to train a dog is get control of his favourites."[211]

She's saying the same thing. So what are a dog's favourites? They can vary from dog to dog, but Donaldson lists:

1. Food.
2. Access to other dogs.
3. Access to outdoors and interesting smells.
4. Attention from people, and
5. Play.

That covers just about everything. So, all you have to do is get control of *everything,* if you want to successfully train your dog with "positive" training. (Don't forget, she's also the one who said you *might/sometimes* get a reliable recall, even after all this extensive training and "control of his favourites"—and also that she had no effective positive reinforcement solutions even for on-lead, where it is *far* easier to control all the dog's favourites.)

So you see that the problem is inherent in the solution that both these authors are suggesting: how exactly do we *prevent the dog from being rewarded* for behaviours we don't want? How exactly do we control *all* the dog's favourite things, all the time? The problem is at once both as simple, and as difficult, as that.

---

[210] Miller, Pat. *The Power of Positive Dog Training*. 2nd ed. Hoboken, NJ: Howell Book House, 2008.
[211] Donaldson, Jean. *The Culture Clash*. 2nd ed. Berkeley, CA: James & Kenneth Publishers, 2005.

PART 4 – Investigating reward

> How exactly do we control *all* the dog's favourite things, all the time?

This is the reason Zak George recommends keeping your dog on a lead *all the time*, whether outside or even *inside* the house, or otherwise putting the dog in a crate:

"The best way to keep control of your dog is by having her attached to you with a leash as often as possible. I'll be honest, *too few people take my advice on this*. They think they can bypass this step, but that's not advisable for the majority of dogs. I'm *not talking about having your dog on leash only when outside of the house*. **I want you to attach her to you often when inside the house, too**… Of course, you can't always supervise your dog this thoroughly. You have to work, run errands, and relax sometimes. In these cases, *it's **critical** that you make sure you have your dog in a controlled setting… **crates** are an excellent way…*"[212] (emphasis added)

I'm not surprised few people take his advice. Who wants to walk around all day with the dog attached to them, or with it locked in a crate? (These are the types of lengths "positive" trainers have to go to in order to have *any* hope of training a well-behaved dog.)

The science that "positive" trainers base their training on is Skinner's work with rats and pigeons, *in cages*. The environment was completely controlled. Or Pryor's work with dolphins—the dolphins were in a completely *controlled, captive* environment. Or Emily Harvey's work in introducing the failed STEP program to the UK Guide Dogs—she had been a Sea Lion trainer. These situations are *nothing* like *the real world*, with all its infinite *variables* and *unexpected* events. Thus their methods are fundamentally flawed. And so, they advocate that you must keep 100% control of the environment just like when training rats or pigeons in cages in the laboratory. They are right—if you have *any hope* of making "positive" training work, that is what you need to do. And you might need to do it for years, or more likely for the dog's entire lifetime. However, remember what Sidman said:

> "Positive" trainers tell you that you must keep control of the environment, 100%.
>
> They are right.
>
> However, it simply isn't possible in *the real world*.

"Nowhere outside the laboratory is such control possible."[213]

---

[212] George, Zak, and Dina Roth Port. *Zak George's Guide to a Well-Behaved Dog: Proven Solutions to the Most Common Training Problems for All Ages, Breeds, and Mixes*. New York: Ten Speed Press, 2019.
[213] Sidman, Murray. *Coercion and Its Fallout*. 2000. United States. Author's Cooperative.

CHAPTER 8 – Does reward work?

Some of the reward-only people right here are going to be jumping up and down (again) shouting, "but, but, but the *Premack Principle!*" Or, "but, but, but *Life Rewards!*" Nope, they don't fix the problem either. You still have to control the dog's environment *totally*, in order to prevent it from rewarding your dog with *even bigger "life rewards"* for the behaviours you are trying to avoid.

The Premack Principle basically means that you get the dog to do something it doesn't want to do (obey) in order to be allowed to do something which it does. Access to the second behaviour rewards the first, and is *contingent* upon it. Or, in technical jargon, a high-probability behaviour (what the dog wants to do) is used to reinforce a low-probability behaviour (just about anything else). So, you reward the less likely behaviour by giving access to the more likely behaviour.

For example, imagine going to take your dog for a walk. Your dog might be going crazy at the front door, so you have it sit before you open the door, and the door doesn't get opened unless it sits. Your dog will learn that sitting gets you to open the door and commence the walk (a "life reward"). Access to the walk is *contingent* on the dog sitting. And, in this case, you actually do have complete control of the environment—the door isn't going to open randomly by itself. Provided you are 100% consistent (more about consistency soon), this can work over time. However, if you aren't 100% consistent, then you have the undesirable behaviour on a variable reinforcement schedule, and you'll never fix it.

Really, for dog training purposes (or any other purposes, for that matter), the Premack principle is just positive reinforcement training. The only difference is in the nature of the rewards. Instead of treats, it is "go play," or "open the door and let's go for a walk"—what are sometimes referred to as "non-appetitive" rewards. It is basically just semantics—a fancy-sounding term that doesn't mean anything particularly useful at all. There is nothing magical about it. So you make watching TV contingent on your child eating their Brussel Sprouts? Hardly revolutionary. You don't let your dog out the door until it sits? Big deal.

> Premack is basically just a fancy-sounding term that in practice doesn't mean anything particularly useful at all.

You use something the dog likes doing to reward something it doesn't particularly want to do, *which is how positive reinforcement always works anyway* even in basic training like teaching your dog to sit. The dog likes eating treats, so sitting is rewarded because it gets access to the treat. The high probability behaviour (treat eating) rewards the low probability behaviour (sitting).

So, if your dog obeys your recall command, you let it go and chase the cat (a "life reward")! That's the reward for coming to you. However, in the *real world*, a smart dog (in fact, any dog at all) sooner or later learns that it can get to chase the cat *without bothering to come to you at all*, and so once again we're back to square one—competing rewards. Even Donaldson, after utilising this "life reward" or Premack technique, and claiming that "Perfection of this exercise lends tremendous distraction-proofing to your recall", noted that when you test your dog with a friend dragging a furry toy:

> "There's a small chance he'll do it [obey], but a better chance he won't." [214]

Hardly terribly effective, and that's not even with a real, live animal. And that's one of the limitations of Premack—you can't use many of the really big rewards that your dog *really* wants to do (such as chase wildlife, chase cars, go and harass bicyclists, attack other dogs) and that you're trying to gain control of—in this way. Are you going to let your dog go and chase cats, or cars, or across the street to chase dogs, just so you can reward it for coming back to you, or sitting, first? Hardly. So the environmental rewards are still bigger than your training reward.

Also, many of the examples dog trainers use of the "Premack Principle" aren't really Premack at all. In the Premack principle, the high-probability behaviour is made *contingent* on the low-probability behaviour. I.e., if you *don't* eat your vegies, you're *not* getting to watch TV. But if your dog is running around playing in a paddock already, or swimming in a pond, and you call it to you, then when it comes you say "go play!" or "go swim!" as a reward for coming, that activity was *not contingent* on the recall. The dog was already engaging in it. I could delve into this in greater detail, but there really isn't much point.

The only way this can work is again if you *have control of everything the dog wants, at all times*. So, when setting off for a walk, you have control of the door—if the dog doesn't do what you want, the door doesn't open. Easy. It's no different from getting a dog to sit for its meal. You control the meal, so the dog doesn't get it unless it sits patiently and waits for you to put the bowl down.

In *these cases*, there is indeed zero need for any correction or punishment (other than negative punishment—pulling the bowl away). The issue, however, is that **the "positive" trainers and ideologically driven academics act as if all cases are as straightforward as these controlled situations. *They very much are not.***

> You only have to fail in keeping control of *everything, once*, for your dog to be massively rewarded for disobedience.

The problem is that you only have to fail in keeping control of *everything, once*, for your dog to be massively rewarded for disobedience. If this happens a few times, you're sunk (in fact, once is

---

[214] Donaldson, Jean. *The Culture Clash*. 2nd ed. Berkeley, CA: James & Kenneth Publishers, 2005.

all it takes). Your dog is now on a variable schedule of reinforcement by that expert trainer of Donaldson's—the environment—using massive *jackpot* rewards that far outweigh yours. As Emma Parson's noted (in relation to walking on the lead, but the principle is identical):

> "If he succeeds even once in a while, you are inadvertently employing what trainers call "a variable ratio schedule of reinforcement," a method of occasional rewards that can make a behaviour stronger and much more difficult to eradicate."[215]

And Pryor likewise notes:

> "Once a simple behavior has been learned, a long and unpredictable schedule can in fact maintain behavior that you *don't* want, with *incredible power*."[216] (emphasis added)

You should certainly do your best to control things so that the environment doesn't get a chance to reward unwanted behaviour as much as practicable, but for nearly every dog owner I can guarantee it will happen at some point—and probably reasonably often if they give their dog much freedom at all. Even something as simple as a guest coming to your house and patting your dog when it jumps up. Again, we just have competing rewards—you doing your training to reward the dog for not jumping, or not rewarding it for jumping, and other random people training your dog to jump up.

One of the big differences between say a highly motivated competition dog trainer (maybe agility like Susan Garrett) at the top of their game, and the general dog owner, is that the successful ones *very* tightly control *every aspect* of their dog's lives *at all times*. Only in that way can they prevent the environment from rewarding their dog. No stone is left unturned. The average dog owner has neither the time, nor the skill, nor the inclination to go down this path, to *micromanage every aspect of their dog's life* for extended periods of time (years), and nor do their kids or partners.

> The most successful competition trainers *very* tightly control *every aspect* of their dog's lives *at all times*.

I have helped even high-level obedience competition trainers with dog problems they simply couldn't fix. And fixed them quickly and easily (as I could have for

---

[215] Parsons, Emma. *Click to Calm – Healing the Aggressive Dog.* USA. Sunshine Books, 2005.
[216] Pryor, Karen. *On my mind – Reflections on Animal Behavior and Learning.* Waltham, MA., USA. 2014.

PART 4 – Investigating reward

Pryor's terrier's lack of recall). If these people couldn't do it with "positive" methods, what makes anyone think the general run-of-the-mill pet owner can?

So, the biggest reason why "positive" training doesn't work in *the real world*? *Competing environmental rewards*, and therefore the absolute need to completely control the environment. Therefore, because the environmental rewards are usually much bigger than our training rewards ("life rewards" notwithstanding), and because it is impossible to perfectly control the environment, so-called "positive" training *simply does not work in the real world*.

Stopping unwanted behaviour with reward

Let's now turn from considering positive reinforcement's failure in relation to teaching behaviours (which is supposedly its forté!) such as a good recall or walking nicely on the lead, to using it for *stopping* unwanted behaviours. One of the criticisms (myths) discussed relating to punishment in an earlier chapter was that punishment supposedly only *stops* behaviours, and can never teach the dog *what to do*. As was demonstrated, this is incorrect.

Using negative reinforcement "suggesting" methods (such as teaching a pup to sit) clearly shows the dog what to do, as do many other negative reinforcement methods. Also, a consequence when the dog doesn't sit clearly communicates to the properly trained dog that it should have, and should now, sit. Or if the dog is in the kitchen when you're cooking dinner, you can easily tell it "out" and then chase it out if it doesn't respond. You are clearly showing it what to do. So punishment can definitely be used to show a dog what to do.

But even if punishment was only good for teaching dogs what *not to do*, that is still a vitally important tool. So much of what all dog owners need when dog training is exactly that—*stopping unwanted behaviours*. Behaviours such as chewing, jumping, barking, nipping, stealing food, pulling on the lead, getting in the rubbish bin, attacking the cat, chasing cars, chasing wildlife, etc., etc.

On the other hand, one of the main shortcomings of "positive" training is that it doesn't tell the dog (or child) what *not* to do. It doesn't directly *stop* behaviour. Only punishment does that. For this reason, the "positive" strategies for stopping behaviours are, like they are for many things, convoluted, difficult, time consuming, and ineffective.

Yet Dunbar claims:

> "lure-reward training is *always so much more effective, especially for eliminating noncompliance and misbehaviour*"[217] (emphasis added)

Once again, the reality is diametrically opposed to Dunbar's claims. In fact, if reward is your only tool, you are at a significant disadvantage, or more likely completely doomed to fail.

The problem, once again, is competing rewards. We can try the "positive" trainers strategy of ignoring our dog for jumping on us, but what happens when it jumps on the kids or the guests who don't do the right things? The dog is still being rewarded for those behaviours. We can try all sorts of complicated procedures of teaching "look at me" and "leave it" so it doesn't steal food off the kitchen bench, but if we aren't *"perfectly* controlling" the environment (to use Zak George's words), and the dog even *occasionally* manages to steal a steak off the bench, it is still *massively rewarded "with incredible power" on a variable reinforcement schedule* for the behaviour we are trying to stop (like we heard from Parsons and Pryor).

Let's look at a couple of examples of reward attempting to *stop* behaviours.

Bin Diving

Getting into the rubbish bin, or stealing food off the kitchen bench, or chewing things they shouldn't, are all basically the same problem. So everything said here will apply to all of these, and more.

Consider again this quote from Karen Pryor, who stated that punishment "almost never really works":

> "My Border terrier, as a young dog, became fond of digging in the wastebaskets and spreading the contents around. I didn't want to punish her, but I also didn't want to constantly empty the wastebaskets.
>
> "I filled a spray bottle with water and added a few drops of vanilla extract: a strong but pleasant scent to me. Then I gritted my teeth and sprayed the dog in the face. She was dismayed and ran. I sprayed the wastebaskets with the scented water. She stayed away from the wastebaskets from then on."[218]

---

[217] Dunbar, Ian, Dr. *Barking Up The Right Tree - The science and practice of positive dog training*. Novato, CA: New World Library, 2023.

[218] Pryor, Karen. *Don't Shoot the Dog!: The New Art of Teaching and Training*. Rev. ed. New York: Bantam Books, 1999.

PART 4 – Investigating reward

Pryor didn't seem to be able to find a non-punishment way of fixing it. And again, even though she stated that punishment "almost never really works", here we have an example of a big name in the "positive" training world (described by Jean Donaldson as "the mother of modern dog training"), using punishment very effectively. The hypocrisy is staggering. Yet Donaldson also states, in direct opposition to these various examples of Pryor's, which all worked quickly, easily, and permanently, that:

> "verbal reprimands, swatting, spanking, hitting with rolled up newspapers, throwing things, *spraying with water or citronella*, shaking by the scruff of the neck and the ubiquitous leash jerk… typically have to be used over and over."[219] (emphasis added)

Not if they are used properly, they don't, as we have seen *over and over*. And Pryor attests to that fact right here. That's one of the great things about correction or punishment—when you use it properly you *don't use it very much at all*.

> When we use punishment properly, we don't need to use it very much at all.
>
> So, although we will talk about punishment a lot in this book, it is something that when done properly is used minimally.

So, although we will talk about punishment a lot in this book, it is something that when done properly is used minimally.

In contrast, as we read earlier, here is Zak George's completely ineffective "positive" method to deal with the same problem:

> "If you catch your dog getting into the garbage, telling her "No, don't do that" in a calm tone and then removing access to the trash is very logical to me... So after you remove the garbage and get your dog to sit and focus her attention on you, then reward the positive behaviour."[220]

Like much of all "positive" trainer's advice, this is about as ineffective a waste of time as I can imagine. It almost can't even be put into words just how completely pathetic this advice is. Pardon the pun, but it's total garbage! So you "remove the garbage", and then once you've done that you "get your dog to sit and focus on you", and then you reward it! Unbelievable.

---

[219] Donaldson, Jean. *The Culture Clash*. 2nd ed. Berkeley, CA: James & Kenneth Publishers, 2005.
[220] George, Zak, and Dina Roth Port. *Dog Training Revolution: The Complete Guide to Raising the Perfect Pet with Love*. New York: Ten Speed Press, 2016.

And keep in mind that Dr. John Ciribassi, DVM, DACVB, the past president of the *American Veterinary Society of Animal Behaviour* (which thinks it is qualified to tell everyone how we should be training dogs), says:

> "His techniques for managing common training problems are spot-on."[221]

The problems with George's "method" (if it deserves to be called that) are numerous. Firstly, calmy telling your dog "No, don't do that", means that, unlike Pryor's squirt bottle, it fails to act as a punishment. The goodies in the rubbish bin are rewarding, and so the consequence must *outweigh* that if the training is to be effective. So, there's no effective deterrent. And dogs don't understand English, no matter how much Dunbar wants it to be so, or how good Zak George's grammar is.

Secondly, by the time you've put the garbage out, then gone back to your dog and made it sit and focus on you and rewarded it, this is completely disconnected from the behaviour. It is absolutely pointless. I can guarantee the next time the rubbish bin is back within reach, George's dog will be back in it. However, this type of strategy is typical in the "positive" world. Let's look at this in a bit more detail.

Rewarding alternative behaviours

There are a few main ways the reward-only ideologues attempt to use positive reinforcement to stop behaviours (as George attempted above). Rewarding what they call "incompatible" behaviours is one way. Rewarding anything else at all is another strategy. And ignoring bad behaviour is another one.

- Reward incompatible behaviours.
- Reward any other behaviours.
- Ignore bad behaviour.

For example, Sidman completely erroneously claims:

> "…we do not have to punish in order to prevent or stop people from acting badly. We can accomplish the same with positive reinforcers… One way to stop people from doing something without punishing them is to *give them positive reinforcers for doing something else.* This is probably the most

---

[221] George, Zak, and Dina Roth Port. *Zak George's Guide to a Well-Behaved Dog: Proven Solutions to the Most Common Training Problems for All Ages, Breeds, and Mixes*. New York: Ten Speed Press, 2019.

practical noncoercive technique of behavior management."[222] (emphasis added)

The strategy of *giving them positive reinforcers for doing something else* (like sitting after Zak George removed the bin) basically looks as follows: you pat your dog when it sits instead of when it jumps; you give your dog treats when it isn't jumping on the kitchen bench to get the juicy steak; you give it treats when it isn't going berserk on the lead trying to kill another dog; etc. etc. These inane strategies fail for the same simple reason—competing rewards. So you pat your dog when it *isn't* jumping on you. Great idea. But perhaps it still enjoys jumping on you and biting your sleeve, so it is still also being intrinsically rewarded for that. So it continues to do both, and gets rewarded for both.

Karen Pryor's statement below highlights this inherent problem, when she used this strategy of teaching Apo (a captive dolphin) to press a lever to get fish instead of harassing its human handlers. The dolphin couldn't be in two places at once:

> "Fortunately, *Apo preferred lever pressing to swimmer harassment*, so the behaviour was eliminated."[223] (emphasis added)

It was indeed fortunate, because if Apo had instead preferred harassing the human, the strategy would have failed. And this is the reason such methods are rarely effective. Dogs (or kids, or students) often enjoy the things we don't want them to do—like play with other dogs, kill the chooks, chase cats, etc.—far more than any alternative we can provide.

Consider some more examples: you give your dog treats when it is in the kitchen when it *isn't* stealing food, or when it comes into the kitchen and sits politely.

> *Both the desired behaviour, and the undesired behaviour, are being rewarded.*
>
> And now your dog is also constantly pestering you for treats every time you are in the kitchen!

Great. But it still gets rewarded massively when it does manage to steal a nice juicy steak, so why would it stop? Now it gets a juicy steak *and* treats for random other stuff. Fantastic. *Both are being rewarded.* And, now your dog is also constantly pestering you for treats every time you're in the kitchen! You are rewarding it for being in the kitchen.

Or, imagine you teach your dog "leave it" by dropping food on the kitchen floor and then covering it with your foot when he tries to get it, as the "positive" trainers suggest. When he leaves it, you give him a treat. But

---

[222] Sidman, Murray. *Coercion and Its Fallout*. 2000. United States. Author's Cooperative.
[223] Pryor, Karen. *Don't Shoot the Dog!: The New Art of Teaching and Training*. Rev. ed. New York: Bantam Books, 1999.

those times when he actually steals food off the bench still get rewarded—why would he stop?

Or even worse, you try to use treats or toys to *distract* your dog *while* it is being reactive. Now you are *directly* rewarding the behaviour!

I had a great example of this a few weeks ago. I turned up to help a woman with her aggressive dog. I rang the doorbell and immediately some savage barking erupted from inside the house. This was closely followed by a dog sliding around the corner of the hallway and racing to smash into the closed wire mesh security door, barking furiously. The owner rapidly appeared behind the dog with treat bag in hand, before waving a treat in front of the dog in an attempt to distract it (as she had learnt from a "positive" trainer). The dog mostly ignored the treat waving in front of its nose, but occasionally turned away from the door to snatch the treat with hardly a pause in its ferocious barking. This is a perfect example of what so many "positive" trainers recommend, and of what you definitely should NOT do. The owners nearly always recognize the stupidity of what they're doing, but assume these so-called "experts" must know what they're talking about.

> The owners nearly always recognize the stupidity of what they're doing, but assume these so-called "experts" must know what they're talking about.

Let's give a human example. You are driving in the car with your kids, and they're being highly annoying. So, you try a reward-based approach and say, "if you sit quietly for the rest of the drive, we will get McDonalds!" They comply. Great—reward fixed the problem, at least short term!

But guess what? Your kids aren't stupid. Next time, your son nudges his sister and says, "Want to go to McDonalds? Every time we misbehave, mum offers us McDonalds to get us to shut up—watch this!!!" So, the reward can work short term, but makes the behaviour worse long term.

> Reward can work short term, but makes the behaviour worse long term.

Don't waste your time attempting these pointless and often counter-productive ideas. Here is a great example of this "reward an alternative" madness if your dog is biting you, again from Zak George:

PART 4 – Investigating reward

> "If your dog is getting bitey, grab one of those soft, room temperature dog treats from a sealed bag or container. At this point, your dog will likely stop biting and go into a sit. Great! You've got her where you want her."[224]

Ah, no, Zak. She has got *you* exactly where she wants *you*. She bites you, and you bring out the treats. Not the greatest of strategies.

Zak George's other strategy is to "redirect her attention to a toy to play tug". So every time your dog bites you, it's time to play tug. Guess how your dog now learns to initiate a game of tug? **A basic understanding of dog psychology, or the psychology of learning in general, or just some** *plain common sense***, is sorely lacking.**

So many things that *seem* to work, or perhaps even do work *short term*, are actually making things worse *long term*. Yes, your dog stops at the time when she bites you and you bring out a treat. But long term you are simply *rewarding bad behaviour*. The foolishness of this type of "positive" training is difficult to comprehend. And yet, millions upon millions of viewers and readers are being deceived by just this sort of insanity.

So, rewarding alternative behaviours is rarely a good answer, because either you are directly rewarding the behaviours themselves, or the behaviours themselves are still being rewarded by the environment at other times anyway.

> Rewarding alternative behaviours is rarely a good answer.
>
> Either you are directly rewarding the behaviours themselves, or the behaviours themselves are still being rewarded by the environment at other times anyway.

On the other hand, if you say to your kids, "if you don't stop yelling and screaming, you won't be going to your friend's house tomorrow" (punishment), the behaviour will stop (provided the consequence is big enough to outweigh how much fun they are having yelling and screaming, and provided they know that you will follow through), *and* the behaviour will be less likely to occur in the future.

If your dog is biting your arm, and you want some real training for your dog that actually works (and has additional benefits), here it is. Simple, quick, and effective: *quietly* and *calmly* tell your dog "No", once, while it is biting you. If it doesn't stop, give it a firm smack on the nose (or head area—the same area dogs, or wolves, correct each other) while it's still *in the act* of biting you. If it still doesn't

---

[224] George, Zak, and Dina Roth Port. *Zak George's Guide to a Well-Behaved Dog: Proven Solutions to the Most Common Training Problems for All Ages, Breeds, and Mixes*. New York: Ten Speed Press, 2019.

CHAPTER 8 – Does reward work?

stop, you weren't firm enough. Get firmer. Don't be frightened to give your dog a good firm whack (without getting angry or flustered). When it stops, calmly give it instant praise that it did the right thing, "Good girl" (without the "positive" trainers high-pitched baby-voice). Just a couple of repetitions and problem solved. Quick and effective, and without any long-term negative side effects (providing you are firm enough), only positive ones.

Unlike George's method for stopping biting which, even if it works (and it won't):

> "…it can feel like *it takes forever* to resolve this… it can *take some time* to stop… any time you try to override an instinctive behaviour, you must be *extremely patient…*"[225] (emphasis added)

Seems like that's a common denominator to all his methods. That, and requiring "*perfect* control of the environment" (emphasis his) as I discussed earlier. And he isn't alone. The same applies to all the "positive" methods. And yet George claims in his book that he would show us "a faster way".

One of the problems with using punishment for play biting, according to Dunbar as we heard earlier, is that it might actually work. He says, that if you use punishment:

> "At worst, your puppy will stop play-biting altogether, and hence will not develop bite inhibition."[226]

So, while admitting punishment might work (it does, even though elsewhere he claims how ineffective it is), for some bizarre reason he says this is bad because you need to let your pup keep biting for months while it learns how to do it softly. So, when your Great Dane pup is biting your baby and young kids (and you), you have to accept this behaviour *for months* as you painfully follow Dunbar's completely crazy process.

(Note also, that treat training itself encourages your pup to bite or nibble at human hands, and steal food from your kids, as the pup is always looking for treats.)

> Treat training itself encourages dogs to nibble at human hands, and even to snatch your kid's food from their hands.

Dunbar's process basically involves giving your pup what he calls "verbal feedback" like the following (using his ESL—English as a Second Language—approach):

---

[225] George, Zak, and Dina Roth Port. *Zak George's Guide to a Well-Behaved Dog: Proven Solutions to the Most Common Training Problems for All Ages, Breeds, and Mixes*. New York: Ten Speed Press, 2019.

[226] Dunbar, Ian. *The Good Little Dog Book*. 3rd ed. Berkeley, CA: James & Kenneth Publishers, 2003.

PART 4 – Investigating reward

"…verbally express your displeasure, "Ouch!!! That hurt! You miserable worm.""[227]

Then he continues on:

"For added effect, you may accentuate your injured feelings by sobbing."[228]

Sobbing?!!! And no, he isn't joking—he's serious. But wait, there's more:

"If you feel your puppy still ignores your feedback, call it a "Jerk!", leave the room, and shut the door."

Figure 11 – Dunbar's "scientific" approach

(Stilwell's advice for mouthing in *"Train your dog positively"* is basically the same—get up and walk away if your pup or dog is nipping, or give your dog more exercise—neither of which will work.)

---

[227] Dunbar, Ian. *The Good Little Dog Book*. 3rd ed. Berkeley, CA: James & Kenneth Publishers, 2003.
[228] As above.

CHAPTER 8 – Does reward work?

And yet, on one of Dunbar's older training videos (SIRIUS Puppy Training), he advises a young girl in the puppy class that simply saying, "Ouch that hurt," won't work:

> "If she [the pup] hurts you, I want you to *scream* "OFF!" [here Dunbar does in fact scream OFF aggressively at the top of his voice] It's no good saying, "Oww, that hurt," or, "oh, Sheena's ripping my jumper," *Sheena won't learn anything from that, ok?* So, if Sheena ever puts her mouth on you, you say "OUCH!" [once again Dunbar screams OUCH *very* loudly and aggressively]"

Jean Donaldson likewise recommends this type of screaming at your dog:

> "Screech "OUCH!" even if it didn't hurt and abruptly end the game."[229]

If you scream at your pup like Dunbar does in this example (and on the video the pup cringes to the floor when he screams at it), or screech like Donaldson recommends, *it is pure and simple punishment*, and it might well be effective if the pup is relatively soft natured. However, it's hardly the approach I would suggest taking—we are far better keeping things quiet and calm rather than *screaming* and *screeching* wildly and aggressively like Dunbar demonstrates and Donaldson suggests. And, you can hardly call screaming and screeching "positive" training.

In fact, this training video of Dunbar's contains many, many examples of straight up punishment—strong leash jerks, grabbing the pup by the cheeks and screaming extremely loudly and aggressively in its face, smacking pups on the end of the nose, and so on. And yet, according to Dunbar's website, Jean Donaldson says:

> "SIRIUS Puppy Training is the video which changed dog training. It remains *the best video on positive methods* and early socialization available today." Jean Donaldson, author of The Culture Clash."[230] (emphasis added)

And the website claims:

> "The original puppy training video and still…"the leader of the pack." Voted BEST VIDEO (every year) by the Association of Pet Dog Trainers."[231]

---

[229] Donaldson, Jean. *The Culture Clash*. 2nd ed. Berkeley, CA: James & Kenneth Publishers, 2005.
[230] *SIRIUS Puppy Training Classic.* Accessed May 27, 2024. https://store.payloadz.com/details/183683-movies-and-videos-educational-sirius-puppy-training-classic.html.
[231] As above.

PART 4 – Investigating reward

Despite these claims, there's a great deal that is anything but "positive"—it is replete with punishment-based methods, and mostly poor ones at that. Of course, there's nothing wrong with a video that demonstrates punishment-based approaches (because in many cases these are indeed useful and necessary), *provided* it isn't claimed by supposedly "positive" trainers that it is "the best video *on positive methods*", and provided the approaches used are actually good ones. Dunbar's video generally fails on both counts.

Dunbar's next piece of advice, if his "ouch that hurt, you jerk" English as a Second Language verbal feedback approach fails (I would say *when* that fails, just like he said to the young girl while advising the use of yelling as punishment), is to "schedule a home consultation with a Certified Pet Dog Trainer." He means, of course, a trainer certified in his *Association of Professional Dog Trainers* courses—I wonder what different approach they'll have to offer?!

> If you're looking for a dog trainer, avoid any that are "qualified" through these reward-only companies like APDT or Karen Pryor or Victoria Stilwell and so on, and equally avoid so-called "veterinary behaviourists" who will mostly just want to drug your dog.

If you are looking for a dog trainer, I suggest avoiding any that are "qualified" through these companies like APDT or Karen Pryor or Victoria Stilwell and so on, otherwise these are the types of ineffective methods you will encounter. And equally avoid so-called "veterinary behaviourists" who will have nothing better to offer, and will mostly just want to drug your dog.

So, rewarding alternative behaviours, or this crazy "verbal feedback using English as a Second Language" nonsense, is pointless at best and counter-productive at worst. Yet, if we follow the pattern of how the pup's mother would train it, with a simple "No" command followed by a smack on the nose for non-compliance, we will fix the behaviour very rapidly in only a couple of applications, and with only positive "fallout".

On extinction—and consistency

Aside from *rewarding other behaviours* in a pointless and often counter-productive attempt to stop undesired behaviour, the other "positive" approach is to *ignore* the undesired behaviour. It is the second part of the mindless mantra everyone who has ever attended one of the reward-based puppy schools has heard: "reward good behaviour and *ignore bad behaviour*". As a strategy it is generally useless. In scientific terms this is referred to as "extinction". The idea is that if a behaviour is not rewarded it will fade away and disappear.

CHAPTER 8 – Does reward work?

In *theory* (i.e. with rats in a laboratory, etc.) not rewarding a behaviour can work (sort of). However, in *practice* it usually just makes things *worse*. First, this is due to the almost inevitable *inconsistency* in the *real world* in failing to ignore the behaviour *every single time, always*. And second, due to the competing environmental (or intrinsic) rewards that exist in the *real world* but don't exist in the laboratory or with captive dolphins.

Imagine your dog is outside wanting to be let in. It's jumping on the glass sliding door, or whining, or barking, or all three. So, we ignore it—"reward good behaviour, ignore bad behaviour". Only when the pup is *not* doing any of these things do we let it in. Great. The dog doesn't get rewarded for jumping, barking, whining, and only for calm behaviour at the door.

However, in the *real world*, what if you slip up occasionally? What if you're running late for work and just *have* to let the pup inside, and it's barking and scratching? What if you *need* to go outside, and the dog just happens to be engaging in one of these behaviours at the time? What if your 5-year-old child opens the door for the pup when it's doing these things? Or your parents who are staying with you?

Ian Dunbar levels the following accusation at punishment training by saying:

> "…punishment must be consistent, immediate, and instructive. Most people are rarely consistent or attentive 24/7, so it is next to impossible for people to punish effectively."[232]

I deal later with the absolute falsity of this statement, because one of the great advantages of punishment is that it very often does *not* have to be consistent (see *Chapter Fourteen – Myths about principles of using punishment*). Rather, it is "positive" strategies like "extinction" that actually require this type of *absolute consistency* if they have any hope of working. Zak George, the "positive" YouTube influencer, says (correctly this time):

> "One of the most essential aspects of ["positive"] dog training is *extreme consistency. I can't stress this enough.*"[233] (emphasis added)

---

[232] Dunbar, Ian, Dr. *Barking Up The Right Tree - The science and practice of positive dog training*. Novato, CA: New World Library, 2023.

[233] George, Zak, and Dina Roth Port. *Zak George's Guide to a Well-Behaved Dog: Proven Solutions to the Most Common Training Problems for All Ages, Breeds, and Mixes*. New York: Ten Speed Press, 2019.

PART 4 – Investigating reward

But as Dunbar says—in his misplaced criticism of punishment—this "extreme consistency" is "next to impossible". And so, the many "positive" strategies that depend on great consistency simply *do not work in the real world*.

> Strategies that depend on great consistency, such as many "positive" strategies, simply *do not work in the real world.*

Ignoring is one of those strategies that *sort of* works in theory, in the laboratory. Or in a *perfectly* controlled environment (remember Zak George saying you need to have "*perfect* control" of the environment?). When using a Skinner Box, the scientist only has to change a setting and the treats stop coming. The reward stops completely and never comes back on—it is 100% perfectly consistent. The scientist has "perfect control" of the environment. And guess what? The behaviour in question fades away and disappears over time (assuming it isn't an intrinsically motivated behaviour—in other words, the dog just likes doing it for its own sake. The Breland's, mentioned earlier in their criticism of operant conditioning, talk about these common problematic situations).

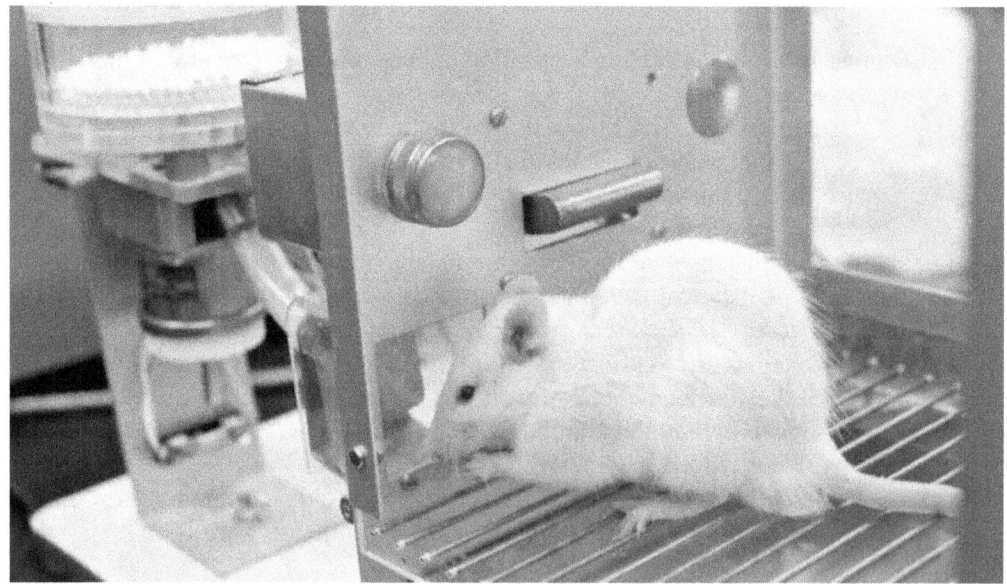

Figure 12 - The Skinner Box—perfect control of the environment

But what happens if you aren't 100% consistent? If you even *occasionally* slip up in one of the ways outlined above? If you don't have "perfect control" of the dog's environment? Now you simply have the undesired behaviour on a *variable schedule of reinforcement*. Therefore, **not only won't the behaviour improve, but it will get** *worse*. As we heard from Sidman:

CHAPTER 8 – Does reward work?

"Nowhere outside the laboratory is such control possible."[234]

And from Emma Parsons:

> "If he succeeds even once in a while, you are inadvertently employing what trainers call "*a variable ratio schedule of reinforcement*," a method of occasional rewards that can make a behaviour stronger and much more difficult to eradicate."[235] (emphasis added)

And Pryor:

> "Once a simple behavior has been learned, a long and unpredictable schedule can in fact maintain behavior that you *don't* want, *with incredible power*."[236] (emphasis added)

If you're a single person, with no one else to derail your training, and you start out with this strategy right from the beginning before habits have started to form, with a pup not much inclined to these behaviours anyway (genetics plays a huge role), you *might* just achieve your goal. But it rarely works.

My advice is to definitely try not to reward bad behaviour. Don't let your pup in when it scratches or barks, don't pat it when it jumps up, etc. But due to the fact that you, or others, are unlikely to be 100% consistent, the pup *will* be rewarded for the undesired behaviour on a variable schedule of reinforcement. Therefore, other strategies are likely to be required. And this certainly applies with established habits.

> Due to the fact that you, or others, are unlikely to be 100% consistent, the pup *will* be rewarded for the poor behaviour on a variable schedule of reinforcement.
>
> And so other strategies are likely to be required.

What about ignoring other types of behaviours? What if the dog is chewing the couch? I generally think of two categories of behaviours—those where the dog is seeking something from us, and those where it isn't. When the dog jumps on us or scratches at the back door, it wants something from us—perhaps attention, or the door to be opened. We hold the cards—we have control over what the dog wants. And so, we just don't give that to the dog. As I said, if you are 100% consistent it might even work, particularly if you start right from a young age with some pups.

---

[234] Sidman, Murray. *Coercion and Its Fallout*. 2000. United States. Author's Cooperative.
[235] Parsons, Emma. *Click to Calm – Healing the Aggressive Dog*. USA. Sunshine Books, 2005.
[236] Pryor, Karen. *On my mind – Reflections on Animal Behavior and Learning*. Waltham, MA., USA. 2014.

Most other behaviours are *not* dependant on us. When the dog is destroying the couch, it doesn't want anything from us—the behaviour is *intrinsically* rewarding. The dog simply enjoys chewing the couch for its own sake. Ignoring *that* is not a strategy at all (regardless if we attempt rewarding "alternative" behaviours), so we can safely debunk the idea of "reward good behaviour and ignore bad behaviour". We need to intervene directly.

Barking is similar. Some dogs just bark due to a genetic inclination to do so. It isn't necessarily because of a lack of exercise, boredom, or anything else like the "positive" crowd would have you believe. For example, many Kelpies are bred to bark when working sheep, and will keep it up all day while working. Are they bored? No. Lacking in exercise? No. It's just an intrinsic behaviour. Terriers often like digging. Scent hounds loving sticking their nose on the ground, and so on.

Ignoring any of these behaviours will fail, as it will for things like jumping on us. Therefore, we would all be wise to completely reject the mainstay of positive training "reward good behaviour, ignore bad behaviour". If common sense isn't enough for you, and you want some science (the "positive" trainers say they are big on science, all the while referencing junk pseudo-science by the likes of Meghan Herron or Zazie Todd and friends, and ignoring everything else), then consider the study by Azrin and Holz:

> "Typically, extinction will not produce a reduction of responses to the operant level because of the phenomenon of "spontaneous recovery" of responses at the start of each session (Skinner, 1938). The extinction procedure also resembles satiation in not maintaining response reduction after the procedure is terminated. Just as responding resumes completely after the subject is again deprived of the food reinforcement, so does reconditioning produce *complete and often immediate recovery once a few reinforcements have been obtained*...When extinction is introduced after intermittent reinforcement, **thousands of responses may occur prior to any substantial response reduction**... As a method of eliminating responses, **punishment appears to be potentially more effective** than either stimulus change, satiation, or extinction."[237] (emphasis added)

In other words, if you have the "patience of Job" as Dunbar described it, and ignore your dog jumping and scratching and barking at your back door for the "thousands of responses" we read about in the above study that are necessary to see "any substantial response reduction", you may see some improvement. But then, if you, or someone else, slip up even a few times, you will have "complete and often immediate recovery" of the unwanted behaviour. And now the behaviour is on a

---

[237] Holz, W. C., & Azrin, N. H. (1963). *A comparison of several procedures for eliminating behavior*. Journal of Experimental Analysis of Behavior, 6(3), 399–406. DOI: 10.1901/jeab.1963.6-399

really long variable schedule of reinforcement. Karen Pryor recognizes the problem this creates:

> "The longer the variable schedule, the more powerfully it maintains behavior. Long schedules work against you, however, if you are trying to eliminate behavior. Unreinforced, any behavior will tend to die down by itself; but ***if it is reinforced from time to time, however sporadically... the behavior, instead of being extinguished, may actually be strongly maintained by a long, variable schedule.***"[238] (emphasis added)

And yet, regarding teaching dogs not to jump on people, Victoria Stilwell's advice is:

> "The best way to stop your dog jumping on you is to ignore her while she is in the act of jumping."[239]

Ah, no, Victoria, it isn't. What you actually have there is a completely ineffective extinction-based approach, which in the real world not only *won't work*, but will simply make the behaviour *worse*. All I can say is, like most "positive" strategies, good luck with that. When you attempt to ignore bad behaviour you generally just put it on an increasingly longer variable schedule of reinforcement, and thereby make it worse and more persistent.

> All I can say is, like most "positive" strategies, good luck with that!

Confinement

When it comes to keeping your dog from straying, what effective methods do the "positive" crowd suggest? Here is an example we heard earlier, again from Karen Pryor:

> "The same principle is at work in the Invisible Fence systems for keeping a dog on your property. A radio wire is strung around the area in which you want to confine the dog. The dog wears a collar with a receiver in it. If the dog gets too near the line, the collar shocks it. However, a few feet before that point, the collar gives a warning buzz... if the setup is properly installed, a trained dog can be effectively confined and will never receive an actual shock. I used such a fence when my terrier and I lived in a house in the woods. An actual fence would have been a perpetual invitation to try

---

[238] Pryor, Karen. *Don't Shoot the Dog!: The New Art of Teaching and Training.* Rev. ed. New York: Bantam Books, 1999.
[239] Stilwell, Victoria. *Train your dog positively.* New York. Ten Speed Press, 2013.

to dig under it or escape through an open gate; the conditioned warning signal and the Invisible Fence were far more secure."[240]

So, another example of the effectiveness, and practical usefulness, of punishment-based methods. Pryor offers no "positive reinforcement" alternatives for achieving the same outcome. Again, one of the most influential "positive" trainers of all time resorting to highly effective means of punishment.

> Reward has no solutions because there are no solutions from reward.
>
> The "incredibly powerful" competing environmental rewards—on a variable schedule of reinforcement—are simply far superior.

Why not try Dunbar's so-called "autoshaping" method to reward the dog for staying around home? Well, because it won't work for starters—the dog receives much greater rewards for running off and exploring the bush. Again, reward has no solutions because there are no solutions from reward. The "incredibly powerful" (as Pryor rightly described them) competing environmental rewards—on a variable schedule of reinforcement—are simply far superior.

But it worked for my dog!

Quite often we'll hear this argument that so-and-so trained their dog with "positive" methods, and that this therefore proves that "positive" training works. However, there are potentially one or more of the following problems with this argument:

1. The dog wasn't actually *well-trained* at all.
2. The owner had actually used punishment, but just didn't *realize, remember,* or *admit it.*
3. The dog is one of those dogs that *less obvious punishment is sufficient.*
4. The owner is one of the top, most obsessed competition trainers in the world.

Let's start at the top. Is the dog that is claimed to be trained by "positive" methods *actually well-trained?* For example, Pryor's terrier that couldn't be let off the lead. Can we call that a *trained* dog? Of course not. And so, if anyone was to claim that Karen Pryor's dogs are *trained* without punishment, and therefore "positive"

---

[240] Pryor, Karen. *Don't Shoot the Dog!: The New Art of Teaching and Training*. Rev. ed. New York: Bantam Books, 1999.

CHAPTER 8 – Does reward work?

training works, they would be wrong. The training has failed. So that's the first question—is the dog that is claimed to be trained by positive methods actually *trained?* Have you seen it tested? In many cases, the answer will be *no*. The dog is simply *managed*. The owner just has a very low standard as to what "trained" means.

Secondly, in many cases, what appears to be (or is claimed to be) a pure-reward trained dog isn't anything of the sort. The owners have either *not realized, forgotten, or simply don't want to admit*, the punishment they've used.

Even Leslie Nelson with her "Really Reliable Recall" recommends that you "talk in a disappointed voice" to your dog if it doesn't listen. This is, frankly, quite laughable on a number of levels. But, they can't claim to have trained without resort to punishment.

I had a client here the other morning for her first sheepdog training lesson. When the time came for her to go home, she opened the doors to her car to allow it to cool down a bit (it was a pretty warm day). Then she asked her dog Ruby to hop up into the back seat. Ruby went to jump in the front instead, and the owner corrected it—"uh, uh, Ruby" in a voice that said "no, don't do that", and a tug on the lead. Ruby went to do it again, again the owner corrected her. Then Ruby jumped up into the back seat.

That right there is correction or punishment (both physical and vocal), something the reward-only crowd say we should never use, and that will without exception cause your dog all sorts of unwanted trauma. Remember Susan Garrett, a multiple national and world agility champion, stating:

> "…there is *always* fallout with *any form* of punishment—even something as benign as a time-out."[241] (emphasis added)

I doubt there are any dog owners in the world who don't do this sort of thing. So can anyone actually claim to have never, ever used punishment on their dog? The answer is no. And, of course, anyone who uses punishment- and force-based harnesses or head collars cannot claim to be "positive" or "force free". They are relying on punishment and force, as the Karen Pryor Academy and Victoria Stilwell do, or as Jean Donaldson did when holding the dog's front end off the ground with a noose around its nose.

What about if you have your dog on a long line in order to prevent it being rewarded by the environment, and it takes off to chase a kangaroo? Assuming you are still holding the long line, what is going to happen? The dog will hit the end of

---

[241] Garrett, Susan. *Punishment: Pros and Cons.* Susan Garrett's Dog Training Blog, 2010, https://susangarrettdogagility.com/2010/04/punishment-pros-and-cons/

PART 4 – Investigating reward

the long line, and thereby be punished. You might not have *meant* to punish the dog, but it was punished all the same, simply through your efforts to "control the environment". If that happens even just once, then you cannot claim that positive reinforcement alone achieved the result.

With a softer natured dog, that punishment of hitting the end of the long line might be all it takes to cure it of kangaroo chasing permanently. Punishment is, after all, *very effective*, and one sufficient application is often all it takes.

Years later, such an owner might claim "I never used punishment on my dog", but they would be wrong. The may not have *thought* of it as punishment, because they didn't do it deliberately, or they may simply have forgotten the incident. But punishment had certainly played its part in training their dog.

And, sometimes, when treat training *appears* to work, keep in mind that some trainers who *claim* to be "positive" trainers (and this applies to many competition trainers), actually resort to other techniques when no one is watching. It isn't at all uncommon. Given that it has become "politically incorrect" to use punishment, they might not admit it openly. And if they also teach classes or workshops, they won't talk about it. But deep down they realize how much more effective it is, and how ineffective "positive" methods are, and how much faster and easier it is (despite Garrett's completely false claim that her "positive" methods are the fastest and most effective available).

> Maybe they have a secret weapon, and maybe that is why you're confused when the methods they teach seem to work for them, but don't work for you.

So, when others see the results these trainers achieve, they assume it all came from positive reinforcement. Maybe not. Maybe they have a secret weapon, and maybe that is why you're confused when the methods they teach seem to work for them, but don't work for you. You think their dogs are trained with "positive" methods, but such is not the case at all.

I am sure many fans of Karen Pryor either are unaware, or completely ignore, the various examples she gives of where she used punishment-based methods with great success. All they hear is "punishment bad, reward good". Yet Pryor's own terrier was not a "positive" trained dog. Pryor had very effectively used an e-collar containment system, and a water spray bottle, as punishments. And those are only the ones she tells us about. If only she had followed such principles for recall, she could have succeeded there also. Pryor also says it's okay to use punishment at times when needed (i.e. to save the Christmas turkey from getting stolen off the table) but just claims that this won't have any long-term training effects.

So that's the first two points – is the "positive" dog actually trained? And has the claimed "positive" dog actually had effective punishment-based methods used in its training, even though the owner either doesn't *realize, remember,* or *admit* it?

The third point is that—with some dogs—much less *obvious* forms of punishment can be successful. This ties back into point two, that owners may simply not *realize* the times punishment has occurred, or how successful it has been. What is punishment to some dogs is—from our perspective—very mild, and therefore easily overlooked as "punishment". But to the dog itself it certainly was punishment.

Consider Pryor's two contrasting dogs. Her timid Misha was fine off the lead. Why? Because for it the environment was a very scary place, and so didn't outweigh the treat training to the same extent it did with her terrier, and because just telling Misha off was a sufficient enough punishment to create obedience.

I have a couple of contrasting examples like this here at home at the moment. A very strong, highly driven young Turkish Kangal livestock guardian dog (Nikki) who has required some strong and obvious punishment. And then a more sensitive Maremma (as some of them can be), for whom a stern voice is strong punishment and *might* be all that is ever required.

Figure 13 - The author's wife with young Nikita (Nikki)

PART 4 – Investigating reward

They are both trained following exactly the same principles, and they are both trained with punishment—it is just different *forms* of punishment and different *levels* of punishment. But what we need to realize is that the different level of punishment is from *our* perspective—from *their* perspective the levels of punishment are similar either way—an e-collar or strong smack for the Kangal, or a stern voice for the Maremma. In fact, the stern voice is probably stronger punishment for the Maremma than the seemingly much stronger punishment is for the mentally tougher Kangal. But the "positive" crowd would cringe at the thought of the smack or e-collar for the Kangal, while largely ignoring a sometimes raised voice for the Maremma.

There are some dogs who never really need much in the way of deliberate training at all (as pets). There are some dogs who are pretty quiet and never really jump on the kids or bite them. With some dogs, you don't actually need to do much training for a general pet. With those dogs, perhaps "positive" training could appear to have worked. But so does not doing much actual training at all.

Every dog is different. And positive reinforcement training will be more or less effective with some dogs than others. If your dog is highly food motivated, then treat training will work better (if you keep it food deprived, even better still, as you will learn in the next section).

If your dog isn't much motivated to chase animals, doesn't like playing with other dogs, doesn't like other people, or perhaps is frightened by most or all of these things, then the environment might not offer much in the way of "environmental rewards" to outweigh your training reward. And so, reward-based training will get better results than for other dogs. As we heard from Pryor, she could let her extremely timid mini poodle of the leash, but even with all her positive reinforcement training experience could still not let her terrier off the lead. Remember what Jamie Dunbar said about methods:

> "The world is full of self-proclaimed dog trainers who promote methods that sound reasonable, but they just don't work. Again—the proof is in the pudding. If your dog's behavior improves, the training methods are good. If your dog's behavior doesn't improve, the methods are not good."[242]

I have had any number of examples of this over the years. Recently I had a family call me out for their young Golden Retriever. They had an older Golden Retriever which they said they had never really done any training with, and he was basically the perfect pet. They had assumed this was what all Golden Retrievers were like! The young dog, however, had all the usual problems—pulling strongly on the lead,

---

[242] Dunbar, Ian. *Six Simple Steps to Solve Your Dog's Behavior Problems*. Dunbar Academy. Accessed Feb 2024. https://www.dunbaracademy.com/courses/six-simple-steps.

destroying things around the house, jumping and nipping the kids and guests, and so on. So one dog didn't need much training at all—in such a case someone might claim "positive" training worked (in fact, as they said, they really did hardly any deliberate training at all – but I am sure he had been told off occasionally). The other one certainly did need some effective, deliberate training.

The problem is that most dog trainers and authors of dog training books fall into a similar category to these owners—as do most of those commenting on social media advocating for "positive" methods and attacking everything else—they've only owned a small number of dogs and think they're experts.

> Most "dog trainers"—and even dog training book authors—have only owned one or two dogs and then think they know how to train dogs.

Another example I had was an older couple who called me out for a young adult German Shepherd. They told me that this was their ninth Shepherd, and that they had trained the first eight themselves. They admitted to believing they were good trainers, until they got the ninth! This time they failed to teach it much at all—it had all the general problems—toilet training, lead pulling and reactivity, destructive chewing, barking, jumping, nipping, wouldn't come when called, etc. They had simply been lucky in the genetic lottery with their first eight (or quite probably they had succumbed to the "positive" propaganda with their ninth!).

The fourth situation where "positive" training can *appear* to work, is with a *very* highly skilled (in the positive reinforcement sense), *obsessed* trainer, whose life revolves around training their dog (and even then, as we've seen, Karen Pryor herself couldn't make it work with her own dog).

If all you do is train dogs because you want to win the world championships (in certain dog sports), and you keep your dog's life *very tightly controlled at all times* in order to achieve that, and you enjoy putting *hours of training* into your dog *every day for months or years* in order to get results, and you have *forty years' experience*, and you *carefully select the right dog*, then maybe, just maybe, with *some* dogs, you might make "positive" training *appear* to work in that scenario. However, keep in mind that we don't actually know what methods really get used behind the scenes, either deliberately or not.

But as Dunbar said:

> "…the owner's not you, they don't have forty years' experience, they're not champion of the world—they're an owner. But they bought a Malinois, you know, and so it's got to be easy or else they can't do it. It's got to be quick, or else they probably won't. Dogs aren't their life… the sooner we as trainers get used to that, instead of holier-than-thouing them that you

shouldn't have got a dog, the quicker we can help people out. And it's got to be effective."[243]

So, can "positive" training *ever* work? The answer is that—in the *real world*—it never can. "Positive" training fails. Where it

> Can "positive" training *ever* work?
>
> No, never.
>
> Not in the *real world*.

appears to work, we never need look far to find that one of the above situations applies. Either the training wasn't actually successful; or the owner didn't *realize, remember*, or *admit* when punishment had occurred; or the dog was susceptible to very low levels of punishment and so again, the owner didn't realize or remember that this had occurred; or that a carefully selected dog's entire life was so tightly controlled by an obsessed and highly skilled trainer that it worked better than it would for any other dogs in actual *real world* scenarios.

In captive situations, positive reinforcement training is much more successful than in the *real world*. The more captive and controlled the situation, the better it works (remember Zak George's advice to keep your dog on a lead or in its crate). Consider performing dolphins, or captive animals in zoos, and so on. But in the *real world*, it fails. And that's without even considering the various other significant downsides to positive reinforcement training we're about to discover in the next chapter.

Conclusion – Does reward work?

So, we've seen directly from the mouths of some of the most influential "positive" trainers in the world (those described as "the mother of modern dog training", "the godfather of modern dog training", one of the most influential YouTube dog trainers in the world currently, and "the very best dog book" Dr. Dunbar has ever read, alongside other well-known figures), that "positive" methods often fail to get the necessary results. And they fail in both the absolute sense, and even more so in the practical sense, in the areas that are most important such as walking on the lead, recall, and stopping undesirable behaviours.

We also saw the three main reasons why this occurs:

---

[243] Balabanov, Ivan. *Training Without Conflict Podcast Episode Four: Dr. Ian Dunbar*. YouTube, April 30, 2021. Accessed Feb 2024. https://www.youtube.com/watch?v=EF-INPvEhA8

CHAPTER 8 – Does reward work?

1. **Competing rewards**. The environment and all that this entails is working against you, using rewards that far outweigh yours. And, in Jean Donaldson's words, it is an expert trainer. Your dog is rewarded enormously on an "incredibly powerful" variable schedule of reinforcement for its undesirable behaviour.

2. **You can't perfectly control the environment**. And the environment is a very good trainer, rewarding your dog for all the things you don't want it to do. Perhaps, if you keep your dog's life *very* tightly controlled (on a lead attached to you even *inside* the house, or in a crate at all times) like an obsessed "positive" competition trainer might, and never let it have much freedom (certainly not free access to your kids, husband, wife, guests, paddocks, etc., that will work against your training), then *maybe* with *some* dogs you can make "positive" training *appear* to work.

3. **The huge amount of knowledge, time, skill, and effort required**. The "positive" methods are confusing, convoluted and complicated. By the "positive" trainer's own admissions they are extremely slow and difficult (even while they claim how fast and easy they are). Unless you're a mad keen dog trainer who is willingly to try to understand these complicated processes, and is just dying to spend hours every day training your dog for years (and even then most such trainers don't succeed), then it simply isn't going to happen.

Sure, you can teach a dog advanced tricks with clicker training (positive reinforcement). And if trick training is your interest, then I suggest you learn clicker training. I still teach it to owners today. Yes, you can teach dolphins to perform in a captive environment. You can teach a chook to dance. You can even teach your pup to behave brilliantly *when it wants a treat*. But even Dunbar admits that:

> "By its very nature, clicker training is a slow, gradual, and involved process... unnecessarily complicated... it does require a sizable knowledge base of animal learning theory to know when to click...." [244]

Therefore, even though "positive" training might *appear* to work in a *tiny* percentage of cases, the conclusion is obvious (and *real world* experience proves likewise): in the most important areas of a pet's life, it fails.

So, having seen that positive reinforcement training fails, and why, let's look now at some of the other negative side effects of so-called "positive" training. Yes, there's more!

---

[244] Dunbar, Ian, Dr. *Barking Up The Right Tree - The science and practice of positive dog training*. Novato, CA: New World Library, 2023.

PART 4 – Investigating reward

9

More downsides and side-effects of "positive" training

Up to this point we've clearly seen—including straight from the mouths of the most influential so-called "positive" trainers in the world—the ineffectiveness, and certainly the immense impracticality, of using "positive" methods in order to train a reliable, obedient pet (or working dog for that matter). We've also seen that many of the "positive" methods are actually counter-productive and make things worse. This already should be enough to put an end to the *dark ages of the "positive" training cult*. Either the recommended "positive" methods don't work, make things worse, or are so painfully slow and require such 100% "relentless consistency" and "*perfect*" control" of the environment for such a long time, that it's the same thing—practically speaking for the majority of the most important things *"positive" training fails*.

> That right there should be enough to put an end to *the dark ages of the "positive" and "force free" ideological cult*.

But the downsides of "positive" training don't end there. There are various other side-effects and factors that are rarely mentioned. In fact, Dr. Michael Perone (Professor of Behaviour Analysis and co-ordinator of the Behaviour Analysis Program at West Virginia University, with a Ph.D. in Psychology) wrote an article titled, *"Negative Effects of Positive Reinforcement"*. He says:

"The purpose of this article is to cause you to **worry about the broad endorsement of positive reinforcement** that can be found throughout the literature of behavior analysis... At issue is whether it is free of the negative effects commonly attributed to the methods of behavioral control known as "aversive" ...I believe that **much of what has been said about aversive control is mistaken, or at least misleading**. Aversive control, in and of itself, is not necessarily bad; sometimes it is good. And, more to the point, the alternative—**positive reinforcement—is not necessarily good; sometimes it is bad.** Aversive control is an inherent part of our world, an inevitable feature of behavioral control, in both natural contingencies and contrived ones. When I say that aversive control is inevitable, I mean just that: *Even the procedures that we regard as prototypes of positive reinforcement have elements of negative reinforcement or punishment imbedded within them.*"[245] (emphasis added)

And later in the same article:

"There are plenty of bad things to say about positive reinforcement. In fact, many of them were said by Skinner himself."

In this chapter, we'll delve more deeply into the "bad things" about positive reinforcement. We'll talk about how, as Perone mentions, in reality *there is really no such thing as pure positive reinforcement training*, and we'll also discuss how "positive" training:

- Can both directly and indirectly result in aggression
- Can cause learned helplessness
- Can both directly and indirectly contribute to fear and anxiety
- Is dependant on highly coercive strategies of food deprivation, addiction, and extended and stringent restrictions on freedom
- Is often highly manipulative and dishonest
- And, finally, how a "positive" training approach contributes to abuse, rehoming, and euthanasia

As nice as it appears on the surface, "positive" training is certainly not what it is sold to the largely unsuspecting public as being.

---

[245] Perone, Michael. *Negative Effects of Positive Reinforcement.* The Behavior Analyst 26, no. 1 (Spring 2003): 1-14.

CHAPTER 9 – More downsides and side-effects of "positive" training

Positive reinforcement methods contain punishment

The first thing to realize with so called "positive" methods is that they contain punishment *by their very nature*. The simple fact is that most of these "aversives" are internal (although some are external). As mentioned in *Chapter Three – Some problems with operant conditioning*, positive reinforcement contains punishment every time it is withheld. As Perone said:

> "Even the procedures that we regard as prototypes of positive reinforcement have elements of negative reinforcement or punishment imbedded within them."[246]

For example, he says (and note that this comes from *real world* experience, not the laboratory):

> "Outside the laboratory, I cannot help but be impressed with the *propensity of people to respond to the negative side of positive contingencies*. Consider college students. In my large undergraduate courses, I have tried a variety of contingencies to encourage class attendance. Early on, I simply scored attendance and gave the score a weighting of 10% of the course grade. There were lots of complaints. The students clearly saw this system as punitive: Each absence represents a loss of points towards the course grade. So I switched to a system to positively reinforce attendance. When students come to class on time, they earn a point above and beyond the points needed to earn a perfect score in the course. Thus, a student with perfect attendance, and a perfect course performance, would earn 103% of the so-called maximum. A student who never came to class, but otherwise performed flawlessly, would earn 100%. If course points function as reinforcers, then this surely is a positive contingency. But the students reacted pretty much the same as before. *They saw this as another form of punishment: With each absence, I was denying them a bonus point. Of course, the students are right. Whenever a reinforcer is contingent on behavior, it must be denied in the absence of that behavior.*"[247] (emphasis added)

Positive reinforcement trainer Jane Killion also recognizes that positive reinforcement schedules contain punishment, when she says:

---

[246] Perone, Michael. *Negative Effects of Positive Reinforcement*. The Behavior Analyst 26, no. 1 (Spring 2003): 1-14.
[247] As above.

PART 4 – Investigating reward

> "The operant dog looks for reinforcers as confirmation that he is doing the right thing. *If a reinforcer is not forthcoming, he assumes he is doing something "wrong"*. Think about this statement carefully, for it is of profound significance. The **absence of reinforcement is actually punishment to the operant dog**. In contrast, a dog that has been trained to heel using leash checks thinks he is in good shape so long as he is not getting leash checks."[248] (emphasis added)

I second Killion's emphasis when she says, "think about this statement carefully, for it is of profound significance". The positive reinforcement trained dog (or person) is happy when the treats are flowing, but stressed and worried when they aren't (which is *most of the time* in a variable schedule) because he "assumes he is doing something wrong". Just consider the obedience competition dogs doing heel work panting with their tongue hanging out, their mad junkie eyes, and the signs of obvious stress, desperately seeking their next fix. Just like the gambling addict who is unhappy every time they pull the lever and don't get a win.

So, you can see that positive reinforcement schedules contain punishment. It's just *mental punishment* rather than physical. And, from this point of view, they contain *a lot more instances of long-lasting and ongoing punishment* than sound punishment-based training. On the other hand, the punishment trained dog is only occasionally punished (and rarely once trained), and is rewarded perfectly consistently (not on an addictive variable schedule) every time it avoids the punishment ("he thinks he is in good shape so long as he is not getting leash checks"), which is the vast majority of the time. Karen Pryor fails to understand this when she criticizes the use of commands instead of "cues", by saying:

> Positive reinforcement methods contain *a lot more instances of long-lasting and ongoing mental punishment* than sound punishment-based training.

> "In contrast to a command, which is a veiled threat, a cue is a promise: if you understand what I'm saying, and you carry it out correctly, you will definitely win."[249]

Pryor is missing the fact that, as Perone and Killion point out, the opposite is also true— "if you *don't* understand what I'm saying, or you *do not* carry it out correctly, you will definitely *lose*." And even if the subject does carry it out correctly, it will still *not* "definitely win", because with variable schedules of

---

[248] Killion, Jane. *When Pigs Fly!: Training Success with Impossible Dogs*. Wenatchee, WA: Dogwise Publishing, 2007.
[249] Pryor, Karen. *Reaching the Animal Mind: Clicker Training and What It Teaches Us About All Animals*. New York: Scribner, 2009.

CHAPTER 9 – More downsides and side-effects of "positive" training

reinforcement only a small percentage of successful responses actually "win". *Most* do not.

In fact, there's no such thing as pure so-called "positive" or "force free" training. As soon as you withhold reward you are punishing the animal and using aversives (despite what Kay Lawrence claims about "avoiding every atom of punishment"). And when you withhold food to increase the effectiveness of treat training, you are now using an aversive (hunger), and *negative* reinforcement, not positive reinforcement. And as soon as you put a lead on a dog, you are using force to prevent it from being rewarded by the environment. "Force free", "positive", etc., are all just euphemistic marketing terms—in other words, they are all just lies.

> "Force free", "positive", etc., are all just euphemistic marketing terms – in other words, they are all just lies.

Karen Pryor admits "positive" trainers use "aversives" (besides the aversive possibilities of the training itself, which she references repeatedly), but she tries to make a pointless distinction (which she does a number of times in her books) between these "aversives" and "punishment":

> "Removing punishment from your tool kit is *not* the same as removing all aversives from your learner's life. Many non-clicker trainers leap to that conclusion, point out that it's impossible to remove all aversives—dogs wear leashes and get shut in crates, clicker trainers use the Gentle Leader [which many dogs find highly aversive]—and conclude that clicker training is not punishment-free. But that initial assumption is wrong. While all punishment is aversive, not all aversives are punishment." [250]

What she is saying is that unless you use aversives *deliberately* to change behaviour (when, by her definition, it now becomes punishment), aversives are fine. Effectively, Pryor is saying: "We don't use punishment, we just use things the dog doesn't like and that function due to force and punishment or negative reinforcement (like head collars), but we pretend we aren't using them to teach them anything, so that's fine". So, when Pryor locks a dog in a crate for long enough so that it becomes "restless and hungry"[251] before a performance, she's using an aversive, but, according to her, that's fine because she isn't using it as a punishment.

So, aside from these factors that there's a lot about so-called "positive" methods that aren't actually "positive" at all, and that positive reinforcement has negative reinforcement and punishment "embedded within it" as Dr. Perone noted, let's look

---

[250] Pryor, Karen. *On my mind – Reflections on Animal Behavior and Learning.* Waltham, MA., USA. 2014.
[251] As above.

at some of the other downsides that are inherent to, or result from, an emphasis on positive reinforcement.

Aggression

Despite claims to the contrary, we saw that when using punishment, aggression is highly unlikely to be the outcome unless the punishment is *non-contingent* and *continuous*. That is, unless it is simply abuse. Conversely, however, in various situations *positive reinforcement* schedules themselves can cause aggression:

> "A variety of species exhibit aggression when exposed to intermittent schedules of positive reinforcement."[252]

We only have to consider the angry gambling addict when he loses (a variable schedule of positive reinforcement), or of how we might get angry when we put our money into a drink vending machine and no drink comes out (a continuous schedule of positive reinforcement, when the reinforcement is withheld). And gambling addicts (Jean Donaldson calls them "reinforcer junkies") is what "positive" training aims to turn all dogs into. Karen Pryor notes:

> "If there's any group of people who can teach us about reinforcement, it's casino owners."[253]

> "Positive" training (if it is at all successful) turns all dogs into gambling addicts.

So-called "positive" training seeks—*at its very core*—to turn all dogs into gambling addicts. The better the training is carried out, the more successful it is in achieving this.

Karen Pryor gives a good example of positive reinforcement causing aggression in her book (and also, once again, of her resorting to punishment!), of where she withheld reward because the elephant she was training didn't retrieve the frisbee close enough, and so the elephant hit her with its trunk:

> "When I didn't fall for that, she whopped me on the arm. When Jim and I both yelled at her for that (a sign of disapproval, which elephants respect) she started fetching nicely..."[254]

---

[252] Looney, Thomas A. *Attack during time-out from a fixed-interval schedule in pigeons.* Bulletin of the Psychonomic Society 19, no. 6 (1982): 369-371.
[253] Pryor, Karen. *On My Mind – Reflections on Animal Behavior and Learning.* Waltham, MA. Sunshine Books. 2014.
[254] Pryor, Karen. *Don't Shoot the Dog!: The New Art of Teaching and Training.* Rev. ed. New York: Bantam Books, 1999.

CHAPTER 9 – More downsides and side-effects of "positive" training

So, the elephant didn't bring the frisbee close enough, Pryor withheld the reward, and the elephant responded aggressively by hitting Pryor with its trunk—just like we might kick the drink vending machine if our treat isn't forthcoming—to which Pryor responded with punishment. Pryor also notes:

> "Extinction is a very unpleasant process for any organism. The experience often arouses anger, even in lab animals." [255]

> "Want an example? Guess what a dolphin does when you accidently frustrate it by suddenly *not reinforcing* behavior you used to pay for, a mistake I sometimes made in my early training days. The *dolphin gets mad, of course*. Behaviorists have a name for it: *extinction-induced aggression.*"[256] (emphasis added)

You can see a more dramatic and well-known example of this where the killer whale Tilikum killed his positive-reinforcement trainer when the trainer likewise withheld reward. Or think about the tantrums thrown by spoilt children when parents don't give them what they want. So, if you think "positive" training can't cause aggression, think again. It is, in fact, *far* more likely to do so than punishment. It's interesting to note the huge increase in dog bites that has coincided with the shift into "positive" training (in England from 1998 to 2018 dog bites *tripled*[257]), as you'll hear more about soon (see *Rehoming and Euthanasia*).

> "Positive" training is *far* more likely to cause aggression than punishment.

So, not reinforcing behaviour (which is integral to any variable schedule) can cause aggression. And if you're training (or parenting) with positive reinforcement, there is *no way you can avoid* using these intermittent schedules of reinforcement—they are the backbone of so-called "positive" training.

Aside from *initiating* and *causing* aggression, "positive" approaches also extremely commonly *exacerbate* it. Take the example of a pup with food aggression. The first thing most owners do when their pup growls at them when they go near it when it's eating, is to back off. They've been told at puppy school or YouTube or by their vet that you should never smack your dog or tell it off, and instead to "reward good behaviour and ignore bad behaviour". Therefore, they have *no other useful strategy* when their pup growls at them other than to move away. This simply and directly

---

[255] Pryor, Karen. *On my mind – Reflections on Animal Behavior and Learning.* Waltham, MA., USA. 2014.
[256] As above.
[257] Medical Xpress. *Adult Hospital Admissions for Dog Bites Triple in 20 Years.* Medical Xpress, January 19, 2021, https://medicalxpress.com/news/2021-01-adult-hospital-admissions-dog-triple.html.

rewards the aggression, and so it escalates from there. If, instead, the owner had instantly given the pup a good firm smack, the problem would have been solved instantly and permanently. And it therefore wouldn't have had a chance to escalate into something far worse over time.

A fear of using punishment—thanks to the reward-only ideologues—results in rewarding, and therefore increasing, aggression.

Other recommended reward-based strategies also directly reward the aggression. Common advice around using treats or toys in an attempt to *distract* the dog from the aggression are madness. You are simply rewarding the aggression.

Of course, if we use reward correctly (definitely *not* like the "positive" advocates recommend), these things won't occur. And that simply means that for many things we *should not use reward at all*, because it either simply will not work or it will make the problem worse. And, we should be prepared to use punishment *instantly when required* (as Pryor did with the elephant) to nip a much worse problem in the bud.

> For many things we *should not use reward at all*, because it either simply will not work, or it will make the problem worse.

The final reason why "positive" training often results in aggression is related to dominance. When people ill-advisedly avoid any forms of correction or punishment, the dog will often be the more dominant party (depending on its temperament—some dogs are naturally more dominant or submissive by nature). And the dominant dog in a relationship is far more likely to assert its wants (for food, the couch, etc.), through aggression, just like dogs and wolves (and other animals) do with each other.

So, despite the reward-only ideologues claims, it is actually "positive" training that is *far more* likely to cause, maintain, and escalate aggression than is punishment.

Learned Helplessness

As we learnt earlier, learned helplessness *never* occurs with contingent punishment—i.e. punishment that is connected to behaviour. Only *non-contingent* punishment that is also continuous can cause learned helplessness, and even that is rare. Conversely, *positive reinforcement* can cause learned helplessness. As we heard from "positive" trainer and world agility champion Susan Garrett earlier:

CHAPTER 9 – More downsides and side-effects of "positive" training

"I have seen many "shut down" dogs with handlers that would never so much as raise their voices to their dogs."[258]

She's correct. In fact, Karen Pryor gives various examples of *herself* inducing this sort of learned helplessness:

"If two or three expected *reinforcers fail to materialize*, the animal may simply *give up and quit on you*. You can see this clearly on the video of my fish learning to swim through a hoop. When three tries "didn't work," the fish not only *quit trying*, he had an *emotional collapse*, lying on the bottom of the tank *in visible distress*."[259] (emphasis added)

Did you get that? "Quit trying", "emotional collapse" and "visible distress", all thanks to "positive" training. And:

"The fourth rat, however, had gotten stuck on one side, and was in a classic extinction curve, *giving up all behavior, curled up in a corner in despair*."[260] (emphasis added)

Or, in an example where positive-reinforcement training by Pryor caused both aggression (redirected) *and* learned helplessness:

"Mali [the dolphin] has not yet experienced failing to be reinforced for something that had previously paid off. I have carelessly put her into extinction. *Mali is furious*. After the fourth unreinforced jump, she breaches twice (away from me, not on me), then retires to the side of the tank that is farthest away from me and turns her back. *There she stays, ignoring people and refusing all fish for two days*."[261]

So Pryor gives various examples of where, during *the normal course* of positive-reinforcement training, even by one of the world's most experienced positive-reinforcement animal trainers like herself, learned helplessness is caused. Just think how much more easily and regularly this happens (like Susan Garrett noted) with less experienced trainers.

In a study titled: *"Learned helplessness, depression, and positive reinforcement"*, the authors also recognized this fact:

---

[258] Garrett, Susan. *Punishment: Pros and Cons.* Susan Garrett's Dog Training Blog, 2010, https://susangarrettdogagility.com/2010/04/punishment-pros-and-cons/
[259] Pryor, Karen. *On my mind – Reflections on Animal Behavior and Learning.* Waltham, MA., USA. 2014.
[260] As above.
[261] Pryor, Karen. *Reaching the Animal Mind.* New York. Scribner. 2009.

"...helplessness could be induced using positive reinforcement"[262]

In the above study, unlike in Pryor's examples, the positive reinforcement was non-contingent. This is exactly what teachers who give rewards to everyone regardless of merit are doing—these are non-contingent rewards—and they are demotivators that can cause learned helplessness. Unfortunately, this is common place in schools nowadays. Also, consider welfare benefits for people who don't really need them—once again, these are non-contingent rewards with the potential to cause learned helplessness. Things that on the surface give people warm and fuzzy feelings about how nice and kind they are, but which actually, in the long run, cause more harm than good to the recipients and to the broader society.

> This is exactly what teachers who give rewards to everyone regardless of merit are doing – these are non-contingent rewards – and they are demotivators.

So, positive reinforcement can, and does, cause learned helplessness, both when it is non-contingent and also during *the normal course of positive reinforcement training*. As Susan Garrett points out, learned helplessness being caused by trainers who are only using positive reinforcement is not uncommon. On the other hand, it *never* occurs during the normal course of sound, correction or punishment training. Once again, the *exact opposite* to what the positive ideologues like to claim.

Fear / Anxiety

Fear and anxiety can easily be, and often are, created and/or exacerbated by reward, again either directly or indirectly.

For example, severe separation anxiety is almost always caused by reward. When the dog is anxious and stressed when locked outside, or in a crate, etc., the owners often reward that with attention, or by letting the dog back inside, or letting it out of the crate. Or, in the case of a recent client, they had a treat dispenser with a built-in camera which that they could control via their phone while they were at work—when the dog was looking stressed and anxious they would have the machine spit out a treat in an attempt to calm the dog. In any of these situations the behaviour (and the emotion that goes with it) is rewarded and shaped, and escalates over time.

Or, take the example of a dog my wife and I owned with a severe thunder phobia. When I met my wife she had two dogs. One was Reggie, a little black Chihuahua

---

[262] O'rourke, T., Tryon, W., & Raps, C. (1980). *Learned helplessness, depression, and positive reinforcement.* Cognitive Therapy and Research, 4, 201-209.

CHAPTER 9 – More downsides and side-effects of "positive" training

cross (the other was Tippy, a white Maltese cross). When a thunderstorm occurred, Reggie would be shivering and shaking under my wife's jumper, who attempted to console her with pats and cuddles. Zak George would approve, as he did the same with his Border Collie:

> "…while Venus wasn't much of a cuddler, she would be the best cuddler in the world if she heard thunder… I'd reassure her and pet her softly."[263]

and Victoria Stilwell:

> "If you are present during the storm, spend time with your dog in the safe haven or give her attention if she comes to seek comfort. *Far from reinforcing the fearful behavior*, your presence will help your dog cope—as long as you remain calm."[264] (emphasis added)

Despite Victoria's claims to the contrary, this is indeed simply *"reinforcing the fearful behaviour"* as my wife was doing, and making the behaviour worse (it had been a problem for years at that point, so it certainly wasn't improving things). Of course, my wife could be excused as she makes no claims to being a dog trainer—but Victoria Stilwell and Zak George on the other hand? They have no excuse. Their strategy is definitely and directly "reinforcing the fearful behaviour." It took a bit of convincing my wife, but from then on whenever Reggie started to get frightened and look to my wife for safety, we would simply reprimand her and send her back to her bed—"Ahh, on your bed". *This correction-based approach soon had her much calmer and less stressed.* All the trembling and shaking and drooling disappeared, and although she didn't learn to love thunder (she was an exceptionally timid dog), very soon she would lie calmly in her bed without the previous levels of distress.

Other strategies for thunder phobias, etc., can be used. For example, desensitization (often in conjunction with counter-conditioning). We play a high-quality recording of storms very softly at first, and slowly build that up. However, generally we need a *very* high-quality stereo for it to be at all realistic, and many people simply do not have that. The dog knows the difference compared to the real thing. And also there is a good risk that if part way through the training (which can be *very* slow) we get a real storm, all that work is ruined. So, in theory it can work, but in the *real world* it often isn't a particularly successful strategy, particularly not on its own.

We also do not want dogs to escape the things they are afraid of—to run and hide—because this also simply acts to reinforce the fear. George (aside from

---

[263] George, Zak, and Dina Roth Port. *Zak George's Guide to a Well-Behaved Dog: Proven Solutions to the Most Common Training Problems for All Ages, Breeds, and Mixes*. New York: Ten Speed Press, 2019.
[264] Stilwell, Victoria. *Train your dog positively*. New York. Ten Speed Press, 2013.

PART 4 – Investigating reward

directly "reinforcing the fearful behaviour" when his dog Venus was frightened by reassuring and patting her) also made this elementary mistake by allowing this very thing to also happen:

> "Other times during storms, Venus loved to nestle in the bathroom or bedroom closet… I considered that very valuable desensitization time as well." [265]

Victoria Stilwell advocates much the same:

> "The most important thing you can do for your thunder phobic dog is to provide her with a "bolt hole"—a place she can escape to in the event of a storm."[266]

Again, this is the complete opposite to what a trainer should be doing. Remember the trigger warnings?

> "Knowing avoidance is a symptom of the problem, experts pointed to the *ironic effects* of trigger warnings. Several studies show that providing explicit alerts to students that upcoming class content might induce unpleasant reactions offers little to no benefit, and *often backfire (only making the anxious more anxious)*."[267]

Very much the same applies here (and, indeed, with most fear-based issues). Do Stilwell or George really think that your dog going and hiding in the closet or "bolt hole" is "valuable desensitization" and that it's going to help things? I'm really not sure what planet these "positive" trainers are living on.

An obvious example is comparing tough, resilient farm dogs to their often nervous, anxious city cousins who largely live their lives insulated from the real world, and for whom many owners unwisely attempt to avoid "every atom of punishment". The "school of hard knocks" makes for stronger, calmer, more resilient animals, as it does for people, while molly coddling dogs (or children) does the opposite. As Alison Gopnik (Professor of Psychology - California University) notes:

---

[265] George, Zak, and Dina Roth Port. *Zak George's Guide to a Well-Behaved Dog: Proven Solutions to the Most Common Training Problems for All Ages, Breeds, and Mixes*. New York: Ten Speed Press, 2019.
[266] Stilwell, Victoria. *Train your dog positively.* New York. Ten Speed Press, 2013.
[267] Kashdan, T. (2024, January 29). *Rethinking Trigger Warnings.* Retrieved from https://toddkashdan.medium.com/rethinking-trigger-warnings-998d9ef9a89

CHAPTER 9 – More downsides and side-effects of "positive" training

> "by shielding children from every possible risk, we may lead them to react with exaggerated fear to situations that aren't risky at all and isolate them from the adult skills that they will one day have to master."[268]

What we see nowadays is an epidemic of anxiety thanks to the "positive" or "gentle" approaches, which simply cause and then exacerbate the very problems they are supposed to avoid. And then, when parents or dog trainers see this anxiety (created by these misguided approaches) they tend to double-down even further on the "positive" methods, *trying even harder to avoid any form of punishment or correction*. Remember Cindy Benson's poor anxious dog for whom a "stern look" or a "sigh" were problematic, and the proposed solution was to try harder to avoid even this? Instead, what these parents or trainers should do is throw out the "gentle" parenting—or "positive" training—approaches altogether. You don't get out of a hole you've created by digging harder.

Rather than causing fear or anxiety, correction or punishment-based procedures can quickly and dramatically cure it. This is seen all the time in horse training (you can check out the video of me working with one of my horses, Madness, on www.realworlddogtraining.com.au). A good dog-related example is a sheepdog I had many years ago, named Bianca. I bought her at about 18 months old. She had a lot of natural sheep working ability, but she was naturally *very* timid. She wouldn't come within about 20 metres of anyone when I got her. I used a strong negative reinforcement approach very similar to that with the horse in the video, that substantially cured her basically in the space of ten minutes.

I had her on a long line of about ten metres, and a noisy chaff feed bag that I flapped around. This frightened her, and she would attempt to run away. But I would draw her in with the line, up to, and then under, the flapping bag, right to my feet. Immediately when she reached me, I would drop the bag to one side and give her a big pat. After a short break, I would repeat the exercise. After just a few repetitions she would come running instantly and straight under or through the flapping bag to get to me. Most of her fear of frightening objects basically disappeared instantly, as did all her fear of me. As we read earlier about "trigger warnings", an approach like this has the effect of:

> "…modifying their assumptions about danger in the world, and building up their sense of agency and tolerance of distress…"[269]

Bianca had learnt to brave out her fear—by being brave she learnt she had power to stop the frightening object. She bonded to me instantly and strongly from that point

---

[268] Lukianoff, Greg, and Jonathan Haidt. *Your Child Is More Resilient Than You Think*. Reason, November 18, 2018. https://reason.com/2018/11/18/your-child-is-more-resilient-t/
[269] Kashdan, T. (2024, January 29). *Rethinking Trigger Warnings*. Retrieved from https://toddkashdan.medium.com/rethinking-trigger-warnings-998d9ef9a89

on, and was like a completely different dog. (I went on to work her in quite a few sheepdog trials, and did a lot of work with her on large sheep and cattle stations.) I use variations of this approach regularly.

Now, this isn't to say we can magically cure every last vestige of every iteration of fear and anxiety in every dog instantly. We *can* cause *drastic* transformations very quickly in many cases, and these can certainly appear like magic to the uninitiated. However, genetics always plays a role, as do some other factors, and so in some cases there is a limit to the changes we can make. Nevertheless, we can nearly always make substantial improvements.

So, contrary to claims that punishment causes fear and anxiety, it is much more likely to be the other way around. Common strategies used by "positive" trainers at best simply don't work, and at worst, regularly make things worse. On the other hand, a sound use of negative reinforcement or punishment can very rapidly, and highly successfully, in many cases cure or vastly improve fear and anxiety issues. And, when used from the start, prevent such problems ever arising.

Hunger

Aside from the potential for "positive" training to cause aggression and learned helplessness, and to greatly contribute to fear and anxiety, there are some other major downsides. Hunger is the first of what I term the **"Big Three Necessities of Positive Reinforcement Coercion"**. The other two, to be discussed next, are *addiction*, and then the *extended, stringent restrictions of freedom* that are required if "positive" training has *any* hope of being even *partially* successful.

Let's deal with hunger first. As everyone knows, treat training doesn't work particularly well *when your dog isn't hungry*. If common sense isn't enough, and you need some science to back that statement up, here it is:

> "…satiation reduces the very effectiveness of food as a reinforcer of responses…satiation reduced responding immediately…"[270]

In other words, when the dog is full it immediately stops responding to food. No kidding? I bet no pet owner ever figured that one out. So, what do you do if you want "positive" training to work? Why, keep your dog hungry, of course. Veterinarian Leon F. Whitney, in his book *"The Natural Method of Dog Training"* that Dunbar referenced in the start of this book, states that:

---

[270] Holz, W. C., & Azrin, N. H. (1963). *A comparison of several procedures for eliminating behavior.* Journal of Experimental Analysis of Behavior, 6(3), 399–406. https://doi.org/10.1901/jeab.1963.6-399

CHAPTER 9 – More downsides and side-effects of "positive" training

> "There's no question about the incentive of hunger… Hunger is the best incentive for our training method."[271]

Whitney also states that 36 hours hungry is the optimal time to treat train your dog:

> "It is a curious fact that animals of all sorts train most easily when they are 36-hours-hungry, more easily than if they have fasted for 24 hours or 48 hours…. The way it works means you have a training session every other day. If you feed your dog Monday morning, he'll be ready for a lesson Tuesday evening."[272]

So, Whitney is advising that you only feed your dog every 36 hours, and then train it just before you give it its once-every-day-and-a-half meal, in order that its hunger is at its peak. Or Karen Pryor:

> "…I found a lively Jack Russell, ascertained that he had never been exposed to any agility training, *had him crated* for a couple of hours *so he'd be hungry and restless*, and brought him up on stage." [273] (emphasis added)

What is really happening here is that the training is *negative* reinforcement not *positive* reinforcement. These trainers create something uncomfortable (hunger) and then remove that (or go part way to removing that) when the dog does what they want. Instead of some pressure on the dog's rump when teaching it to sit (how awful—using "force" based methods—just think about all the "disastrous" fallout we are going to cause!), they now have an uncomfortable hunger the dog wants to remove. So, trainers who think they're using "positive reinforcement" only, are often relying on "negative reinforcement" instead. Steven R. Lindsay (author of the three volume "Handbook of Applied Dog Behaviour and Training") writes:

> "Thus, from this perspective, working for food may be interpreted as escape-avoidance behaviour aimed at reducing or terminating the aversive condition of starvation."[274]

Remember how I talked about the confusing overlap between the four quadrants of operant conditioning at the start of this book? So, what trainers think of as *positive* reinforcement can in fact be, and often is, *negative* reinforcement. When research

---

[271] Whitney, Leon F., Dr. *The Natural Method of Dog Training*. New York: Tandem Books, 1969.
[272] As above.
[273] Pryor, Karen. *On My Mind – Reflections on Animal Behavior and Learning*. Waltham, MA. Sunshine Books. 2014.
[274] Lindsay, Steven R. *Handbook of Applied Dog Behavior and Training, Volume 1: Adaptation and Learning*. Ames, IA: Blackwell Science, 2000.

PART 4 – Investigating reward

is done using supposed positive reinforcement with food, the same always applies. This is a typical statement you find in basically all "positive reinforcement" studies:

> "Three male, White Carneaux pigeons were used as subjects. They were ***deprived of food*** until their weights were 75 to 80 per cent of their free-feeding levels."[275] (emphasis added)

> The experimental science that so-called "positive" trainers rely on is based on food deprivation.

So all this type of science that the "positive" trainers rely on is based on food deprivation. Jean Donaldson, the strong reward-only advocate (who sometimes likens punishment methods to torture during the Spanish inquisition[276]), also advocates depriving the dog of food like Whitney (and Dunbar, as you will see shortly):

> "Reinforcers tend to become more potent after a period of *deprivation*: *hungry dogs work harder* for food than satiated dogs… When trainers deliberately use *deprivation* to increase motivation it is an example of what's known as an *"establishing operation"*. **Food is established as a reinforcer through *deprivation***. When trainers really need maximum motivation, they will *"close the economy,"* i.e., *not provide any nutritive reinforcement for free…* Establishing operations are like pretraining measures to make sure training will be effective."[277] (emphasis added)

Just like Pryor's "establishing operation" of locking the Jack Russell up in the crate so it became "restless and hungry" before her performance. And:

> "Take control of the goodies the dog wants in life—stop handing them out for free. Nothing is "free" anyway…"[278]

> In Dunbar's "Dog Behaviour Problem Flow Chart" the first step is always to *stop feeding your dog*.

Likewise, in Dunbar's "Dog Behaviour Problem Flow Chart" his first recommendation is always to *stop feeding your dog*. You stop feeding your dog so it's hungry, and then treat training works better. No kidding.

---

[275] Holz, W. C., & Azrin, N. H. (1961). *Discriminative properties of punishment*. Journal of the Experimental Analysis of Behavior, 4(2), 225-230. Retrieved from https://onlinelibrary.wiley.com/doi/pdf/10.1901/jeab.1961.4-225
[276] Donaldson, Jean. *The Culture Clash*. 2nd ed. Berkeley, CA: James & Kenneth Publishers, 2005.
[277] As above.
[278] As above.

CHAPTER 9 – More downsides and side-effects of "positive" training

If your treat training isn't going too well, you probably forgot to deprive your dog of food.

And, once again, we're now talking about *negative* reinforcement. It's just that the uncomfortable pressure is *internal* rather than external. In marine animal parks deprivation is also integral to training and performing. If the dolphins or seals or killer whales aren't hungry, they won't bother working for their fish. If you want to learn more, watch the documentary "Blackfish" about the killer whale Tilikum. In the documentary you will learn about the deprivation techniques that were employed as a matter of course, as they are by all serious "positive" trainers like Dunbar, etc.

And remember, training by reward can take *years*, and *thousands* upon *thousands* of reinforced trials, so you might need to keep your dog hungry for *a long while* (likely its entire life in order to keep it under some semblance of control) all while the so-called "positive" trainers claim to be "non-coercive".

In fact, in relation to keeping your dog hungry *for life*, Dr. Dunbar advises that you should simply *never feed your dog a proper meal*. In his method, your dog only ever gets food as reward contingent on behaviour, doled out throughout the day (this is the same as Donaldson's "close the economy" or her recommendation that you should "take control of the goodies the dog wants" because "nothing is free anyway"). So, nothing much different there—just the same coercive deprivation tactics. But then, in an amazing twist in an attempt to justify the imposition of this lifelong, never-ending hunger, Dunbar tries to argue that if you're *not* doing this, if you have the temerity to feed your dog *an actual meal*, then you're abusing your dog! If I didn't have the exact quote, you probably wouldn't believe me—"you couldn't make this stuff up!" He says of people who psychologically torture their dogs by feeding them a real meal:

> Training by reward can take *years*, and *thousands* upon *thousands* of reinforced trials, so you might need to keep your dog hungry for *a long while* (quite likely its whole life in order to keep it under some semblance of control) all while the treat trainers claim to be "non-coercive".

> "… it's *psychologically devasting* for the dog that its day now revolves around a single time 5pm and one bowl of food… what you've now *stolen from the dog* is its raison d'être, its *reason for being*." [279] (emphasis added)

Wow! So according to Dunbar what you're actually doing if you feed your dog a meal is terrible psychological abuse. It's "psychologically devasting" for your poor dog, and you've now "stolen" its entire "reason for being"! If you have to read that a couple of times to make sure you're reading it right, I'm not surprised.

---

[279] Dunbar, Ian. *Six Simple Steps to Solve Your Dog's Behavior Problems*. Dunbar Academy. Accessed Feb 2024. https://www.dunbaracademy.com/courses/six-simple-steps.

PART 4 – Investigating reward

As you see, Dunbar advocates withholding food unless your dog does something you want to reward, *for life*. Imagine rearing a child like that—you don't get *any* food unless you do what I want; then, and only then, you can have *one little piece of potato* if you do your homework, *half a sausage* if you clean your bedroom, a *spoonful of peas* if you sit quietly. **But this is the real face of "positive" training.** The coercion is more covert than overt, but for that reason all the more underhanded and manipulative, and hypocritical.

> This is the real face of so-called "positive" training.
>
> The coercion is more covert than overt, but for that reason all the more underhanded and manipulative, and hypocritical.

And lest anyone misunderstand, this is not some fringe opinion. Rather it is *integral* to nearly all "positive" trainers, including the most influential and best known. You may have heard it referred to as "Nothing In Life is Free" or something similar—they "close the economy" to use Donaldson's words, or simply never feed the dog a meal as Dunbar recommends.

And keep in mind that if a so-called "positive" trainer claims they don't do this, then their training will be even *less* effective than the already pitiful results we've heard about.

It wouldn't be such an issue if these people weren't having so much influence among the general public, or up-and-coming animal trainers, or even guide dog trainers and so on, who *think* they are doing the humane thing by their dogs by avoiding correction and punishment, when in actual fact the opposite is the case.

Daniel Abdelnoor, the New Zealand trainer who goes by the name "Doggy Dan" (and bases his approach largely on Jan Fennell's so-called "Amichien Bonding"), advertises as using "no treats, no tricks and no force" (in reality he uses all three). He recommends you apply the same *deprivation* approach, only this time with *affection*. He doesn't describe it in these terms, but rather uses "dominance" related terms such as:

> "Doing EVERYTHING ON YOUR TERMS helps establish that you're in charge, and you're the one calling the shots." [280] (emphasis his)

He calls it "Golden Rule #3: Everything on your own terms", and describes it as a "crucial rule".

---

[280] Doggy Dan. (2024). *How to Improve Dog Obedience: Doggy Dan 5 Golden Rules — Rule #3*. Retrieved from https://theonlinedogtrainer.com/improve-dog-obedience-doggy-dan-golden-rule-three/

CHAPTER 9 – More downsides and side-effects of "positive" training

Much like the "Nothing in Life is Free", "Closing the economy" and "take control of the goodies the dog wants" deprivation tactics of the treat trainers, he does the same with regards to "pats, cuddles and affection" (some of the "goodies" your dog wants). So, he recommends *only* giving your dog attention if you make it work for it first, by commanding "here", or "sit", etc. You never pat your dog or give it any affection if it simply approaches wanting a pat. After all, according to Jean Donaldson, "nothing in life is free anyway".

> Much like the "Nothing in Life is Free" and "Closing the economy" deprivation tactics of the treat trainers, Daniel Abdelnoor does the same with regards to "pats, cuddles and affection".

The default position of so-called "positive" training is that your dog doesn't get what it wants *without begging you for it*. It has to beg for every morsel of food for the treat trainers, and every scrap of attention for "Doggy Dan". With balanced training, it's the opposite: the default position is the dog can have free food, and a pat whenever it wants it and you feel like giving it, and various other "goodies" likewise. And yet the so-called "positive" trainers claim they are the ones who are "non-coercive"!

So this deprivation aspect is another major downside of "positive" training. If you want it to work (as well as possible, which is not very well), then, regardless of the reward you're using, you need to use *deprivation* in order to increase your dog's motivation. Conversely, when using correction or punishment, you usually do not.

Another related strategy is to use "high value" rewards. Skinner notes that treats high in sugar and salt are highly reinforcing:

> "Salt and sugar are critical requirements... as a result, the human species, like other species, is powerfully reinforced by sugar, salt..."[281]

This is, of course, true. Karen Pryor approvingly notes her social worker friend's use of candy, when *clicker training adults and kids* in her "positive" family therapy sessions (yes, you read that right also):

> "On the table is a big bowl of treats for the dogs and another, much bigger bowl of wrapped candy of many different sorts, for the people."[282]

Certainly your treats will be more effective the higher sugar, salt, and fat content they have. It's no different to McDonald's. Which would work better? Rewarding your kids by saying, "If you clean up your room, you can have some *nice healthy*

---
[281] Skinner, B.F. *About Behaviorism.* United States. Vintage Books. 1976.
[282] Pryor, Karen. *On My Mind – Reflections on Animal Behavior and Learning.* Waltham, MA. Sunshine Books. 2014.

vegetables as a reward"? Or, by saying, "You can have McDonalds, or some candy"? Not only are such treats more reinforcing *per se*, as Skinner notes, but they are also more *addictive*. And if we can get our dog (or kids) *addicted*, then our treats are more effective than they would otherwise be (see the next section). Heroin based treats would work brilliantly! In that case, we could actually get "positive" training to work because it would outweigh everything else.

So, if you want to try to make "positive" training do something it generally just doesn't do (*stop* behaviours, or gain *obedience* under distraction), then make sure your dog *stays hungry*, or is *deprived* of affection, or both, and use the most reinforcing, most addictive treats you can—high fat, salt and sugar content will work much better. Trainers attempting to use "positive" methods, especially highly competitive competition trainers, will use every trick in the book, all the while avoiding punishment (or claiming they do) and attacking anyone who uses it because it is "bad". They will even starve dogs for days in an attempt to make so called "positive" training work, in their rabid desire to avoid using correction.

Just to be clear, all this is not to say you should *never* use treats in training. I don't use them myself training my dogs for reasons I explained earlier, but up to a point they're fine. You can teach most pups the *meaning* of "sit" using treats without having to "close the economy". And *if you then move on to correction-based approaches* (like Whitney did) you won't need variable schedules of reinforcement (addiction) to gain improved response (or at least not for long). And the deprivation, if any, will also be relatively short-lived. However, if you *don't* move on to correction, then in order to have *any* hope of actually *gaining obedience under distraction* with "positive" training, you certainly *will* have to resort to these *Big Three Necessities of Positive Reinforcement Coercion*: Deprivation, Addiction, and Extended Stringent Restriction of Freedom, ***probably for life.***

> If you *don't* move on to correction, then in order to have *any* hope of actually *gaining obedience under distraction* with reward-only, you certainly *will* have to resort to the *Big Three Necessities of Positive Reinforcement Coercion*.

So that's the first necessity of *positive reinforcement coercion*—deprivation (usually of food, but can also be affection or anything else the dog likes as Donaldson said—"take control of the goodies your dog wants in life"). Let's now move on to the second necessity of "positive" training—addiction.

Addiction

As we've seen, the psychology of positive reinforcement training, in particular the psychology of variable reinforcement schedules, is the psychology of addiction—that is, the psychology of gambling. This is the second element, after hunger, in the

CHAPTER 9 – More downsides and side-effects of "positive" training

Big Three Necessities of Positive Reinforcement Coercion. As Victoria Stilwell notes:

> "…intermittent reinforcement actually makes your dog respond more quickly and reliably because this learning is *based on the same concept that makes a casino slot machine so addictive*."[283] (emphasis added)

If you've ever seen competition obedience dogs doing heel work—or some half-crazed agility dogs trained by positive reinforcement methods—then you've probably noticed the crazed eyes, the mouth hanging open, the slobbering tongue, and that half-demented look that the dog has. You are looking at highly-stressed positive reinforcement addicts seeking their next fix. As Donaldson notes:

> You are looking at highly-stressed positive reinforcement addicts seeking their next fix.

> "Trainers love reinforcer-junkie dogs."[284]

Figure 14 - "Reinforcer Junkie" dogs—addiction in action

---

[283] Stilwell, Victoria. *Train your dog positively.* New York. Ten Speed Press. 2013.
[284] Donaldson, Jean. *The Culture Clash.* 2nd ed. Berkeley, CA: James & Kenneth Publishers, 2005.

PART 4 – Investigating reward

The way these "junkies" are trained is exactly the way slot machines are programmed (and now computer games for kids). They are carefully set to create *the greatest possible addiction*. And when people become addicted, they will sit like mindless zombies playing these games for hours and hours on end, day after day, desperately seeking their next fix. Jane Killion says:

> "We must scientifically build an *ongoing state* of *excited anticipation* that will *keep your dog on the hook* and vibrating with eagerness to see what you will do next."[285] (emphasis added)

In other words, "scientifically" get your dog addicted ("keep your dog on the hook") and turn it into a "reinforcer-junkie", all to get some basic obedience (what Zak George described as "advanced behaviours"). Karen Pryor says:

> "If there's any group of people who can teach us about reinforcement, it's casino owners."[286]

Some dogs, like some people, are more prone to addiction. Sheepdog breeds are a good example—they often get obsessed and addicted relatively quickly and easily. They've been bred with that sort of nature. You'll notice many positive reinforcement trainers who compete with dogs often use border collies (as in the above picture). They're very susceptible to addiction, and variable reinforcement schedules work better on them than many other breeds.

If we can select a dog that is naturally prone to addiction, and then create an addiction for treats (or toys, or whatever), we can improve the effectiveness of reward-based training. But do you think an addict is in a calm, happy, relaxed state? Do those obedience champions doing their heel work look calm and relaxed? Hardly. Compare that to the video of one of my dogs, King, walking relaxed, calm and happy at my heels on my website (see *www.realworlddogtraining.com.au*). (Horror of horrors, he was trained using negative reinforcement and punishment in conjunction with praise.)

> Do you think an addict is in a calm, happy, relaxed state?

And it's particularly problematic when we have to use addiction *all the time* just to create some level of obedience in day-to-day life. Remember? Close the economy. Nothing in Life is Free. Deprive them of food or affection and anything else they

---

[285] Killion, Jane. *When Pigs Fly!: Training Success with Impossible Dogs*. Wenatchee, WA: Dogwise Publishing, 2007.
[286] Pryor, Karen. *On My Mind – Reflections on Animal Behavior and Learning*. Waltham, MA. Sunshine Books. 2014.

CHAPTER 9 – More downsides and side-effects of "positive" training

enjoy. And then offer it back on a variable schedule of reinforcement, in order to create addiction for *everything* in life we want the dog to do.

If we use variable schedules of reinforcement to create an addiction related to *all behaviours*, because we've determined to use positive-reinforcement in an attempt to train them *all*, then we have a dog that is *perpetually* striving to have its addiction fed. Obviously, this isn't a great place to be psychologically. It certainly isn't relaxing, as anyone who has had an addiction can attest to.

> Close the economy. Nothing in Life is Free. Deprive them of food and/or affection and anything else they enjoy.
>
> And then offer it back on a variable schedule of reinforcement in order to create addiction for everything we want them to do.

The dog is, in effect, living constantly in the presence of the slot machines (its "positive" trainer), to get nearly everything it wants or needs.

And if you are one of these trainers, don't kid yourself that your "reinforcer junkie" dog "loves" you. It only "loves" you in a "toxic dependency" sort of way, like a gambling addict loves slot machines (as killer whale trainer Carol Ray eventually realized as we heard earlier).

Also, as mentioned in relation to hunger, when you create an addiction through the use of variable schedules of reinforcement, you are no longer in positive reinforcement territory, but negative reinforcement. The "reinforcer-junkie" dog is *driven* not only to ease his hunger, but also to *assuage his addiction by getting his fix*, just like the drug addict.

This is the world of so-called "positive" training. When using correction or punishment approaches (just like dogs or wolves do) we have none of these issues. We do not need to deprive our dog of food or affection (at least not very often or for very long), and we do not need to rely on addiction to create (and then maintain) some level of compliance. And, as we will learn next, we do not need to so tightly restrict our dog's freedoms, and certainly not for *anywhere near* as long.

Stringent and extended restriction of freedom

The third integral aspect of *The Big Three Necessities of Positive Reinforcement Coercion*, after deprivation and addiction, is the *extremely tight* control under which it is necessary to keep the dog *at all times*. As we've learnt, this is in order to prevent the dog from ever being rewarded by the far greater environmental rewards. Remember Pat Miller:

PART 4 – Investigating reward

> "All you have to do is figure out how to *prevent your dog from being rewarded for the behaviours you don't want* and rewarded consistently and generously for the behaviours you do want." [287] (emphasis added)

And Jean Donaldson:

> "Like most reinforcers, these are always more potent after periods of *deprivation*. All you need to do to train a dog is *get control of his favourites*."[288] (emphasis added)

And Zak George:

> "I'll be honest, *too few people take my advice on this*. They think they can bypass this step, but that's not advisable for the majority of dogs. I'm *not talking about having your dog on leash only when outside of the house. I want you to **attach her to you often when inside the house, too**…* Of course, you can't always supervise your dog this thoroughly. You have to work, run errands, and relax sometimes. In these cases, **it's critical** *that you make sure you have your dog* **in a controlled setting**… **crates** *are an* excellent way…"[289] (emphasis added)

Dunbar advises something similar. So, in order to get control of *all* your dog's favourites, to prevent your dog from ever being rewarded for undesirable behaviour, you need to *greatly restrict its freedom* not only while it's being trained, but also after, lest its behaviour rapidly deteriorate. And, as we've learnt from the mouths of these very trainers, reaching the place where your dog is "well-trained" is possibly never going to happen. Or at least, it's going to be a very slow process, perhaps taking years. Who on earth is going to do that? On the other hand, with sound, *real world* training, we can have a dog trained in a *fraction* of the time to a *much* higher standard, and then we can give it *far greater freedom*.

> With sound, *real world* training we can have a dog trained in a *fraction* of the time to a *much* higher standard, and then we can give it *far greater freedom*!

If you have a dog that is highly reactive and aggressive when you take it for a walk, Victoria Stilwell's advice is:

---

[287] Miller, Pat. *The Power of Positive Dog Training*. 2nd ed. Hoboken, NJ: Howell Book House, 2008.
[288] Donaldson, Jean. *The Culture Clash*. 2nd ed. Berkeley, CA: James & Kenneth Publishers, 2005.
[289] George, Zak, and Dina Roth Port. *Zak George's Guide to a Well-Behaved Dog: Proven Solutions to the Most Common Training Problems for All Ages, Breeds, and Mixes*. New York: Ten Speed Press, 2019.

CHAPTER 9 – More downsides and side-effects of "positive" training

> "The secret to successfully treating aggression is to *never put your dog in a situation* where he goes beyond his stress threshold."[290] (emphasis added)

So how do you walk your dog? You don't. At least, not for a very, very long time, while you *very* slowly and with a *huge* amount of time and effort attempt to work on the reactivity, utilising all your friend's dogs as training partners, using these trainer's highly ineffective positive-reinforcement approaches. So, once again, you are going to have to *severely restrict your dog's freedom* for a *very extended period of time*. On the other hand, with a sound use of negative-reinforcement we can have the issue completely fixed in a week, with only positive side-effects.

To quote Dunbar again (who, remember, calls us "trainers from the dark side"):

> "What is the bigger abuse? Kelly... she's always said, "What is greater and more abusive in a dog? To solve the problem quickly using something that could be a little scary or physically painful, or not solve the problem and live with it?"—giving the dog grief every day even if you're just in a huff, denying the dog your joy and companionship because you haven't trained the dog?"[291]

Kelly (Dunbar's ex-wife) is 100% correct. Do we subject our dogs to the three essential strategies of positive-reinforcement coercion (deprivation, addiction, and stringent and extended restriction of freedom) that these reward-only ideologues advocate, during the slow, time-consuming, complicated, difficult training processes, which in most cases will still not get reliable results? Or, do we instead use fast, effective training that actually works, and which has only positive "fallout"?

> Deprivation, addiction, and stringent and extended restriction of freedom, or quick, effective training?

And yet, in Dunbar's latest 2023 book, he states:

> "A lot of dog trainers ask me why I discuss aversive punishment at all, since *I never recommend its use*."[292]

So, he agrees with Kelly that it is a "bigger abuse" to "not solve a problem and live with it" rather than to "solve the problem quickly using something that could be a little scary", but also says he "never recommends its use". Personally, I find the use

---

[290] Stilwell, Victoria. *Train your dog positively.* New York. Ten Speed Press, 2013.
[291] Balabanov, Ivan. *Training Without Conflict Podcast Episode Four: Dr. Ian Dunbar.* YouTube, April 30, 2021. Accessed Feb 2024. https://www.youtube.com/watch?v=EF-INPvEhA8
[292] Dunbar, Ian, Dr. *Barking Up The Right Tree - The science and practice of positive dog training.* Novato, CA: New World Library, 2023.

of something "a little scary" that almost instantly gives a dog great freedom from that point forwards, far more desirable than "managing" its behaviour (i.e. never walking it, or never letting it off lead, etc.—the solution of choice for many "positive" trainers) or messing around with positive reinforcement-based methods for years that will never achieve success.

Of course, Dunbar also claims that:

> "…when food lures and rewards are used effectively to teach ESL, any aversive stimulus becomes unnecessary." [293]

This is blatantly untrue. Food lures and rewards are virtually never effective enough (ESL notwithstanding!) that "aversive stimulus becomes unnecessary". He also claims:

> "…the most effective punishment bar none is verbal instruction and guidance—even a single word can prevent, reduce, or eliminate so many undesired behaviors." [294]

And:

> "Single-word instructions can be used extremely effectively to teach dogs to cease and desist and instruct them what to do instead. And poof! The undesirable behavior is history. So easy, so simple, and so amazingly effective." [295]

Just like that! A single-word instruction and poof! Behaviour gone. Wow. I'd love to see the science (or *real world* evidence) he has to back up *this* preposterous claim. I can hardly believe I am even writing about it, it is so patently absurd. We have seen throughout this book just what absolute rubbish these two statements, and those like them, are.

In the *real world*—regardless of Dunbar's anti-reality claims—if you want "positive" training to work you are going to have to severely restrict your dog's freedom for a very long time, if not for life. As Kelly Dunbar asked, "what is the bigger abuse"?

So those are the Big Three Necessities of Positive Reinforcement Coercion – *Deprivation, Addiction,* and *Extended Stringent Restriction of Freedom*, all of

---

[293] Dunbar, Ian, Dr. *Barking Up The Right Tree - The science and practice of positive dog training*. Novato, CA: New World Library, 2023.
[294] As above.
[295] As above.

CHAPTER 9 – More downsides and side-effects of "positive" training

which are *essential* if you rely solely on positive reinforcement methods. And none of which are necessary with sound training methods.

Manipulation and dishonesty

Another downside with the "positive" approach, particularly in regard to training or handling people (like the "never correct your child or student" approach, because *punishment never works it only makes things worse*), is the amount of complicated manipulation involved. It's all very underhanded and deceitful, and will simply backfire. Kids (and adults) aren't as stupid as the people recommending these methods. As we heard from Perone:

> "Outside the laboratory, I cannot help but be impressed with the *propensity of people to respond to the negative side of positive contingencies*."[296]

Kids and adults alike know when they're being manipulated, even if the coercion is coated in sugary-sweet gold stars and fake smiles. Karen Pryor notes:

> Kids and adults alike know when they are being manipulated, even if the coercion is coated in sugary-sweet gold stars and fake smiles.

> "…the little girl turned to me and said, "I get praised for reading. I get praised for being nice to my little sister, too." I nodded. She added, "Of course I'd read, anyway." For that child, praise was a known manipulator… Children can usually tell when praise is intentional, artificial, and designed with some end point in mind—when it's fake—and when it is spontaneous and genuine."[297]

Case in point—remember Pryor's own example of "effecting change in a hard case" (which was actually a case of her effective use of punishment and negative reinforcement, not positive reinforcement)? Remember Pryor's "pleasant smile" during the process, and big "thankyou" and fake smile when the young visitor finally put her wet clothes in the washing machine? I'm certain, for that child also, that Pryor's "big thankyou" and fake smile were both "known manipulators" as Pryor elsewhere rightly describes such tactics. It simply leads to distrust.

---

[296] Perone, Michael. *Negative Effects of Positive Reinforcement*. The Behavior Analyst 26, no. 1 (Spring 2003): 1-14.
[297] Pryor, Karen. *On My Mind – Reflections on Animal Behavior and Learning*. Waltham, MA. Sunshine books. 2014.

Rather than a simple, honest ultimatum – "I don't like that behaviour, if you do it again this is going to be the consequence", the "positive" psychologists instead try to invent all kinds of circuitous—and highly ineffective—ways of arriving at the same place. Pippa Mattinson, with her "Total Recall" that we read about earlier, approvingly calls it the "cunning use of rewards"[298]. "Cunning" describes it perfectly.

I'm sure, from a human perspective, we all know those types that, rather than being upfront and honest, instead have that big fake smile, and all the while we know they're attempting to manipulate us. Do kids trust their parents more or less when they're honest and direct, or when they simply try to manipulate them in a more "cunning" way? Or their teachers? Or their bosses? I know what I would prefer.

> It is disingenuous, and therefore dishonest, and therefore leads to distrust.

It seems to me that simple and honest communication, where good behaviour is rewarded and poor behaviour is corrected or punished as appropriate, is not only far simpler and far more effective, but also far more open, transparent, and honest, and will lead to far more trust. And as long as we are fair, there are no downsides.

Anger and abuse

Anger and abuse are often consequences of forcing "positive" methods on people, and preventing them from using—or not teaching them—common sense training that actually works in *the real world*. Because the "positive" methods usually don't work (and certainly not with any ease or practicality as we've continually seen), it often ends up with dog owners becoming increasingly frustrated with their pets (or parents with their kids).

After your thousands of dollars' worth of couch is destroyed, the carpet is ruined, the back yard is demolished, the neighbours hate you because of the barking (and you're getting short on sleep!), your dog keeps nipping you and the kids, walking it is a nightmare, and it will never come when called, people, understandably, can become just a little bit frustrated and angry!

In fact, I helped one family who told me that their two little dogs had cost them at least $20,000 in damage to various objects in their house (carpets, couches, tables, curtains, etc.). They had tried all the usual rubbish—giving their dog a toy instead when it chewed the couch (just rewarding the behaviour), etc. And yet, with some real training, the problems were solved in weeks.

---

[298] Mattinson, Pippa. *Total Recall: Perfect Response Training for Puppies and Adult Dogs*. Shrewsbury: Quiller Publishing, 2013 reprint.

CHAPTER 9 – More downsides and side-effects of "positive" training

And add to that wasting a lot of time trying all the YouTube and puppy school methods, or even working with "positive" trainers for a year or more, and who can blame them? Only a fool would keep trying the same things that aren't working. Sooner or later, the punishment will probably come out. But if people haven't been taught how to use it correctly, and they use it out of frustration or anger, the results usually won't be great.

Zak George even admits that many qualified "positive" trainers resort to using correction or punishment (as we've repeatedly seen):

> "It's important to find an authentic positive trainer. Almost all dog trainers claim to be positive trainers, but most of them are not. Their general game plan may be to pay lip service to positive training methods, but if they don't get instant results, they'll often view this as justification to use an overabundance of force in lieu of real teaching."[299]

Why exactly aren't most of these "positive trainers" actually "*authentic* positive trainers" to use his words? If it really is so quick and easy as he, and Dunbar, and Garrett, and so on, all try to make out? Remember, Zak George claims reward-only methods are faster:

> "I will offer you a better option—one that will *teach your dog faster* without the use of these tools or other harsh corrections…"[300]

And from Susan Garrett:

> "The fastest and most effective way to bring out the best in your dog. No matter what age, breed, or experience."[301]

We've seen how misleading these sorts of claims are, and from some of their very own mouths. The reason the trainers Zak George refers to end up resorting to correction or punishment is *because the "positive" methods don't work*. It's that simple. If the reward methods were actually faster and more effective as claimed, the trainers claiming to be "positive" wouldn't feel the need to resort to other methods.

> If the reward methods were actually faster and more effective as claimed, the trainers claiming to be "positive" wouldn't feel the need to resort to other methods!

---

[299] George, Zak, and Dina Roth Port. *Dog Training Revolution: The Complete Guide to Raising the Perfect Pet with Love*. New York: Ten Speed Press, 2016.
[300] As above.
[301] Susan Garrett, 2024. The fastest and most effective way to bring out the best in your dog. No matter what age, breed, or experience .https://dogsthat.com/hstd-joinnow/

PART 4 – Investigating reward

Yet we see exactly this even from the likes of Karen Pryor, or Jean Donaldson, or Pat Miller, or Victoria Stilwell, all resorting to punishment when their methods fail (and they fail a *lot* more than they care to admit). Or when their methods have nothing effective to offer. According to Zak George, therefore, none of these are "authentic positive trainers". So where might we fight one, I wonder?!

Honest trainers will use methods that work. Only *dishonest* ones will continue taking money from clients while *peddling reward-based methods that simply don't work*, or that take months or years to work where the problem can actually be solved in a single session. It's a great business model, I'll say that for it. Of course, Ian Dunbar claims differently:

> "Food lure/reward training is just so incredibly efficient and effective that most dogs are well-trained within minutes. By advocating more time-consuming and laborious techniques, the trainer is able to keep clients for longer and extract more money!"[302]

As we've seen, this is laughable. As with so many of Dunbar's claims, the exact opposite is again the case.

Also, much like George's accusation, I can tell you that many trainers who *claim* to use only positive methods are simply lying. Behind the scenes, many high-level obedience or agility competitors (or others) use correction or punishment. Why? Because they know how ineffective "positive" training is, and punishment gives them an edge over their well-meaning but gullible competitors who try to avoid it. They would never openly admit to it for two reasons: they fear the criticism of their peers, and they want to retain their secret advantage.

So, by advocating methods that really *don't work*, or are highly *impractical*, in many cases the end result is highly frustrated, and even angry, owners. This sooner or later tends to lead to some sort of punishment as the owner lashes out in frustration. The problem is that this is then rarely done *calmly and methodically* following *sound methods of training* (which are rarely taught nowadays), and therefore it often doesn't work.

If, instead, the owners had been taught some simple, effective, correction-based methods of training from the start, all these problems could have been averted. A simple, calm, deliberate smack on the nose when the pup is swinging off the curtains the first time or two it does it, and the learning of an effective "no" command, and most of these problems are solved before they even really get going. Or, they're cured once they are.

---

[302] Dunbar, Ian. *The Good Little Dog Book*. 3rd ed. Berkeley, CA: James & Kenneth Publishers, 2003.

CHAPTER 9 – More downsides and side-effects of "positive" training

But the earlier they're dealt with, the easier it is on the dog and on the owner, *and the less correction or punishment that will be required.* Even Karen Pryor recognizes this fact:

> "Punishment has the best chance of halting a behavior in its tracks if the behavior is caught early, so that it has not become an established habit"[303]

So rather than a *last resort*, effective approaches such as these should be taught as the *first resort*.

(This is discussed further in *Chapter Thirteen* in relation to the LIMA, LIFE, and PROTECT training frameworks.)

> The earlier problems are dealt with, the easier it is on the dog and on the owner.
>
> And *the less correction or punishment that will be required* in the long run.
>
> Therefore, rather than as a *last resort*, effective approaches such as these should be taught as the *first resort*.

So, owners often become frustrated or even angry when restricted from using effective training methods, which then sometimes leads to abuse. Even if it doesn't, poorly trained dogs suffer a diminished quality of life: they are rarely or even never walked, never experience the freedom of being off leash due to an unreliable recall (unless they escape and can't be caught), and are generally "managed" rather than being involved. These are all direct side-effects of "positive" training.

Let's now look more deeply at where this so often also leads—to rehoming and euthanasia.

Rehoming and Euthanasia

Even if the owners don't end up lashing out at their dog in frustration, or aren't keen on keeping it confined for the rest of its life (management), their other option is to *get rid of the dog*. It's either rehomed to become someone else's problem, or put to death. *All for the want of some simple discipline.* A huge number of dog's deaths can be laid squarely at the feet of the "positive" trainers.

> A huge number of dog's deaths can be laid squarely at the feet of the "positive" trainers.

As Lindsay so rightly points out:

> "Besides misrepresenting and confusing the facts, excessive moralizing about the use of punishment and other aversive training procedures may

---

[303] Pryor, Karen. *Don't Shoot the Dog!: The New Art of Teaching and Training.* Rev. ed. New York: Bantam Books, 1999.

> have a very undesirable effect on the dog-owning public, *making responsible dog owners feel guilty about exercising the necessary aversive prerogatives needed to establish constructive limits and boundaries over a dog's behaviour*. Many of the basic facts of life that all dogs must learn to accept (if they are to become successful and welcome companions) are won through the mediation of directive training, combining a balanced application of behaviour modification—not just positive reinforcement. Instead of grinding away on a very dull ax [positive reinforcement], *a dog's welfare is better served by teaching the owner when punishment is necessary and how to use it effectively and humanely*."[304] (emphasis added)

I couldn't agree more. If every puppy school taught some good, sound, *balanced* training, instead of the pointless and often counter-productive "positive" rubbish, a far greater number of dogs would be living happy lives in their original homes with their happy owners, rather than having been surrendered or destroyed.

> "Can't thank Tully enough! We decided to try the New Puppy Consult & highly recommend anyone with a new puppy to give it a go! In 1 hour Tully had stopped her from trying to get in the house, shown us how to walk her properly as she wasn't keen on the lead (now loves her walk) and stopped her from jumping. I was amazed at the difference it made to our gorgeous girl in just 1 hour. Don't put it off give it a go as you won't be disappointed."
>
> - Jodi Lewis

So one of the major side-effects of an emphasis on "positive" training is a lot of very poorly trained dogs, and the resultant huge number of dogs being put down. Remember Zak George's comment:

> "What's tragic is that so many of these issues are part of the reason our shelters are overflowing with unwanted dogs. **Millions of people get rid of their pets, often because they just can't deal with certain disruptive behaviors**... a study in the Journal of Applied Animal Welfare Science found that **65 percent of people who relinquished their dogs reported some behavioural issues as a reason**. Sadly, **hundreds of thousands of such shelter dogs are euthanized each year**."

---

[304] Lindsay, Steven R. *Handbook of Applied Dog Behavior and Training, Volume 1: Adaptation and Learning.* Ames, IA: Blackwell Science, 2000.

CHAPTER 9 – More downsides and side-effects of "positive" training

He's right. He's just completely wrong about the cause. ***He, and those like him, are the cause***. The cause is the push into ineffective "positive" training, and the demonisation of correction and punishment.

> The cause is the push into ineffective reward-only training and the demonisation of correction and punishment.

Victoria Stilwell likewise notes the large rise in behavioural problems:

> "My *ever-growing workload* as a dog trainer and behaviour consultant attests to there being *more problems than ever*."[305]

She then makes the astounding claim that this is due to:

> "…this *epidemic of punishment-based dog training*… [and] increase in owner-surrendered pets at rescue shelters due to behavior problems has led to steadily increasing euthanasia rates. The number of dog bites and attacks, especially on children, are increasing…"[306] (emphasis added)

Her reasoning and logic here (when she blames this mythical "epidemic of punishment-based dog training") is diametrically opposed to reality, as any thinking person will quickly realize. In the past, prior to the current "positive" revolution, punishment-based training was basically 100% of trainers. Practically every single obedience club in the world relied on correction-based methods (such as choker chains). Those have become largely a thing of the past.

The only thing that *has* increased is the reliance and faith in "positive" methods, and the demonisation of punishment. Almost every "puppy class" teaches the ineffective "positive" strategies, as do the *vast* majority of obedience classes. The use of punishment in dog training has *greatly reduced*. No rational, honest person could argue otherwise.

Claiming that there's some "epidemic of punishment-based training" is a lie.

The only correlation that can realistically be drawn is that, as the popularity of so-called "positive" methods has *increased*, and the use of punishment-based approaches has *decreased*, the number of

> Claiming that there's an "epidemic of punishment-based training" is a lie.
>
> As the popularity of reward-based methods has *increased*, and the use of punishment-based approaches has *decreased*, the number of behavioural problems and pets being euthanised has significantly *increased*.

---

[305] Stilwell, Victoria. *Train your dog positively*. New York. Ten Speed Press. 2013.
[306] Stilwell, Victoria. *Train your dog positively*. New York. Ten Speed Press. 2013.

behavioural problems and pets being euthanised has risen *significantly*.

Not only that, but as Stilwell mentioned, there has been a huge increase in dog bites over a similar period. According to the US Department of Health and Human Services, from 1993 to 2008, dog bites increased by 86%[307]. And in England, where reward-only training is even more prevalent than in the US, between 1998-2018 dog bites *tripled*[308] (200%).

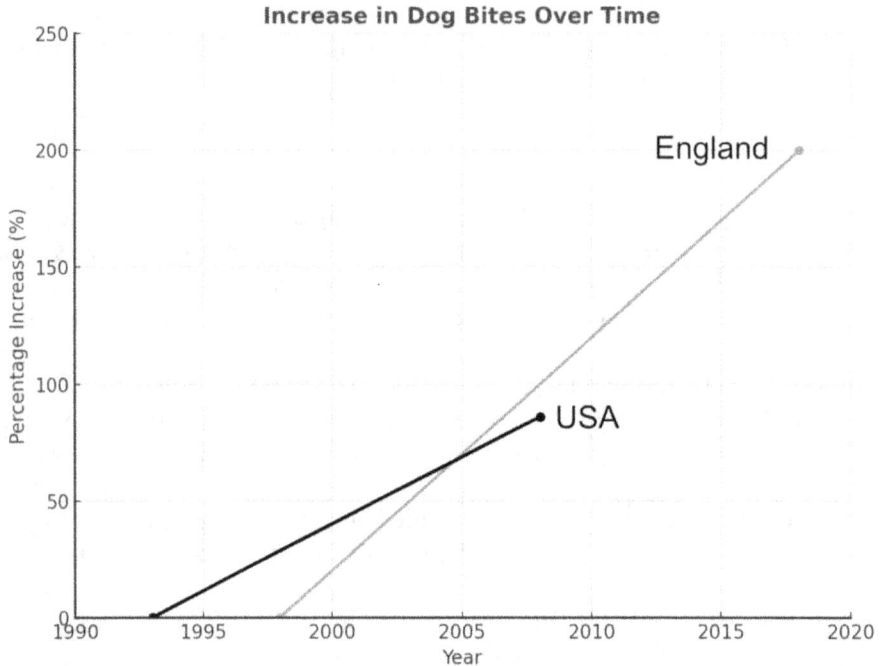

Figure 15 - Increase in dog bites as "positive" training has increased

What has also greatly increased over this time? Reward-based training. What has greatly decreased over this time? The use of punishment in training. Coincidence? Definitely not. We've seen how punishment rarely ever causes aggression, and that it's actually the positive methods that do so. And we've seen how punishment very effectively cures aggression, while positive approaches fail. A belief in, and reliance on, "positive" methods is what is *actually* responsible for the epidemic of behavioural problems leading to rehoming and euthanasia.

---

[307] Agency for Healthcare Research and Quality. "Hospital Admissions for Dog Bites Increase 86 Percent Over a 16-Year Period: AHRQ News and Numbers, December 1, 2010." Last reviewed December 2010. Agency for Healthcare Research and Quality, Rockville, MD. https://archive.ahrq.gov/news/newsroom/news-and-numbers/120110.html.
[308] Medical Xpress, "Adult Hospital Admissions for Dog Bites Triple in 20 Years," Medical Xpress, January 19, 2021, https://medicalxpress.com/news/2021-01-adult-hospital-admissions-dog-triple.html.

CHAPTER 9 – More downsides and side-effects of "positive" training

It is the epidemic of so-called "positive" dog training that is to blame.

Murray Sidman (his fundamentally flawed theories are discussed in *Chapter Eleven – Murray Sidman's "Coercion and its fallout"*) promotes pretty much the same fallacy in relation to school discipline and dropouts. He says:

> "The current *discipline* and *dropout crises* are th*e inevitable outcome* of a history of educational *coercion* [which he defines as negative reinforcement or punishment]"[309] (emphasis added)

And yet, in a completely contradictory statement in the very next paragraph (adding to the huge volume of contradictions that come from the reward-only ideologues), he says:

> "This is not to say that coercion in the educational system is the sole cause of dropouts. *Coercion has always been practiced in schools*, yet dropouts have **become an acute problem only within recent memory**. We have no evidence that our schools are more coercive today than in the past, or that today's coercion is more severe. Clearly, other factors are also involved."[310] (emphasis added)

So, while claiming that problems with discipline and dropout in schools are the "inevitable outcome" of a usage of negative reinforcement and punishment, he then notes that this hasn't increased! So, obviously, there is no correlation between the two. In fact, any rational person should realize that what this has actually coincided with is the increase in the demonisation of punishment, and in the *reduction* in its use in schools, not the opposite. Far from too much punishment (clear boundaries and sufficient consequences for breaking them) being the cause of a lack of discipline and dropouts, it's evidently the exact opposite.

So, whether regarding the epidemic of badly behaved dogs, students, or children, it is ludicrous to claim that this is the result of punishment. And yet here we have **Sidman, Stilwell, and George—and many, many others—all trying to claim a causal relationship from what is clearly an *inverse* relationship.** That is some mighty impressive mental gymnastics. The rise in reward-based approaches has coincided with a significant rise in problems such as increasing anxiety, a lack of discipline in schools and homes, as well as *higher rates of dog bites, rehoming, and euthanasia*—and yet they claim that the blame lies with punishment!

> The rise in reward-based approaches has coincided with significant increases in problem behaviours – and yet they claim the blame lies with punishment!

---

[309] Sidman, Murray. *Coercion and Its Fallout*. 2000. United States. Author's Cooperative.
[310] As above.

PART 4 – Investigating reward

Figure 16 – Stilwell, Sidman and George—when ideology trumps evidence

Conclusion – More downsides and side-effects of "positive" training

So, as you can see, there are many significant downsides to the use of reward, and particularly of reward-*only* approaches. We've seen that positive reinforcement can cause aggression, and often does. We've seen that even when used by some of the top trainers in the world during their normal training, positive reinforcement can and does cause learned helplessness—including "emotional collapse", "visible distress", and refusing food for days as we heard from Pryor. And we've seen that positive reinforcement can exacerbate fear and anxiety, while the sound use of negative reinforcement or punishment can be very successful in *curing* fear and anxiety.

We've also seen that if you want to gain obedience under distraction using reward, you will *have* to use methods of *deprivation* like 36 hours feeding, "close the economy", "take control of all the goodies your dog wants", "nothing in life is free", no free "pats, cuddles and affection", simply never feed your dog a meal, and so on. Whereas with a balanced approach none of this is necessary. You will also have to rely strongly on the psychology of addiction, and turn your dog or student or child into a "reinforcer-junkie". Again, with a balanced approach, this is not

CHAPTER 9 – More downsides and side-effects of "positive" training

necessary. And, you will have to impose stringent and extended restrictions on your dog's freedom, both inside the home (such as attached to you on a lead at all times, or in a crate) and outside, during the long training process (and probably for life). And quite probably you will still never get the results you need in order to be able to give your dog much freedom.

Keep in mind also, that if you're considering (or attempting) so-called "positive" parenting or school teaching, that all the same principles apply. You really cannot (or should not!) apply these positive reinforcement essentials of deprivation, addiction and severe restrictions on freedom to your children or students—and *yet without them* "positive" methods are doomed to even more certain failure, as we've seen.

> Without *deprivation, addiction,* and *extended stringent restriction of freedom,* "positive" methods are doomed to even *more* certain failure.
>
> Therefore, they are even *less* effective with *children, students,* or *people in general* than they are with animals.

This is the reason that positive reinforcement-based methods are *even less effective* with children, students, or society in general, than they are with animals.

So, we've now seen just how slow and ineffective "positive" training is in the most important areas for pet owners, and conversely how quick and effective punishment can be. We've also seen that the reward-only claims about the "fallout" from punishment were unfounded, and that in fact there are many positive side-effects from punishment (for dogs and humans). And now we've seen that there is much that is "bad" about the "positive" methods (and there is more that hasn't been covered).

Let's now turn our attention to the "science". The "positive" advocates always claim they have "THE SCIENCE" on their side—let's debunk that myth also.

Part 5

THE SCIENTIFIC BASIS – FACT AND FALLACY

Chapter 10 – The scientific basis—fact and fallacy

Chapter 11 – Murray Sidman's "Coercion and its fallout"

Chapter 12 – Some real-life junk "science"

Chapter 13 – LIMA, LIFE and PROTECT

10

The scientific basis—fact and fallacy

The "positive" ideologues constantly parrot the claim that they have "science" on their side. "The Science proves punishment doesn't work", "The Science proves reward-based methods are superior", "The Science proves we never need to use punishment", "modern science-based methods," etc., etc., etc.

Laurie Luck, a faculty member of the *Karen Pryor Academy for Animal Training and Behaviour,* professional member of Dunbar's *Association of Professional Dog Trainers,* and who was a past president of the *Certification Council for Professional Dog Trainers* (an offshoot of APDT), parrots the same rubbish when she says that:

> "I train with positive reinforcement *because science says* it's efficient and effective."[311] (emphasis added)

YouTube influencer Zak George is notorious for it. As the saying goes, "if you can't convince them with evidence, then baffle them with science." He likes to provide a long list of scientific "studies", and to imply that "The Science" is *irrefutable*. Perhaps he banks on the fact that hardly anyone will

> If you can't convince them with evidence, then baffle them with "science".

---

[311] Luck, Laurie. *Karen Pryor Archives.* Smart Dog University, May 7, 2024. https://smartdoguniversity.com/tag/karen-pryor/.

PART 5 – The scientific basis—fact and fallacy

actually analyse the studies—he most likely hopes they will just be impressed by the list he provides. Or, at best, that people might glance at the *conclusions* on the studies (which often don't even correlate to the "evidence" provided), without understanding the major flaws the studies have, and without looking at all the other studies he doesn't reference which refute those he does. A number of concrete examples of this will be given soon, when two such often-quoted studies are dissected in some detail.

At best, we can say "*this* study proves" or "*this* study supports". When people say "THE SCIENCE proves," they are basically claiming that *all* the science agrees, and that science is perfect. When I say, "REAL science says," then I'm admitting that there is science out there that disagrees. I'm just claiming it's wrong, and that it's poor science. And I've presented plenty of evidence to support that claim, and will present more here.

There is so-called "science" that says punishment doesn't work, and there's real science that says it does. And the real science says it often works highly effectively, permanently, and quickly. And, if the *real world* evidence I have presented, and the evidence people see all around them, doesn't convince them, then anyone can experiment for themselves to prove it works.

There is very definitely good and bad science. You can probably find some "science" to support just about any crazy position you have, on just about anything. You only have to consider the "science" that says there are 63, or 78, or whatever it is nowadays, different "genders". If you believe *that* insane "science" then this book probably isn't for you (unless you're open-minded enough to reconsider your views). But it *is* a great example of "junk" science.

Modern "science" is plagued with issues, and many (often all) of these are evident in most of the studies supporting the use of "positive" training. What follows is a quote from Professor John Staddon in his 2022 book "*Science in an Age of Unreason*". Staddon is the James B. Duke Professor of Psychology, and Professor of Biology and Neurobiology, Emeritus, at Duke University, USA. He has authored more than 200 research papers and five books, so he has a very good idea of what he's talking about:

> "Science is in trouble. The tranquil realm of reason has been invaded by man's strongest passions. *Facts have taken a back seat to ideology*, and *political activism masquerades as scholarship*… The corruption of science, unfortunately, harms not only scientists but everyone. It poisons *the social sciences* [which includes behavioural science] and even the humanities, with disastrous results for our entire culture. Can science be saved? The first step is to acknowledge *the rot at the core of our scientific*

CHAPTER 10 – The scientific basis—fact and fallacy

establishment—a rot that is worse than most of us realize."[312] (emphasis added)

Few people understand just how bad that rot really is. In another of his books, "*The Scientific Method – How Science Works, Fails to Work, and Pretends to Work*", he also says:

> "…flawed and even fallacious claims meet a need and get wide circulation… Unfortunately, there is reason for distrust. All too often hypotheses based on correlations ('links') are pushed in the media as conclusions about causes."[313]

> "The first step is to acknowledge *the rot at the core of our scientific establishment – a rot that is worse than most of us realize.*"
>
> - Professor John Staddon

(Or, far, far worse as we saw attempted by Sidman, Stilwell, and George—an *inverse correlation* being pushed as the cause!)

And:

> "Contemporary science, especially biomedicine and social science, is *not in a very healthy state*… If science is to thrive, people must understand the difference between *good science and bad*. They need to be *sceptical* to recognize *claims that are baseless or exaggerated.*"[314] (emphasis added)

I've presented many of the false claims from people who claim they have science on their side, like these:

> "No matter what the behaviour is, responding with punishment simply will not work."[315]

Or:

> "Punishing Unwanted Behaviour Just Makes it Worse."[316]

---

[312] Staddon, John. *Science in an Age of Unreason.* Washington, D.C. Regnery Gateway. 2022.
[313] As above.
[314] Staddon, John. *Scientific Method – How Science Works, Fails to Work, and Pretends to Work.* New York. Routledge. 2018.
[315] Lemon Drops Kids Therapy. *Why Punishing Our Children Doesn't Work.* Lemon Drops Kids Therapy, www.lemondropskidstherapy.com.au/blog/why-punishing-our-children-doesnt-work
[316] Neurodiversified. *Punishing Unwanted Behaviour Just Makes it Worse.* Medium, https://medium.com/neurodiversified/how-punishing-unwanted-behaviour-just-makes-it-worse-baf22793d07b

PART 5 – The scientific basis—fact and fallacy

We've seen throughout this book just what absolute rubbish these claims are. Mazur's assessment in the 6th edition of *Learning and Behaviour* is spot on:

> "The question of whether punishment can be an effective way of controlling behaviour has been settled. This chapter has presented numerous studies, from both inside and outside the laboratory, that have demonstrated that punishment can change behaviour and, in many cases, *change it permanently*."[317] (emphasis added)

How do we know which study is right? I suggest that first we compare their conclusions to *reality*, we compare them to *experience*. But then we have to analyse the studies themselves, and their strong and weak points.

When the study (or a reward-only ideologue) says that punishment doesn't work, we know it is false. How? Because we see all around us that punishment *does* work. I've given plenty of examples to prove all those types of statements false, including from these "positive" trainers themselves. So, we know that any "science" that says punishment doesn't work is *garbage*. From that we can quickly see that other "science" which follows on from this "science" and cites this "science" is also going to be garbage.

When later researchers then cite these flawed studies, they generally don't reference the flaws in those studies—they simply reference the conclusions. And so, in a case like "Chinese Whispers", before long the foundation of those original very flimsy studies has been built upon with further flimsy studies, into a complete *house of cards*. And that house of cards, it is claimed, "proves" many of the claims made about positive reinforcement and punishment. But a strong wind should be sufficient to blow it down, provided people are open to the facts. Hopefully this book will drastically stir up that wind.

> A strong wind should be sufficient to blow down the reward-only house of cards, if people are open to the facts.
>
> Hopefully this book will drastically stir up that wind.

I have personally trained perhaps four hundred sheepdogs to varying levels over the last thirty years. (I only say "perhaps" because I stopped counting after three hundred, so the next hundred is an estimate.) You can look at any of the videos of my dogs working sheep. Check out *www.campaspeworkingdogs.com*. All were trained with a balanced approach that included reward (not treats, not toys), correction, and punishment (never e-collars, although I do use them for other training). I don't think anyone can honestly argue that the methods don't work.

---

[317] Mazur, James E. *Learning and Behavior, 6th Edition*. New Jersey. Pearson Prentice Hall, 2006.

(If you'd like to learn more about the methods and principles used, check out my book *"Working Sheep Dogs – A Practical Guide to Breeding, Training and Handling."* Or, for online sheepdog training instructional videos visit *www.workingsheepdogtraining.com*.)

I have also helped thousands of clients with every behavioural problem (from the mundane to the extreme) and general training issue under the sun. And unlike most "positive" trainers, I don't turn down aggressive dogs or hard cases. Many of these hard cases get referred to me from obedience clubs, various dog clubs, veterinarians, and from other trainers. I work with them all. I know punishment works, conclusively, without a shadow of doubt. I also know it doesn't cause any of the side effects often claimed (and also produces positive side effects), when used properly. And remember that reward, even when used as directed, causes all sorts of problems. So the *real world* evidence proves much of the "science" wrong.

The aim in this section is, firstly, to discuss why "observational science" (like the above) is often superior to laboratory or university "science". Then to talk about the fact that science is done by *people*, and people are prone to all sorts of issues. Then to examine the problems with studies that are often little more than surveys of the general dog owning public (this is very often the type of "science" that the reward-only ideologues quote).

Then we'll consider, in some detail, one of the most influential scientists and authors in creating the reward-only ideology—Murray Sidman—and his wild, unproven, utopian claims.

Then we'll move on to directly consider three regularly cited examples of this type of pseudo-science. Of course, it's impossible to dissect every single scientific study ever done on the subject, or even very many, in a book like this (I personally have studied a very large number of them). But hopefully you will see enough (you probably already have) to question the lie that "science proves that positive training is superior" and that "science proves punishment doesn't work".

Finally, we'll discuss the LIMA and LIFE models or frameworks for training, and I'll offer what I believe is a superior alternative—the PROTECT framework.

Why observational evidence is so often superior to "science"

When science studies physics, or chemistry, or mathematics, it is more often dealing with fairly concrete, provable subjects. (Of course, this isn't always true nowadays, with the crazies who claim mathematics is "racist"—and claim to have "science" to back them up!—but for normal, rational people this applies.)

PART 5 – The scientific basis—fact and fallacy

However, when science attempts to study things like learning or psychology, it is an entirely different scenario. There are so many variables that come into play that can be almost impossible to quantify.

Science tends to attempt to *isolate* certain aspects in order to study them, which certainly has its place. In doing so it usually removes as many other complicating variables as possible. Like Skinner with his rats or pigeons *in cages*. Or Pryor with her *captive* dolphins. The problem is that the *real world* is nothing like that. These studies often fail to *see the forest for the trees*.

When we step back, and look at the subject in the *real world*, we now see how it functions with *all the variables in place*. We see the forest with *all* the trees in place. So when *real world experience* contradicts "science", the onus is on the science to prove itself.

The *real world* is always right, and we should believe it first.

Even Murray Sidman (you'll hear all about him in the next chapter), notes:

> "The proof of such applicability [to the real world] comes not from the basic experiments *but from experience* with informed transformations of scientific principles *into practical technologies*… starting with abstract theory and the **artificially controlled settings** of the intellectual arena, we move on to *practical tests*…"[318] (emphasis added)

He also acknowledges that the *real world* might debunk some of their theoretical "knowledge" when he says:

> "When our *extrapolations to everyday life fail*, we start again; as our extrapolations begin to prove successful, we gain more confidence that we are on the right track… the pioneers of the behavior laboratory, acutely conscious of historical precedent, accepted the assumption that their science, too, would prove applicable to everyday life. *This assumption remains to be tested in many areas of behavioral analysis, and failures in specific areas may still await us*…"[319] (emphasis added)

The fact is, the extrapolations of their theories to everyday life *have* failed. The reward-only experiment *has* proven a failure, whether in the animal world or the human. We see far more undisciplined dogs leading to an increase in dog bites, rehoming, and euthanasia. We see far more undisciplined school children, and plunging standards in the classroom.

---

[318] Sidman, Murray. *Coercion and Its Fallout*. 2000. United States. Author's Cooperative.
[319] Sidman, Murray. *Coercion and Its Fallout*. 2000. United States. Author's Cooperative.

CHAPTER 10 – The scientific basis—fact and fallacy

The positive reinforcement experiment is over; it has failed. It is time to get back to reality and put this utopian fantasy to rest.

When these reward-only advocates say, "punishment doesn't work," and cite some probably long since disproven study, they can easily be refuted with *real world evidence* like Sidman suggests, exactly as discussed in this book. We don't need a study to back this up (although there are plenty, and some are provided here for those who refuse to believe their eyes or to trust in logic). Are you really going to ignore the evidence right in front of your face in preference to some activist with *zero real world* experience, in a university office somewhere, doing a *survey* and claiming, "punishment never works, in fact it only makes things worse"?!

Or, are you going to believe it when Karen Pryor, while likewise saying punishment *almost never really works*, gives so many examples of it working perfectly (and examples of positive reinforcement failing dismally)? Believe your eyes. Believe your experience. We all know punishment works. We all know it doesn't traumatise dogs or kids for life, ***provided it is fair and used properly***. We know that the insanity of "no punishment" is playing its part in destroying our schools and judicial system and society, and creating a generation of anxious, fearful, spoilt children.

> Are you really going to ignore the evidence right in front of your face in preference to some activist with zero real-world experience in a laboratory doing a *survey* and claiming, "punishment never works, in fact it only makes things worse"?!

As Sidman noted, observational evidence trumps "science" when the "science" is proven wrong by the *real world*. Why? Because the *real world* takes into account *all* the variables, whether we understand them and account for them *or not*.

What is wrong with science – the people

The next thing to consider with science is that it is conducted by *people*. Some people have a great deal of honesty and integrity, most don't. Some scientists want to find the truth, others just want to confirm their existing bias, gain as much funding as possible, be popular with their peers, or push their agenda. Even Pryor recognizes this when she says:

> "Science is supposed to be logical and based on proven facts, but scientists are humans, after all."[320]

---

[320] Pryor, Karen. *On My Mind – Reflections on Animal Behavior and Learning.* Waltham, MA. Sunshine Books. 2014.

PART 5 – The scientific basis—fact and fallacy

There's a lot of peer pressure in science. I have experienced this firsthand. About ten years ago I was involved in fighting the state government here in Australia over some entirely misguided dog-welfare related legislation that was proposed. One particular scientist told me privately that although he agreed with my position, he couldn't support it publicly. Why not? Because it would hurt his career.

There is a lot of cowardice like this in science, like there is in all walks of life. People don't like to rock the boat; it's easier, and probably better for your career, just to go along with the crowd. And nowadays, the crowd—the fashion—*is* so-called "positive" training. How popular do you think someone would be at a university who said they wanted to study how punishment is more effective than reward? Do you think they would get funding to start with? Or would they be popular with their peers or friends? It takes courage and sacrifice to go against the flow.

When you consider any institution in society, you see that they *all* have problems. They all have corruption. They all have good and bad. The justice system—major problems, corruption, bias, etc. The political system—major problems, corruption, bias, etc. The health system—major problems, corruption, bias, etc. You name it, every human institution has major issues. Science is no different. People like to quote "science" like it is some infallible, pure, perfect institution. It is nothing of the sort. Like everything, we have to consider it carefully, logically, and rationally, and decide for ourselves what is backed up by real evidence, and what is just hogwash "junk" pseudo-science.

So, when analysing any "science", keep in mind that behind every study is a few more-or-less flawed humans, many lacking in much *real world* experience in the subject they are studying, more-or-less susceptible to greed, peer pressure, confirmation bias, cowardice, and so on.

What is wrong with science – surveys

A large proportion of the pseudo-science used to support "positive" training is based on *surveys*. How many people realize that so-called "science" is often nothing more than a small-sample-size questionnaire of the general public?

In fact, I remember the first time many years ago when I first looked into the actual scientific studies behind some of the claims about dog training. There was a summary of a study in a magazine that supposedly proved positive reinforcement is better, and that we shouldn't use punishment. Reading the magazine article, I knew, based on *real world* experience, that it was completely untrue. I decided to find the original study, because I wondered how on earth they could come to the conclusions they did. I was flabbergasted when I found that it was just a survey of a dozen or so general pet owners, and their opinions. I have since discovered that

CHAPTER 10 – The scientific basis—fact and fallacy

this wasn't a rare exception. That was the start of my healthy scepticism of "science"—and it has only become much stronger and more confirmed ever since.

As Professor John Staddon points out:

> "…social science relies heavily on correlational studies. But as I have already pointed out, possibly *ad nauseam*, surveys, epidemiology, and the like, cannot prove causation."[321]

Figure 17 - Correlation is NOT necessarily causation!

---

[321] Staddon, John. *Scientific Method – How Science Works, Fails to Work, and Pretends to Work*. New York. Routledge. 2018.

PART 5 – The scientific basis—fact and fallacy

Even the *American Veterinary Society of Animal Behavior* admits as much:

> "Survey studies cannot differentiate between causation and correlation…"[322]

And yet, they then go and base much of their recommendations on exactly such studies!

For example, these surveys might ask a group of general dog owners what types of training they believe had the best results—punishment or positive reinforcement? If you asked a group nowadays that have had the current fashion drummed into them, you can predict the results. If you had asked the same question a few years back, before the current *epidemic of "positive" training*, you would have received a completely different result. If you asked my clients, you would get a definitively different answer. Even Sidman (perhaps the king of the reward-only ideologues) notes:

> "When personal opinion and public policy are supported by correlational statistics, skepticism is justified."[323]

And, of course, it also depends on *what* questions are asked, and *how* the questions are asked. And it also depends on whether the respondents answer these questions truthfully or not; in many cases, they may not be willing to admit the method they used due to the current fashions. And often this specific information is not included in the studies, and so the studies cannot be replicated. An article in *The Scientific American* titled *"Science Funding is Broken"*, notes:

> "Many scientific protocols, analysis methods, computational processes and data *are opaque*. When researchers try to crack open these black boxes, they often discover that many top findings *cannot be reproduced*. That is the case for *two out of three top psychology papers*."[324] (emphasis added)

So, the actual processes and methods are not clear in *two out of three top psychology papers*, nor are the actual questions provided in the studies involving surveys, and so the results cannot be checked, and certainly can't be replicated. Even Sidman would recognize this as *junk* "science". We've all seen how wrong political polls (surveys) often get things. The results are *often the exact opposite* of what they predict. Here is a great example of the conclusion of one such survey:

---

[322] American Veterinary Society of Animal Behavior. *Position Statement on Humane Dog Training.* 2021. https://avsab.org/wp-content/uploads/2021/08/AVSAB-Humane-Dog-Training-Position-Statement-2021.pdf
[323] Sidman, Murray. *Coercion and Its Fallout.* 2000. United States. Author's Cooperative.
[324] *Science Funding Is Broken.* Scientific American, April 5, 2024, accessed May 7, 2024, https://www.scientificamerican.com/article/science-funding-is-broken/.

CHAPTER 10 – The scientific basis—fact and fallacy

> "More owners using reward based methods for recall / chasing report a successful outcome of training than those using e-collars."[325]

This is the type of pseudo-science that is then used to claim that reward-based methods are more effective, and that "the science proves". As we've seen, even the top, most influential "positive" trainers admit how much hard work and time it takes to get a good recall, if they ever do. Remember Jean Donaldson's *"might/sometimes"* recall being the best you can hope for? Or Pat Miller using the cap gun? Or Karen Pryor just not letting her dog off the lead? And yet, this study purports to prove the opposite.

Whereas, I can help any client get a truly reliable recall in a week or so. If I was training the dog myself it would be a lot quicker than that, and I would create a great bond and relationship simultaneously. As we heard earlier:

> "Tully was fantastic. After just one session I was confident using the e-collar to train my dog to come to me on command and within a week he was a new dog!"
>
> - Patrick O'Keefe

So, the science is garbage. And I'm more than willing to *challenge any "positive" trainer in the world* to see who can produce a reliable recall faster and more effectively. The results would not even be close. A year or more for a *might/sometimes* recall using reward (including deprivation, addiction, extended stringent restrictions on freedom, and years of reinforcement history), or an hour or a week for me?

And I'd love to do it in the presence of these researchers, and put a very large wager on it. Let's see how confident they are in their "science" and their claims then.

> Let's see how confident they are in their "science" then.

So, having outlined some of the many problems with "science", such as its often political and activist nature, its peer pressure and funding pressure, its flawed methodologies such as small sample size "surveys", and the often biased positions of the scientists, let's move on and discuss Murray Sidman.

---

[325] Blackwell, E.J., Bolster, C., Richards, G. et al. *The use of electronic collars for training domestic dogs: estimated prevalence, reasons and risk factors for use, and owner perceived success as compared to other training methods.* BMC Vet Res 8, 93 (2012). https://doi.org/10.1186/1746-6148-8-93

PART 5 – The scientific basis—fact and fallacy

11

Murray Sidman's "Coercion and Its Fallout"

Murray Sidman was a highly influential American behavioural scientist, who started his work in the 1950's (while Skinner commenced his in the 1930's). In 1989 he published his book "*Coercion and its fallout*", which continued Skinner's theme of the irrational demonisation of punishment.

If there's been any single, largest contributing factor to this myth that punishment has all this "disastrous" and "terrible" fallout (Sidman even likens it to a "*nuclear explosion's side effects*" [326] which is where the term "fallout" originated), and therefore we should never use it, it is this book, in which we find Sidman's extreme ideas. It provided enormous fuel for the reward-only ideologues to use in the demonisation of the perfectly natural and normal phenomenon we call punishment.

Sidman's book contains many wild, unsupported and unreferenced claims; many false and unproven assumptions; and various logical fallacies. The problem is often not with the "science" behind it as such, it is with the *interpretation* and then the *wild, reckless extrapolation* of that science in which Sidman goes *way* beyond the evidence (aside from the various other mistakes in reasoning and logic that he makes).

---

[326] Sidman, Murray. *Coercion and Its Fallout*. 2000. United States. Author's Cooperative.

Punishment doesn't work?

Firstly, as usual, Sidman claims that punishment doesn't work (while at other times admitting its effectiveness). There has been so much evidence (both before and since) that this is simply not true, and yet the reward-only ideologues still insist on parroting this claim, ignoring all evidence to the contrary. As Sidman himself said, the ultimate test is *reality*—what happens in the *real world*:

> "Starting with abstract theory and the artificially controlled settings of the intellectual arena, we *move on to practical tests…*"[327] (emphasis added)

And:

> "…as the investigation proceeds, we test the knowledge gained in the laboratory *by applying it outside…* When our extrapolations to everyday life *fail*, we start again"[328] (emphasis added)

Therefore, even he would have had to recognize that all the *real world* evidence shows that punishment certainly *does work*, and that this must take precedence over any "*abstract theory*" in "*artificially controlled settings of the intellectual arena*". If he had only been a dog or horse trainer (the majority of horse training is negative reinforcement—pressure/release), he would have realized right from the start that, outside the *artificially controlled settings of the intellectual arena* where he operated, punishment works. Just consider any highly trained sheepdog. As said earlier, the *observational evidence* trumps any "science", because science is prone to all sorts of problems. Reality is always right. The *applications outside*, and the *extrapolations to everyday life*, prove that punishment works.

This claim that punishment doesn't work has already been debunked in some detail in earlier chapters, however I will add just one more aspect to that discussion here—Skinner. Skinner claimed that the effects of punishment are only temporary in nature, and he based that largely on one flawed experiment he conducted because it matched his utopian ideology. (If you'd like to learn more about Skinner's utopian fantasies, read his novel *Walden Two*.) The reward-only ideologues have been quoting his flawed conclusions ever since (while ignoring all the contrary evidence).

Skinner's experiment involved two groups of rats trained to press a lever. Then, once that behaviour was established (by positive reinforcement on food deprived rats, so actually negative reinforcement), he introduced a very mild punishment to one group. For the next 10 minutes, instead of providing a treat, the lever instead

---

[327] Sidman, Murray. *Coercion and Its Fallout*. 2000. United States. Author's Cooperative.
[328] As above.

flicked up and tapped the rat's paw. This mild "punishment" reduced the rat's lever pressing. However, when the lever flicking ceased, the food deprived rats slowly went back to pressing the lever again and receiving food. In the end, the rat's response supposedly returned to the same as the unpunished group.

There are a number of problems with this. The first is that the punishment was very mild. I think about punishment on a continuum from very low, to very high, with a number of threshold areas within that. Consider a dog jumping on us. If we give it a *light* smack on the nose, the dog can actually think this is a fun game—in that case the smack will be a reward and will actually increase or maintain the behaviour. The dog may possibly start nipping at our hands and think it's a great game—you tap me on the nose, I nip at your hands—yippee! So it certainly won't work.

> Punishment occurs on a continuum from very low, to very high, with a number of threshold areas within that.

Once we rise above that threshold and get a bit firmer, at some point the dog begins to dislike this, and it begins to act as punishment rather than being either neutral or rewarding. However, it's still probably not enough to outweigh the variably reinforced positive reinforcement the dog is receiving (or has received in the past) for jumping on us at other times, and therefore it still won't be particularly effective.

When we get firmer than that, we come to what I consider the main threshold—almost the balance point where the punishment is about equal to how much the dog wants to do the behaviour. Provided we are firm enough to rise above this balance point or threshold, with just two or three repetitions the dog will stop jumping. It is only at this level or above that the punishment will be particularly effective, and we won't have to do it very much at all—the problem will largely be solved very rapidly.

So, aside from this problem in Skinner's experiment that the punishment used was very mild, the other problem is that Skinner's actual results have been challenged as the following quote from Rocha and Hunziker explains:

> "…the claim that the suppressive effect of punishment is only temporary (Sidman, 1989) is disputable, with ***relevant data suggesting otherwise*** (e.g., Appel, 1963; Azrin, 1960; Azrin et al., 1963; Boe & Church, 1967; Hake et al., 1967; Herman & Azrin, 1964; Holz et al., 1963; Storms et al., 1962; among others). Apparently, this statement was originally made by Skinner (1938, Experiment II), as he showed that the suppressive effect of punishment was not enduring when punishment followed the first few responses following the onset of extinction. Rats receiving or not receiving

> such punishment showed an equal total amount of responses by the end of two sessions.
>
> "These results were contested by Boe and Church (1967), who pointed out *methodological flaws* in the original research. When they replicated Skinner's procedures with *greater experimental control* and strict manipulation of variables, there was **an enduring effect of punishment**."[329] (emphasis added)

In Boe and Church's study, titled "*Permanent effects of punishment during extinction*", they conclude:

> "In both experiments, punishment produced *a permanent reduction* in the number of responses instead of only temporarily depressing response rate."

So Boe & Church demonstrated that there was still a lasting effect—it wasn't just completely temporary; and therefore according to them Skinner's findings were in error.

And, as we often see (and you're about to see some extreme examples shortly from Sidman), we also see Skinner going *way beyond the evidence*. His mild, paw tapping lever didn't permanently cure the food deprived rats trying to get food by pressing the lever (who would have thought?), therefore he claims that punishment is *always* temporary and *never* works. If only he'd consulted a good dog or horse trainer operating out in the *real world* with *real experience*, he might have learnt something. Or, if he hadn't let his utopian ideology blind him to reality.

So, we see an experiment that involved mild punishment (which is indeed likely to be largely ineffective in such a case), that when it stopped, the rats slowly went back to pressing the lever (although according to Boe & Church, not to the original levels). And we see this is then claimed to support the idea that punishment *never* works, and that it is *always* only a temporary suppression (as so many of the reward-only ideologues continue to claim—including scientists like Meghan Herron as you will read more about soon).

And, finally, like most of the others we've heard from throughout this book, Sidman, while on the one hand claiming punishment has only a temporary, suppressive effect, at other times admits it's effectiveness!

> "It turns out that laboratory data strongly support the position that punishment, *although clearly effective in controlling behavior*, has serious drawbacks, and that we desperately need alternatives."[330] (emphasis added)

---

[329] Rocha, C., & Hunziker, M. *A Critical Assessment of Murray Sidman's Approach to Coercion*. Brazilian Journal of Behavior Analysis 17, no. 2 (2021): 188-194.
[330] Sidman, Murray. *Coercion and Its Fallout*. 2000. United States. Author's Cooperative.

CHAPTER 11 – Murray Sidman's "Coercion and Its Fallout"

Actually, no, the data does not support the supposed "serious drawbacks" of punishment. But he is correct here (in opposition to his contradictory claims elsewhere) that punishment is indeed, as he says, *clearly effective in controlling behaviour*. He also recognizes that it works in nature:

> "Certainly, if a species could not make use of environmental cues for reinforcement and punishment, it would not survive for long."[331]

Furthermore, he also recognizes that it is often easier and faster than positive reinforcement methods, and that it has benefits over them, but that its *nuclear-like fallout* disqualifies it from use:

> "Also, some noncoercive [positive reinforcement] methods are *not as easy to apply or as rapid in their action* as a swift, intense punishment or negative reinforcer. What makes the noncoercive [positive reinforcement] alternatives necessary, even though they are *unfamiliar and sometimes difficult to apply*, is the vast catalogue of punishment's side effects—consequences of punishment that cancel out its benefits and are responsible for much that is wrong with our social systems.
>
> "Anyone unaware of these side effects *cannot be expected to consider it reasonable or even desirable to replace coercive with noncoercive control*. I am therefore going to review the full range of side effects in order to show how they negate whatever justification punishment may have."[332] (emphasis added)

So, he's admitting, that were it not for all of the disastrous side-effects he believes accompany punishment, **no one could be expected to think it *reasonable or even desirable* to replace negative reinforcement or punishment methods with the slow and difficult positive reinforcement ones** (which, as we've discovered, mostly don't work anyway).

So Sidman, *consistent in his inconsistency* with so many of the other reward-only ideologues, on the one hand claims that punishment doesn't work (based on gross overgeneralizations from flawed experiments, including those of

> Sidman admits that were it not for all of the disastrous side-effects he believes accompany punishment,
>
> **no one could be expected to think it reasonable or even desirable to replace negative reinforcement or punishment methods with the slow and difficult positive reinforcement ones**
>
> (which mostly don't work anyway).

---

[331] Sidman, Murray. *Coercion and Its Fallout*. 2000. United States. Author's Cooperative.
[332] As above.

Skinner, and on his own selective blindness to many of the opposing studies), and on the other hand admits that it is "clearly effective in controlling behaviour"!

The Inductive Fallacy – Wild, fanciful extrapolation

The next most glaring problem with Sidman's book is his truly *wild, utopian,* and *fanciful* extrapolations (even worse than those we've already heard about). While at times giving various sensible recommendations on what constitutes good science (which rules out most of the reward-only pseudo-science, as we will discover in the next chapter), Sidman proceeds to commit one of the most egregious examples of the inductive fallacy you could ever wish to see. He proceeds from some laboratory experiments with rats and monkeys (which he describes but does not reference— there isn't a single reference in his entire book), to a wild, universal generalisation that punishment (his definition of "coercion") is, if not THE root cause of all of society's problems, then is certainly close to it:

> "Coercion [punishment/negative reinforcement] is not the root of *all* evil, but until we adopt other than coercive ways to control each other's conduct, no method of physically improving our species will keep our *survival timer from running out.*"[333] (emphasis added)

And:

> "Coercive interactions threaten our well being and even our *survival as a species.*"[334] (emphasis added)

And he says that his book:

> "...describes the *disastrous side effects* of coercion, and even warns of catastrophe if we *fail to eliminate* or reduce our coercive practices"[335] (emphasis added)

Wow. So, the use of punishment or negative reinforcement (which are his definition of "coercion"), produces *"disastrous side effects"*, *"threatens our survival as a species"*, and only by adopting exclusively positive reinforcement approaches throughout society (including, as Sidman mentions, in the law enforcement and judicial systems!) can we hope to keep our *"survival timer from running out"*! Even *if* he was correct about the terrible, disastrous side effects resulting from the use of negative reinforcement or punishment, it is still a *massive* leap from that to claiming that they threaten our very survival as a species.

---

[333] Sidman, Murray. *Coercion and Its Fallout.* 2000. United States. Author's Cooperative.
[334] As above.
[335] As above.

CHAPTER 11 – Murray Sidman's "Coercion and Its Fallout"

However, given that there is a lack of these disastrous side effects when punishment is applied correctly (and, in fact, it has positive side-effects), then there are absolutely zero grounds for his preposterous claims. I've already argued the case against so-called "fallout", however I will provide one more quote, this time from Rocha and Hunziker's 2021 *"Critical assessment of Murray Sidman's approach to coercion"*:

> "The claim that aversive control necessarily generates undesirable side effects (Sidman, 1958, 1977, 1989) that may make difficult an individual's adaptation to his environment also has been *contested by experimental data*. Not only in *applied settings*, where the **need to resort to aversive control to assure client's well-being** *has been verified* (Lernan & Vorndran, 2002), but also in laboratory experiments, several studies have *endorsed the utility* **of punishing noncooperative behavior to establish good levels of cooperation between group members** (Fehr, & Gachter, 2000; Sefton et al., 2007). These data support the argument of Critchfield (2014) that **"punishment, rather than being one of society's great ills... might sometimes function as an** *essential adhesive* **that helps to** *hold society together*"" [336]

And remember this:

> "The *lack of undesirable side effects* associated with the use of punishment has also been noted in the applied literature (e.g., Brantner & Doherty, 1983; Harris, 1985; Johnston, 1972; van Oorsouw et al., 2008). Indeed, the use of punishment-based interventions typically has been *related to increases in positive behavior* (e.g., Bostow & Bailey, 1969; Firestone, 1976; van Oorsouw et al., 2008; Risley, 1968). For example, Matson and Taras (1989) *reviewed 382 applied studies* employing different punishment procedures during interventions with individuals with developmental disabilities and concluded that the results reviewed *did not provide evidence supporting the occurrence of undesirable side effects*. Instead, **the majority (93%) reported positive side effects during punishment interventions**, such as increases in social behavior and responsiveness to the environment."[337] (emphasis added)

I'm not going to bother spending much more time countering Sidman's claim that civilization's very survival depends on us adopting exclusively positive reinforcement methods, because it's so ludicrous that there's simply no need (and positive reinforcement usually doesn't work in the important areas anyway, so the

---

[336] Rocha, C., & Hunziker, M. *A Critical Assessment of Murray Sidman's Approach to Coercion*. Brazilian Journal of Behavior Analysis 17, no. 2 (2021): 188-194.
[337] Fontes, R. M., & Shahan, T. A. (2021*). Punishment and its putative fallout: A reappraisal*. Journal of the Experimental Analysis of Behavior, 115(1), 185–203.

whole argument is pointless). The only thing I will do is emphasize the enormous and completely unjustified leap he makes. He jumps from some cruel laboratory experiments all the way to the irrational conclusion that all society is doomed if it doesn't rid itself of all punishment.

One such experiment he describes (known as Sidman Avoidance, which he is particularly well-known for) involved rats that were given a shock every twenty seconds, unless they pressed a lever, which would postpone the shock for thirty seconds. Not surprisingly, the rats learnt to sit close by the lever and press it regularly in order to avoid getting shocked. Sidman even says that "monkeys will go for days without shock", so evidently they were kept in this situation for *days*. Then he seems surprised that they seem a bit messed up mentally, and in another of his wild extrapolations takes this to mean we should never use punishment because it causes pathological side effects. He says:

> "They [the rats and monkeys] behave with a persistence and a compulsiveness that resembles the pathologically rigid and inflexible behaviour we often see around us... that seeming lack of contact with reality is *an inevitable consequence* of avoidance [punishment]... even professional clinicians often fail to appreciate how coercive control [punishment] may generate seemingly pathological conduct." [338]

I know who was out of touch with reality, and it wasn't the rats or monkeys—there is nothing "out of contact with reality" about continually pressing a lever to avoid getting shocked—that is perfectly rational and reality-based; what would be "out of contact with reality" would be to *not* sit close by and continue pressing the lever. So, Sidman psychologically tortures rats and monkeys, and then concludes that a "seeming lack of contact with reality" is "an inevitable consequence" of *any* punishment. And then, he also leaps to the conclusion that any use of negative reinforcement and punishment is the major cause of the downfall of the human species! That just about sums up much of Sidman's work.

Punishment is coercive and reward is not?

So, those are his first two erroneous claims: that punishment doesn't work, and that punishment is nothing less than the downfall of the human species. We can also add his ridiculous claim that "lack of contact with reality is an inevitable consequence of avoidance [punishment]". Sidman's next glaring mistake and unproven assumption is when, in a case of false equivalence, he conflates coercion as being, by his definition, punishment or negative reinforcement. He says:

---

[338] Sidman, Murray. *Coercion and Its Fallout*. 2000. United States. Author's Cooperative.

CHAPTER 11 – Murray Sidman's "Coercion and Its Fallout"

> "By coercion, I refer to our use of punishment and the threat of punishment..."[339]

And, on the other hand, according to him, positive reinforcement is *not* coercive:

> "Control by positive reinforcement is noncoercive; coercion enters the picture when our actions are controlled by negative reinforcement or punishment."[340]

He also interchangeably applies the terms "abuse" and "tyranny" when describing punishment, regardless of the situation or the way it is used. So, by Sidman's definition, every instance of negative reinforcement or punishment is abuse, tyranny, and coercion. In fact, Rocha and Hunziker in their 2021 article, "*A critical assessment of Murray Sidman's approach to coercion*", when writing about his book, state:

> "Its central weakness is the equivalence it established between coercion and aversive control."[341]

The simple fact is that both punishment and positive reinforcement can be coercive, and often are. As Perone was quoted earlier:

> "Aversive control is an inherent part of our world, an inevitable feature of behavioral control, in both natural contingencies and contrived ones. When I say that aversive control is inevitable, I mean just that: *Even the procedures that we regard as prototypes of positive reinforcement have elements of negative reinforcement or punishment imbedded within them.*"[342] (emphasis added)

The so-called "positive" trainers are simply trying to avoid this reality, whether for marketing reasons (to make their training methods sound nice, when they talk about things like "control without coercion") or they're just deceiving themselves. As soon as anyone puts a lead on a dog they are being coercive (and a head collar even more so). If you make a dog sit for its dinner, you are being coercive. If you don't allow your dog to go and visit every dog it wants to, or chase every animal it likes, you are being coercive. The only way you cannot be coercive, is to not own the dog. However, even wild dogs are living under coercion (as are we), as Sidman admits:

---

[339] Sidman, Murray. *Coercion and Its Fallout*. 2000. United States. Author's Cooperative.
[340] As above.
[341] [341] Rocha, C., & Hunziker, M. *A Critical Assessment of Murray Sidman's Approach to Coercion*. Brazilian Journal of Behavior Analysis 17, no. 2 (2021): 188-194.
[342] Perone, Michael. *Negative Effects of Positive Reinforcement*. The Behavior Analyst 26, no. 1 (Spring 2003): 1-14.

> "Because a large segment of society has succeeded in overcoming the external and internal stresses the natural environment imposes, many of us have lost sight of the extent to which *nature coercively shapes our conduct.*"[343] (emphasis added)

He even recognizes that our own body is coercive:

> "We defend against our own body's coercion by supporting a huge, expensive medical establishment..."[344]

So even wild animals live in a coercive environment (they can't just do whatever they like—there is plenty of punishment and negative reinforcement in the environment), so just setting dogs free wouldn't spare them the coercion either: the punishment of being hit by a car might teach them some road sense (if they survive), the punishment of being bitten when pestering an older dog, and so on.

In other places, however, (when he advocates for so-called "errorless learning"), Sidman also admits that positive reinforcement can, in fact, be coercive, due to the withholding of reward constituting punishment (as discussed in *Chapter Nine – More downsides and side-effects of "positive" training*). He says:

> "An effective teacher will never reinforce mistakes; that is a sure way to perpetuate them. Here is where *coercion* comes back into the picture: to *withhold reinforcement is to punish.*"[345] (emphasis added)

> "To withhold reinforcement is to punish"
>
> – Murray Sidman

So even Sidman admits that withholding reinforcement (as happens all the time when using positive reinforcement) is punishment and coercive, and yet elsewhere he claims that positive reinforcement is not coercive!

Also, when dog trainers use deprivation, as the vast majority of so-called "positive" trainers recommend (and which Sidman hypocritically condemns), they are even further along the coercion road (as will be discussed in the next section).

Rocha and Hunziker (quoting Goldiamond) summarize the issue well, when they write:

> "Furthermore, assuming that "...the issue is *never coercion versus no coercion...* the issue is the amount and type of coercion we are willing to

---

[343] Sidman, Murray. *Coercion and Its Fallout.* 2000. United States. Author's Cooperative.
[344] As above.
[345] As above.

accept, and the protections against abuse we set up" (Goldiamond, 1976, p. 23), behavior analysts *may better acknowledge the* **ubiquity, inevitability of coercion in society at large**, therefore **putting forward** *more pragmatic and less utopian* **projects to social change.**" [346] (emphasis added)

So, Sidman's claim that we should redefine "coercion" as negative reinforcement and punishment, but not positive reinforcement, and that we can avoid all coercion if we simply rely solely on positive reinforcement, is not even remotely true, even by his own admission.

Deprivation – Positive reinforcement IS negative reinforcement

Another problem is that positive reinforcement, as we have learnt, is very often negative reinforcement (and therefore coercion, by all of Sidman's definitions). Anytime someone uses any form of deprivation, they are now using *negative* reinforcement. In fact, this means that virtually ALL scientific studies on positive reinforcement (such as Skinner's, and everything following) are, in fact, studies on *negative* reinforcement. Deprivation is standard practice in so-called positive-reinforcement studies as we saw earlier, where the subjects are deprived of food to bring them down to about 75% of their normal weight.

> Virtually ALL scientific studies on positive reinforcement are actually studies on *negative* reinforcement.

Despite this fact, Sidman says that deprivation is a "misuse of positive reinforcement"! And yet, as we learnt in *Chapter Nine – More downsides and side-effects of positive reinforcement*, this use of deprivation is actually an essential mainstay if positive reinforcement is going to have any chance of working, like it is in all so-called "positive reinforcement" experiments. Sidman even recognized these facts while evidently failing to understand their significance and the utter absurdity of his position:

> "Another misuse of positive reinforcement is *deliberately to create the kinds of deprivations that* **make reinforcers effective**... their effectiveness is a product of *negative* reinforcement; they become instruments of *coercion*." [347] (emphasis added)

---

[346] Rocha, C., & Hunziker, M. *A Critical Assessment of Murray Sidman's Approach to Coercion*. Brazilian Journal of Behavior Analysis 17, no. 2 (2021): 188-194.
[347] Sidman, Murray. *Coercion and Its Fallout*. 2000. United States. Author's Cooperative.

Right, so it's "a misuse of positive reinforcement" to do the very thing (deprivation) that *makes them effective?!* The irony is almost painful.

Remember Jean Donaldson recommending deprivation:

> "You have control of your dog's access to **everything he wants in life**: food, the outside world, attention, other dogs, smells on the ground, play opportunities... *Take control of the goodies the dog wants in life—stop handing them out for free. Nothing is "free" anyway...*"[348]

So Sidman obviously would *not* approve—he recognizes Donaldson's and Dunbar's and Pryor's (and all the other "positive" trainers) use of deprivation as "coercive", and therefore tyrannical and abusive. However, when Sidman condemns the use of deprivation (which he also notes "makes reinforcers effective"), he obviously has a serious problem, as Rocha and Hunziker also note:

> "If Sidman's (1989) reasoning about conditions of imposed deprivation of positive reinforcers is correct, near all experiments involving positive reinforcement **(which so frequently *requires* imposed deprivation)** would demand rectification; **they must come to be conceived as experiments on negative reinforcement**."[349] (emphasis added)

So just about every experiment ever done on so-called positive reinforcement is, in fact, an experiment on negative reinforcement (which Sidman says is abuse and tyranny and we shouldn't use), because they all use deprivation. And all the main "positive reinforcement" trainers use this technique of depriving food in order to make their so-called non-coercive "positive" reinforcement training even somewhat effective. When Karen Pryor locks a dog up in crate so that it becomes restless and hungry prior to doing a demonstration on so-called positive reinforcement training, she is actually not doing a demonstration on positive reinforcement at all. She is, in fact, utilising deprivation, that "instrument of coercion" of Sidman's, and demonstrating *negative* reinforcement. And yet, in a very Sidmanesque manner, Pryor included a chapter in her book, *Don't Shoot the Dog,* titled, "*Control without coercion*". Many others, like Donaldson, say similar things, while simultaneously recommending deprivation. Sidman elsewhere notes again:

> "Such techniques [of deprivation] are, of course, completely coercive."[350]

---

[348] Donaldson, Jean. *The Culture Clash*. 2nd ed. Berkeley, CA: James & Kenneth Publishers, 2005.
[349] Rocha, C., & Hunziker, M. *A Critical Assessment of Murray Sidman's Approach to Coercion*. Brazilian Journal of Behavior Analysis 17, no. 2 (2021): 188-194.
[350] Sidman, Murray. *Coercion and Its Fallout*. 2000. United States. Author's Cooperative.

CHAPTER 11 – Murray Sidman's "Coercion and Its Fallout"

However, he again notes deprivation's effectiveness:

> "Deprivation, however, does contribute to the effectiveness of positive reinforcers. We have little interest in food right after a good meal…"[351]

He's wrong, in the sense that deprivation doesn't improve the effectiveness of positive reinforcement—what it does is turn it into *negative* reinforcement; but he's right about deprivation improving the effectiveness—negative reinforcement is always more effective than positive (your treats work better when your dog is hungry). He also notes:

> Negative reinforcement is always more effective than positive.

> "In certain extreme cases, deprivation for a brief time can produce desirable consequences that are **unavailable any other way**."[352] (emphasis added)

So, sometimes, according to Sidman, positive reinforcement might not work and then deprivation is okay. The implication is that if positive reinforcement isn't working (it often doesn't—not just in "extreme cases"—as you learnt in *Chapter Eight – Does reward work?*) then you should use deprivation to improve the effect of the "positive" reinforcement (or, in actual fact, to change it to negative reinforcement), and in that case he is admitting that coercion is acceptable. And don't forget that he also admitted that positive reinforcement is coercion, because it is withheld so as not to reinforce mistakes!

If negative reinforcement is fine when positive reinforcement fails (it often does), then why could the argument not also be made—using his own reasoning—that punishment is also okay when positive reinforcement *and* deprivation fail and need a bit of help, as they so often do?

If that's all a bit confusing, I'm not surprised. The reward-only ideologues amazing inconsistencies and lack of logic *are* utterly confusing, and quite staggering when you consider just how influential they have been, and continue to be.

Destroying the bond

Another issue Sidman makes a great deal about is the certainty of "destroying the bond" if we use punishment. He argues the case that when you use reward, you *become the reward*; and, conversely, when you use punishment, you *become the punishment*:

---

[351] Sidman, Murray. *Coercion and Its Fallout*. 2000. United States. Author's Cooperative.
[352] As above.

"...people who use punishment will *become conditioned punishers themselves*. Others will fear, hate, and avoid them. If we punish other people, *we, too, become punishments*. Our very presence will be punishing. If we simply come near those whom we customarily punish, *we will put a stop to whatever they are doing*. If we just threaten to approach, *they will flee*. All the side effects that shocks generate, we, too, will generate. *Anyone who uses shock becomes a shock*." [353] (emphasis added)

Once again, we see Sidman's truly wild extrapolations. If we use punishment on others, everyone will not only stop whatever they are doing whenever we arrive, but will also *flee* from our presence! Just like in our dance classes when we punished misbehaving dance students, and the entire room emptied in an instant as all the students fled in abject terror never to return (oh, wait, that's not what happened at all...). He also claims that the same thing happens to the *environment* in which the punishment occurs:

"Environments where we are punished become punishing themselves and we react to them as to natural punishers." [354]

There are two main problems with Sidman's reasoning. The first is that no-one only ever uses punishment. We also use reward. So, on the one hand, we become the reward, and on the other, we become the punishment. And the same applies to the environment. So, which is it? It is simply a question of balance. If we use both, *fairly*, we won't have any problems. Like the example I gave of my favourite teacher at high school—he was very strict, and also very encouraging. The *combination* creates respect and a great relationship. I loved his classes and always looked forward to them.

> The combination of *discipline* and *encouragement* creates respect and a great relationship.

And yet, elsewhere, Sidman recognizes that occasional punishment will actually *not* create any problems if used in balance:

"A scolding or a slap by a parent who is usually loving, concerned, and protective **is unlikely to do any damage**. Children and some adults are always testing the limits. With a background of positive reinforcement, a defensive punishment here and there is likely to be treated less as a shock than as a signal that a reasonable boundary has been overstepped. An occasional correction or negative statement is more likely to be listened to and reacted to as a help rather than a punisher if it occurs in a context of mostly positive interactions." [355]

---

[353] Sidman, Murray. *Coercion and Its Fallout*. 2000. United States. Author's Cooperative.
[354] As above.
[355] As above.

And:

> "Punishment may occasionally be necessary to put a quick halt to a dangerous situation. These are not occasions to be concerned about. In a relationship based on strong and frequent positive reinforcement, *an infrequent punishment is not going to cause any long-term damage*."[356] (emphasis added)

So, on the one hand, he claims that punishment and negative reinforcement has all this terrible, *catastrophic, nuclear-like fallout* and are going to *destroy us as a species*, and on the other hand he admits that when it is used in the context of a balanced scenario—which also provides plenty of encouragement and reward—that it won't cause any problems.

He also reasons that somehow it then magically stops really being punishment, and becomes instead "a signal that a reasonable boundary has been overstepped", and is now reacted to "as a help rather than a punisher"! Great. So now we can just think of a "defensive" smack on the nose for our dog that is biting us as not really punishment at all, but simply a "signal that it has crossed a reasonable boundary." And our dog will take that as "a help rather than a punisher," because it occurs in the "context of mostly positive interactions". Ah, no, Murray, no matter how you try to spin it, it's still punishment. But he's right—it won't cause any problems.

So, while saying that if we use punishment "we become the punishment" and everyone will flee from us, Sidman admits that the occasional use of punishment will not destroy the bond (although he doesn't go far enough in understanding that it will actually create a *stronger* bond, when done fairly). And, as I've mentioned, because we also use reward we also "become the reward". So, like most things, it's simply a matter of balance.

The second problem with Sidman's assertion that we (or places) "become the punishment", is that this is not lasting. As I described a pup that is punished by its mother is initially fearful of her in general, but with *repetition* of the situation so that the pup learns the direct connection between its behaviour and the punishment, the fear disappears. Fontes and Shanan, in their 2020 critique of Sidman's ideas titled *"Punishment and Its Putative Fallout: A Reappraisal,"* correctly note:

> "... those stimuli [people, places] are *only effective as conditioned punishers while correlated with unconditioned punishers*, and do not necessarily acquire lasting effects of the unconditioned punishers with which they are associated. Further, the **generalizability** *of conditioned punishment effects is* **reduced with continued training**, contradicting Sidman's argument that more exposure to punishment results in greater

---

[356] Sidman, Murray. *Coercion and Its Fallout*. 2000. United States. Author's Cooperative.

generalization of response suppression. Thus, it appears that **these concerns of Sidman are *not supported*** by empirical evidence. Instead, the "toxicity" of the conditioned punishment side effect seems to be greatly impacted by the ***animal's control of the punishment delivery*** and the ***information conditioned punishers provide about the contingency***."[357] (emphasis added)

In other words, the conditioned punisher (the person or place) only maintains that status while it is continually paired with punishment. Just like a clicker loses its reinforcing value if it isn't continually paired with an unconditioned reinforcer (food) (thus the advice to follow every click with a treat), so too does a person or place rapidly lose any value as a punisher if they don't keep continually punishing. So, when I use punishment to enforce a recall with a dog, once it's trained (which doesn't take very long), that punishment ceases and hardly ever needs to be used again. And, when we add to that the fact that the vast majority of people and places act in both ways (as reward and punishment), it becomes simply a question of balance, and fairness.

Provided the dog understands the rules—*"the information the conditioned punishers provide about the contingency"*—it then gains *"control of the punishment delivery"*, and provided that we use both reward and punishment as appropriate, **none of Sidman's claimed disastrous nuclear fallout is going to rear its ugly head to destroy the earth as we know it**. In fact, it's the exact opposite—it is the fear of using punishment intelligently that is contributing to many of society's, and dog's (and dog owner's, and teacher's, and parent's) problems.

> It is the fear of using punishment intelligently that is contributing to many of society's, and dog's (and dog owner's, and teacher's, and parent's) problems.

Before leaving this topic of "destroying the bond", Sidman also recognized that *positive reinforcement* has the potential to "destroy the bond," when he says:

"Offering a good friend money in return for a favor may well destroy rather than strengthen the friendship." [358]

As may offering them fake, transactional smiles and other "known manipulators," as we heard from Karen Pryor. So, quite often, an attempt to rely solely on positive reinforcement—whether with our dogs or kids or students—can backfire and have many negative side-effects (as discussed in *Chapter Nine* in the section titled *Manipulation and Dishonesty*).

---

[357] Fontes, R. & Shahan, T. *Punishment and Its Putative Fallout: A Reappraisal.* 2020. Journal of Experimental Analysis of Behavior.
[358] Sidman, Murray. *Coercion and Its Fallout.* 2000. United States. Author's Cooperative.

Conclusion – "Coercion and Its Fallout"

In conclusion, it's been very unfortunate that Sidman's illogical, unproven, wildly extrapolated ideas have been so influential in both the human, and animal, spheres. A great deal of damage has been done to both people and animals, which cannot now be undone. However, for the sake of current and future generations, it is high time we got back to a much more balanced approach.

As a society, and as dog trainers, we need to reemphasize the *necessity and effectiveness of negative reinforcement and punishment*, and its proper use, in both human and animal spheres (alongside reward, in its place). We need to put an end to this nonsense that there are all these terrible, disastrous, *nuclear fallout* like side-effects, that actually never occur with a sound, rational, intelligent use of contingent punishment. And we need to realize that there are *many significant problems* with positive reinforcement approaches.

The truth is often the opposite to how it's portrayed by the "positive" ideologues.

PART 5 – The scientific basis—fact and fallacy

12

Some real-life junk "science"

Aggression Survey – Herron, Shofer & Reisner 2009

So, having now looked at some of the general problems with "science", and having debunked influential behavioural scientist Murray Sidman's irrational rantings and Skinner's claims that punishment's effects are only ever temporary, let's now turn our attention to two specific examples of influential "scientific" studies which are regularly quoted by the reward-only ideologues.

The first of the two studies I will look at is one quoted by Zak George (and often by many other reward-only ideologues, including Victoria Stilwell). One of the authors of the study, Meghan Herron, is someone George quotes a number of times in his book and elsewhere (the other two authors of this junk survey are Frances Shofer and Ilana Reisner). The title of the study is:

> "*Survey* of the use and outcome of confrontational and non-confrontational training methods in client-owned dogs showing undesired behaviours."[359] (emphasis added)

---

[359] Herron, Meghan E., Shofer, Frances S., Reisner, Ilana R.,*Survey of the use and outcome of confrontational and non-confrontational training methods in client-owned dogs showing undesired behavoiurs*. 2009

PART 5 – The scientific basis—fact and fallacy

Firstly, note that it is a "survey", which, as I said, so much of the so-called *"science"* for "positive" training is. It was simply a questionnaire to owners of dogs with pre-existing behavioural conditions, mainly focusing on aggression. The exact questionnaire is not provided, which is a common problem as we just learnt:

> "Many scientific protocols, analysis methods, computational processes and data *are opaque*. When researchers try to crack open these black boxes, they often discover that many top findings *cannot be reproduced*. That is the case for *two out of three top psychology papers.*"[360] (emphasis added)

This survey asked the owners (general dog owning public, not experienced dog trainers) what techniques they had used when attempting to fix aggression, and the "results". The study admits that the owners may not have understood the results question properly, and that they may have been unwilling to honestly answer the questions:

> "It is also possible that owners misinterpreted the meaning of the "effect" section of the survey. The terms "positive", "negative", and "no effect" are subjective, and judging a techniques effectiveness based on these options may not be accurate. Next, owner's self-reporting may have led to bias and/or poor answer reliability. For example…some owners may have felt reluctant to admit to a veterinary professional that they have used physically aversive methods on their dogs."[361]

Not surprisingly, in the study (survey), approaches using punishment of any sort more often elicited an aggressive response (note that these were dogs with *pre-existing* aggression issues). Positive reinforcement methods were less likely to elicit the aggressive response.

> We trigger the aggressive behaviour and then punish it, exactly as I have explained in various examples throughout this book.
>
> Simply *avoiding* the behaviour won't fix it.

To start with, that is stating the obvious. If you attempt to deal with an aggressive dog to train it using anything other than treats, you are no doubt going to elicit the aggression in the process of retraining the dog. In fact, this is a *necessary part* of the training. We trigger the aggressive behaviour and then punish it, exactly as explained in various examples throughout this book. Done properly, it is highly effective (although I suggest you do not

---

[360] *Science Funding Is Broken*. Scientific American, April 5, 2024, Accessed May 7, 2024, https://www.scientificamerican.com/article/science-funding-is-broken/.
[361] Herron, Meghan E., Shofer, Frances S., Reisner, Ilana R., *Survey of the use and outcome of confrontational and non-confrontational training methods in client-owned dogs showing undesired behaviours*. 2009

attempt this without proper guidance if you are inexperienced). Simply *avoiding* the behaviour won't fix it.

If you simply bring out treats to "reward alternative behaviours", you may not elicit the aggression (although, as we've seen, in some cases you certainly can). But that says nothing about the *effectiveness* of the reward in fixing the aggression in the presence of whatever triggers it.

Let's just take one type of intervention the owners in the study had attempted—leash corrections. According to the survey 75% of owners surveyed attempted this. The survey's author's state:

> "*Contrary to expectations*, not all owners reporting an aggressive response to a particular intervention felt that the training methods had a "negative" effect on their dog's behaviour…in our study, 63% of owners who used *leash corrections felt they had a positive effect.*"[362] (emphasis added)

So, "contrary to expectations" (these are the times good scientists sit up and take notice!) 63% of owners who used leash corrections on their dog found it was beneficial to some extent. Of course, *the authors do not mention this in their conclusion*. And, obviously wanting to downplay this finding, they hasten to add their own spin:

> "…in our study, 63% of owners who used *leash corrections felt they had a positive effect.*"

> "It is *possible* that the correction *temporarily* inhibited reactive or other undesirable behaviours, thus *appearing* that the behaviour had improved and that the technique had a positive effect. While it may be effective *as a momentary interruption*, correction or punishment alone does not selectively reinforce desirable behaviour and is an inefficient way to train an animal to perform a specific behaviour (Mills, 2002)."[363] (emphasis added)

Why do they make these claims that correction might be "temporary" or effective only as "a momentary interruption" (as in Skinner's debunked experiments) when, as we've heard from Azrin (from studies that were actually repeatable experiments, not surveys):

---

[362] Herron, Meghan E., Shofer, Frances S., Reisner, Ilana R., *Survey of the use and outcome of confrontational and non-confrontational training methods in client-owned dogs showing undesired behaviours.* 2009
[363] As above.

PART 5 – The scientific basis—fact and fallacy

> "...these results show that aggression is eliminated by direct punishment of the aggression..."[364]

And for how long might punishment eliminate the behaviour when done properly? Is it just a "temporary inhibition"?

> "The question of whether punishment can be an effective way of controlling behaviour *has been settled*. This chapter has presented numerous studies, from both inside and outside the laboratory, that have demonstrated that punishment can change behaviour and, in many cases, *change it permanently*."[365] (emphasis added)

So not "temporary" or "interrupted momentarily" as Herron et al. claim, but fixed *permanently*. And:

> "One of the most dramatic characteristics of punishment is the *virtual irreversibility or permanence of the response reduction* once the behaviour has become completely suppressed. Investigators have noted that the punished response does not recover for a long period of time even after the punishment contingency has been removed... How quickly does punishment reduce behaviour? Virtually all studies of punishment have been in complete agreement that *the reduction of responses by punishment is immediate* if the punishment is at all effective. When the data has been presented in terms of number of responses per day, the responses have been *drastically reduced or eliminated on the very first day* in which the punishment was administered."[366] (emphasis added)

> You'll notice the bias when Herron did not choose to reference Azrin's study or Mazur's conclusions, but instead chose to reference the one by fellow reward-only activist Daniel Mills

And yet, biased researchers like Herron, Shofer and Reisner, make these claims, and choose to ignore the science that doesn't suit their narrative. You'll notice the bias when they did not choose to reference Azrin's study, or Mazur's conclusions, but instead chose to reference a "study" by fellow reward-only activist Daniel Mills (we'll move on to debunking a study he was involved in next, and will demonstrate *his* significant bias).

---

[364] Azrin, N. H., Ulrich, R., Wolfe, M., & Dulaney, S. (1969). *Punishment of shock-induced aggression*. Journal of Experimental Analysis of Behavior, 12(6), 1009–1015.
[365] Mazur, James E. *Learning and Behavior, 6th Edition*. New Jersey. Pearson Prentice Hall, 2006.
[366] Azrin, N. H., & Holz, W. C. (1966). *Punishment*. In W. K. Honig (Ed.), Operant Behavior: Areas of Research and Application (pp. 213-270). New York: Appleton-Century-Crofts.

And, interestingly and tellingly, the survey's authors made no such similar disclaimers about the reward-based methods when owners cited a positive response to them. And yet, *at best*, treats *can only ever* produce a *temporary* or *momentary* response in cases like this. Unless the training is *very* long and intensive (and even then probably not) there will be no significant improvement, as we have seen. *Only* punishment-based methods contain even the *possibility* of achieving quick, long-lasting results. Remember Karen Pryor's comments on using positive reinforcement for aggression?

> At best treats *can only ever* produce only a short-lived response in cases like this, unless the training is *very* long and intensive, as we have seen.
>
> *Only* punishment-based methods contain the possibility of achieving quick, long-lasting results.

> "It sometimes takes *a long time* to clicker-train a dog, step by step, to be calm and confident around strange dogs on the street instead of attacking them on sight; but it can be done."[367] (emphasis added)

But either way, the actual *evidence* in this survey says nothing whatsoever about these things. The survey's authors make a claim here that has *no basis on the evidence* presented whatsoever. It is a statement based on their own existing bias, and on the selective citing of other flawed studies, rather than on anything in the survey itself. And that is true of so much of the reward-only biased research. The evidence says 63% of owners who used leash corrections for aggression felt it had a positive effect. That's what the evidence says, nothing more, nothing less. The rest is simply conjecture and bias.

They also then make two highly misleading statements as quoted above when citing Daniel Mills:

> "While it may be effective as a momentary interruption, *correction or punishment alone does not selectively reinforce desirable behaviour* and is an *inefficient way to train an animal to perform a specific behaviour (Mills 2002).*"[368] (emphasis added)

Firstly, why on earth would we want to "selectively reinforce a specific behaviour" in these cases? We are trying to *remove* an undesirable behaviour—aggression—*not* "selectively reinforce" a specific behaviour. And secondly, why even *mention* correction or punishment being "an inefficient way to train an animal to perform a specific behaviour"? Again, the owners weren't trying to "train an animal to

---

[367] Pryor, Karen. *Reaching the Animal Mind.* New York. Scribner. 2009.
[368] Herron, Meghan E., Shofer, Frances S., Reisner, Ilana R.,*Survey of the use and outcome of confrontational and non-confrontational training methods in client-owned dogs showing undesired behaviours.* 2009

perform a specific behaviour"—they were trying to stop aggression. (And, as we've repeatedly seen, correction and punishment can be *very* efficient ways of teaching specific behaviours—depending on the behaviour—such as teaching the Doberman to go to his bed and stay there, or when teaching a dog to get out of the kitchen, or to sit, *or to walk nicely on a lead with zero reactivity*, and so on).

Also, by punishing a behaviour, we are thereby reinforcing the dog's self-control in not doing the behaviour in the future, and therefore not being punished. Avoidance of the punisher is reinforcing. Even Karen Pryor admits this fact:

> "While touching the [electric] fence has been punished, the behavior of avoiding the fence has been reinforced..."[369]

So, Herron et. al. try to explain away the evidence which doesn't fit their narrative. They might be correct about the results being temporary in these cases—it all depends on the exact methods used, although there is likely to be some lasting improvement. But the reward-based methods will certainly be only temporary, and yet they offer no disclaimer for them. They are *stepping outside the evidence* and presenting their bias. And so, we simply have a *self-perpetuating confirmation bias*.

In their conclusion they state that:

> "Owners of dogs aggressive to family members are especially at risk for injury—and their pets at risk of relinquishment or euthanasia—when certain aversive methods are used. Ultimately, reward-based training is less stressful or painful for the dog, and, hence, safer for the owner."[370]

This is complete and utter *spin*. But this is what people like Zak George quote. And studies like this also then get cited by other reward-only junk science, while ignoring the study's extensive flaws. For example, Emily Blackwell and Emma Williams in their 2019 "study" (again, just another survey), cite Herron, Shofer and Reisner's survey (and another of their own) when they say:

> "Owners who utilize positive punishment-based training techniques in an attempt to change aggressive behaviour are likely to place themselves at increased risk of injury (Blackwell et al., 2008; **Herron, Shofer and Reisner, 2009**)... In light of evidence for the increased risk of harm to both the owner and their pet, associated with punishing aggressive dogs, it's not

---

[369] Pryor, Karen. *Don't Shoot the Dog!: The New Art of Teaching and Training*. Rev. ed. New York: Bantam Books, 1999.
[370] Herron, Meghan E., Shofer, Frances S., Reisner, Ilana R.,*Survey of the use and outcome of confrontational and non-confrontational training methods in client-owned dogs showing undesired behaviours.* 2009

CHAPTER 12 – Some real-life junk "science"

clear why people continue to use these techniques or what prevents them from adopting positive reinforcement-based methods."[371]

So, they simply cite Herron's junk survey and biased conclusions as gospel. This is the way the reward-only scientific community works—they simply quote each other's pseudo-science surveys in a circular fashion.

So Herron's conclusion states that owners of aggressive dogs are at risk of injury when certain aversive methods are used. True—dealing with aggressive dogs can be dangerous (and I recommend you don't attempt it unless you're experienced). But that doesn't mean the methods aren't effective (in fact, aversive methods are by far *the most* effective treatment, and realistically the *only* effective treatment).

> Aversive methods are by far *the most* effective treatment for aggression, and realistically the *only* effective treatment.

They also try to link the pets being at risk of euthanasia with the aversive methods, which is simply not true. The dogs were *already aggressive* and at risk of euthanasia. *But if the aversive methods work then that risk is eliminated or reduced.* The number of aggressive dogs I have saved from being euthanised over the years is considerable. So this is another complete red herring, and nothing in the study relates to this conclusion.

Then the study mentions how reward-based training is less stressful for the dog, and hence safer for the owner. In some cases that can be true. In some it isn't (remember Pryor's dolphin having a meltdown and refusing to eat for two days; or the elephant hitting her with its trunk; or Tilikum killing two of his positive reinforcement trainers, or turning dogs into junkies?). But that says nothing about whether it was *effective* or not. Great, let's assume we didn't cause the dog any stress, and perhaps there was no risk to the owner during training. *But have we actually done anything to fix the problem?* If not, then the risk of the dog being euthanised is *increased*, because for every ineffective method owners try that don't work, the dog's likelihood of euthanasia goes up, not down.

> For every ineffective method owners try that don't work, the dog's likelihood of euthanasia goes up.

It is telling that aggression is one issue that very few "positive" trainers are willing to deal with. For example, Laurie Luck (the *Karen Pryor Academy* faculty member and professional member of Dunbar's *Association of Professional Dog Trainers* and past president of the *Certification Council of Professional Dog Trainers*) whom I quoted earlier, simply states:

---

[371] Williams, E. J., & Blackwell, E. *Managing the Risk of Aggressive Dog Behavior: Investigating the Influence of Owner Threat and Efficacy Perceptions.* 2019.

PART 5 – The scientific basis—fact and fallacy

"I don't work with aggressive dogs." [372]

Why not? If these "positive" approaches are so effective, and if the trainers are perfectly safe so long as they don't use "aversive" approaches, as this study claims, then what are these top-level "positive" trainers afraid of? Why won't most "positive" trainers work with aggressive dogs? The simple fact is that their methods *don't work*, and these trainers have *no solutions*.

This, Emily Williams and Emma Blackwell, is what "prevents them [owners] from adopting positive reinforcement-based methods"—*they simply don't work*. And don't forget, as we have seen, that the ineffective strategies which are recommended also take a *very* long time, and are extremely *difficult* and *complicated*, and liable to simply *make things worse*.

So the interpretation and conclusion of this study is useless and misleading, like so much of Meghan Herron and her associates work. But this is what the reward-only ideologues quote, and is what governments and bodies like AVSAB use to push their "reward only" agendas. As Sidman notes:

> "When personal opinion and public policy are supported by correlational statistics, skepticism is justified." [373]

Let's move on now and look at a very influential study in which Daniel Mills (who was referenced in Meghan's study) was part of.

---

[372] Luck, Laurie. *Karen Pryor Archives*. Smart Dog University, May 7, 2024. https://smartdoguniversity.com/tag/karen-pryor/.
[373] Sidman, Murray. *Coercion and Its Fallout*. 2000. United States. Author's Cooperative.

CHAPTER 12 – Some real-life junk "science"

Livestock Chasing – E-collar versus Positive Reinforcement

The second study I will dissect is one by Lucy China, Daniel Mills and Jonathan Cooper. This is a 2020 study titled:

> "Efficacy of Dog Training With and Without Remote Electronic Collars vs. a Focus on Positive Reinforcement."[374]

It also had a sister study based on the same research, with two additional authors Nina Cracknell and Hannah Wright[375].

It was a study mainly looking at dogs with livestock chasing or recall problems. It involved 63 dogs that were split into 3 groups—Group A were to be trained with E-collar based training. Group B with balanced training, however without e-collars. And Group C with Positive Reinforcement. The dogs were given *two fifteen-minute training sessions a day, for four days*, supposedly based on recall around livestock. The conclusion was that the positive reinforcement training was more effective, and therefore there is never a need for any other type of training:

> "Overall, the professional use of a reward-focused training regime, as demonstrated by Control Group 2, was superior to E-collar and Control Group 1 in every measure of efficacy where there was a significant difference... Given... the finding that the use of an E-collar did not create a greater deterrent for disobedience we conclude that an E-collar is unnecessary for effective recall training."[376]

Of course, the "positive" idealogues just love this finding, and quote it regularly. However, there are numerous serious issues that completely negate these conclusions (apart from the fact that *real world* experience proves them false). In reality, the conclusions aren't even in the same *universe* as the truth. Various other scientists even felt strongly enough that they wrote rebuttals to this study, yet it continues to be widely cited by the reward-only ideologues today.

---

[374] China L, Mills DS and Cooper JJ (2020) *Efficacy of Dog Training With and Without Remote Electronic Collars vs. a Focus on Positive Reinforcement.* Front. Vet. Sci. 7:508. doi: 10.3389/fvets.2020.00508

[375] Cooper JJ, Cracknell N, Hardiman J, Wright H, Mills D (2014). *The welfare consequences and efficacy of training pet dogs with remote electronic training collars in comparison to reward based training.* PLoS One.

[376] China L, Mills DS and Cooper JJ (2020) *Efficacy of Dog Training With and Without Remote Electronic Collars vs. a Focus on Positive Reinforcement.* Front. Vet. Sci. 7:508. doi: 10.3389/fvets.2020.00508

PART 5 – The scientific basis—fact and fallacy

Before even getting into the study itself, we need to address the bias and corruption *behind* the study. The first factor is that it seems pretty obvious that the UK government department known as DEFRA (Department for Environment, Food and Rural Affairs) funded the study as a way to justify its intention to ban the use of e-collars. The study's author's note:

> "We would like to thank DEFRA for funding the original study."[377]

Once complete, this was *the only study* DEFRA cited as justification for their stance on banning e-collars, despite there being far superior studies that support the exact opposite.

As an interesting aside, one of the DEFRA ministers responsible for implementing her government's promise to ban e-collars was later caught using one on her own dog! In fact, after enlisting the help of a dog trainer and witnessing the results, she was singing e-collar's praises. An article in *The Times* titled "Thérèse Coffey: Minister uses 'cruel' electric dog collar" states:

> "A rescue dog owned by the family of a cabinet minister is being trained with a collar that gives off electric shocks despite the government promising to ban the "cruel" devices.
> "Thérèse Coffey, who was promoted to the cabinet last week as pensions secretary, **was so impressed by the collar that she agreed to share her views on it** with Theresa Villiers, the environment secretary.
> "Before becoming pensions secretary Ms Coffey was a minister at the department for environment, which promised last year that "electronic training collars which are used for dogs and cats" would be banned."[378] (emphasis added)

So Coffey, once she actually experienced using an e-collar, "was so impressed" that she wanted to share her experience with one of her fellow ministers! Evidently she didn't buy into the effectiveness of "positive" methods, and didn't believe her own department's spin that:

> "We are a nation of animal lovers and the use of punitive shock collars causes harm and suffering to our pets. This ban will improve the welfare of

---

[377] China L, Mills DS and Cooper JJ (2020) *Efficacy of Dog Training With and Without Remote Electronic Collars vs. a Focus on Positive Reinforcement.* Front. Vet. Sci. 7:508. doi: 10.3389/fvets.2020.00508
[378] Webster, Ben. *Thérèse Coffey: Minister Uses 'Cruel' Electric Dog Collar.* The Times, September 17, 2019. https://www.thetimes.com/article/therese-coffey-minister-uses-cruel-electric-dog-collar-wshlqs85d.

CHAPTER 12 – Some real-life junk "science"

animals and I urge pet owners to instead use positive reward training methods."[379]

So that's the first consideration—the government had already promised to ban e-collars when it commissioned this study, which was evidently done simply to justify this policy. The second factor is the activist status of the researchers that received the funding. As Professor John Staddon notes:

> "Science is in trouble. The tranquil realm of reason has been invaded by man's strongest passions. *Facts have taken a back seat to ideology*, and *political activism masquerades as scholarship.*"[380] (emphasis added)

That is exactly what this study, and so many like it, are—nothing but political activism and ideology disguised as science. At least one of the authors of this study, Daniel Mills, had been involved in attempting to have e-collars banned at least 12 years prior to this study for this *very same government department*[381]. And in a sister paper to this one, based on the same research, two of the other co-authors Nina Cracknell and Hannah Wright had also been involved in attempting to have e-collars banned at least since 2009[382]. So, well prior to this study, at least 3 of the 5 (and possibly 5 of the 5) study authors were actually already attempting to have e-collars banned—hardly impartial scientists. Most likely, this is why DEFRA chose them. Sidman notes that we must ask certain questions to determine the validity of the "science":

> This study, and so many like it, are nothing but political activism and ideology disguised as science.

> "Who did the measuring? Did the people who recorded the data have any stake in the results? If so, they might have unconsciously recorded what they wanted to see rather than what actually happened?"[383]

Obviously, in this case, they certainly did have a stake in what they wanted the study to find. So, what we have is a politically motivated and funded study employing activist "scientists" to find a desired result. It readily fits Staddon's description of ideologues engaged in political activism.

---

[379] Webster, Ben. *Thérèse Coffey: Minister Uses 'Cruel' Electric Dog Collar.* The Times, September 17, 2019. https://www.thetimes.com/article/therese-coffey-minister-uses-cruel-electric-dog-collar-wshlqs85d.
[380] Staddon, John. *Science in an Age of Unreason.* Washington, D.C. Regnery Gateway. 2022.
[381] Animal Welfare Division, DEFRA. 2002. *The Consultation on an Animal Welfare Bill: An Analysis of the Replies.*
[382] https://web.archive.org/web/20090821171442/http://www.gopetition.co.uk/online/27913/ signatures-page14.html
[383] Sidman, Murray. *Coercion and Its Fallout.* 2000. United States. Author's Cooperative.

PART 5 – The scientific basis—fact and fallacy

The study concluded that positive reinforcement training was more effective in teaching dogs to recall around livestock, and therefore there's never a need for any other type of training. And it did all this with just *two fifteen-minute* training sessions a day, for *four days*! As we've seen throughout this book, this is completely laughable. I mean, roll around on the ground in uncontrollable hysteric's kind of *laughable*. We've seen all the top "positive" trainers like Jean Donaldson (in the best dog training book Dunbar has ever read) outline and admit the great amount of time and effort necessary to create even a *might/sometimes* recall (often a year or more). And yet this study purportedly shows that *in just two hours* spread across *four days* positive reinforcement achieved the best results.

> And it did all this with just *two fifteen-minute training sessions a day for four days*!
>
> As we've seen throughout this book, this is completely laughable.

We also saw Leslie Nelson in her "Really Reliable Recall" outline the time required. She recommended giving your dog the following training sessions (whilst also claiming it "gets instant results"):

> "Three times a day, for the next two weeks... After the first two weeks, practice once a day or even *4 or 5 times a week for a year*. After that, practice *once or twice a week for life* and you and your dog will have a recall to be proud of."[384] (emphasis added)

Or Jane Killion (author of "Pigs might fly – training success with impossible dogs"), a well-known reward-based trainer, asked her readers:

> "Are you beginning to get a feel for the *enormity of reinforcement* that is necessary to build a recall?"[385] (emphasis added)

And then a few pages later:

> "If you are going to take your dog to a park or trail and let him off leash, *you may need literally years of substantial reinforcement history of coming when called."* [italics TW][386]

---

[384] Nelson, Leslie. *Really Reliable Recall Booklet*. Dogwise. Kindle Edition.
[385] Killion, Jane. *When Pigs Fly!: Training Success with Impossible Dogs*. Wenatchee, WA: Dogwise Publishing, 2007.
[386] As above.

CHAPTER 12 – Some real-life junk "science"

And Zak George:

> "Also keep in mind that *some dogs will never be okay off leash*, and that's fine too. Only you will know when the time is right to let your dog off leash in certain situations, *if ever*."[387]

And remember Karen Pryor, who finds that it is *just too hard* to train a reliable recall with dogs that are prey motivated (as so many dogs are, and those in the study were) like her own Border Terrier? These positive-reinforcement trainers all admit how long and difficult the process is, and that even then with many dogs it will *still fail*. And yet Jonanthan Cooper, Lucy China and Daniel Mills are trying to convince us that, in *just two hours over four days*, positive reinforcement training gets better results than training that also utilizes negative reinforcement or punishment. It is frankly preposterous.

And, once again, like in the case of Herron's aggression study we just looked at, these studies then get cited by other reward-only ideologue pseudo-scientists like Emma Williams and Emily Blackwell. They say things like:

> "Evidence suggests that training using aversive methods is no more effective than positive reinforcement-based training techniques (Cooper, Cracknell, Hardiman, Wright and Mills, 2014) and some studies suggest that they are less effective."[388]

When Emma and Emily say, "Evidence suggests", they are simply referring to this study we are looking at. There is a whole little clique of these reward-only researchers that simply go round and round citing each other's junk "science" (Zazie Todd is another one).

As mentioned earlier, I will challenge any "positive" trainer on earth to a four-day test similar to this, to see who can get better results with dogs with recall under distraction with livestock chasing and worrying problems—provided the dogs are *actually tested off lead at a distance in the presence of free ranging livestock*. And, I want Jonanthan Cooper, Daniel Mills, Lucy China, Hannah Wright, Nina Cracknell, Emily Blackwell, Emma Williams, and

> I will challenge any "positive" trainer on earth to a four-day test similar to this, to see who can get better results with dogs with recall under distraction with livestock chasing and worrying problems.

---

[387] George, Zak, and Dina Roth Port. *Zak George's Guide to a Well-Behaved Dog: Proven Solutions to the Most Common Training Problems for All Ages, Breeds, and Mixes*. New York: Ten Speed Press, 2019.
[388] Williams, E. J., & Blackwell, E. *Managing the Risk of Aggressive Dog Behavior: Investigating the Influence of Owner Threat and Efficacy Perceptions*. 2019.

PART 5 – The scientific basis—fact and fallacy

all the others present to witness the results, and then attempt to justify their study and their conclusions, for all to see.

Now, as if all that's not enough already to entirely discredit this often-cited study, let's delve into the actual flaws within it.

One issue was in the way dogs were allocated to the three groups (positive reinforcement, e-collar, and balanced training without e-collar). You would think the dogs would either be randomly allocated, or an attempt made to allocate a similar selection of dogs to each group. That would make sense in a truly impartial study. That was, however, not the case at all—dogs were supposedly randomly allocated to the two correction-based groups (i.e. one utilizing e-collars and the other utilizing non-e-collar corrections), however were actually *selected specifically* for the positive reinforcement group. Professor Doug Elliffe, who wrote a scathing response to this study, says:

> "…the classic way to minimize this problem is by random allocation of participants to groups. *This was not done*—dogs were randomly allocated to the e-collar group and one control group, but were *recruited separately* at a later date for the second control group whose training emphasized positive reinforcement. The authors claimed that this conferred the advantage of being able to match the behavioural problems for which the dogs had been referred to training. They were unsuccessful in this—dogs with poor recall but *no chasing* (e.g., of sheep) issues were *three times as common* in Group C2 [the positive reinforcement group] than in Group E [the e-collar group] and dogs with a *history of aggression* against other dogs were *twice as common* in Group E [e-collar group] than in Group C2 [positive reinforcement group]… Those differences were *in the direction unfavourable to the e-collar*—Group E appears likely to have been a more difficult group to train than Group C2 in particular."[389] (emphasis added)

In other words, they kept most of the sheep chasers and aggressive dogs out of the positive reinforcement group. Elliffe had *many* other criticisms of the study also. And he's not the only one. Two other scientists, Rebecca Sargisson and Ian McLean, who could also see the serious issues with it, also put out a combined rebuttal pointing out some of the study's various flaws. In their 2021 commentary on the study, they said that, amongst other things,

> "…the conclusions go well beyond the results."[390]

---

[389] Elliffe, Dr. Doug. Commentary on China, L, Mills, D.S., & Cooper, J.J. (2020). *Efficacy of dog training with and without electronic collars vs. a focus on positive reinforcement.*
[390] Sargisson RJ and McLean IG (2021) *Commentary: Efficacy of Dog Training With and Without Remote Electronic Collars vs. a Focus on Positive Reinforcement.*

CHAPTER 12 – Some real-life junk "science"

So, it's exactly like Herron's aggression study in that way. Unfortunately, this is a painfully familiar theme in the reward-only "science". Sargisson and McLean also outline various problems with the statistical analysis, concluding:

> "These statistical anomalies are not minor issues, as they potentially compromise or *even reverse the key results* in the paper."[391] (emphasis added)

Did you get that? Their statistical analysis might even *reverse* the key results. Remember the saying: "Lies, damned lies, and *statistics*"?

Also, Sargisson and McLean note that,

> "…the *problem behaviours were not addressed*… attacks by dogs are unlikely to be initiated when the **owner is within 1m and the dog is on-lead**… [the] *results shed no light on the possible behaviour of the dog off-lead* or when the owner is absent… The problem for which the dogs were referred to trainers was *not directly addressed*, and the *real-life context* in which problem behaviours occur was not replicated in the study."[392] (emphasis added)

Wow. Did you get all that? The actual problems were not even addressed. In a study on livestock chasing and predation, the dogs were *on-lead about 1m from the owner*, and there is nothing in the results about how the dogs would behave *off lead!* Also, as far as I can make out (the exact methodology is *opaque*, as we learnt is the case for *two out of three* "top" psychology papers, but I have seen photos), the sheep were kept contained in tiny pens where they couldn't run or move around, and the dogs were trained outside these pens. This prevented the sheep from running and moving, which would be much more likely to trigger a predatory chasing/hunting instinct than a sheep lying down calmly chewing the cud. And yet, this study is used to justify policy decisions of governments around the world wanting to appease these ignorant and harmful "welfare" activists.

I have attempted to contact Jonathon Cooper a number of times to obtain copies of the video footage from this study, in order to ascertain the exact situation and methods employed, and to discuss his findings, however he has never responded. Others have tried also, without success. In contrast, many other scientists and others were happy (or at least willing) to talk to me and discuss their findings. Even Sidman notes that:

---

[391] Sargisson RJ and McLean IG (2021) *Commentary: Efficacy of Dog Training With and Without Remote Electronic Collars vs. a Focus on Positive Reinforcement.*
[392] As above.

PART 5 – The scientific basis—fact and fallacy

> "A fundamental characteristic of experiments is that they make data and data-gathering techniques *accessible for public evaluation*. An experimenter has to tell exactly what he or she did, and under what conditions, so that informed criticism and repetition of the study by others will be possible." [393]

Cooper et.al., certainly haven't done this. What is also particularly interesting is that, prior to this study, these very same authors did what they referred to as a "preliminary study" (this is not something I've seen addressed anywhere). This so-called "preliminary" study was with 8 dogs, all trained with e-collars, for the same problems. Only this time, the trainers were free to use whatever methods they choose while using the e-collars. And, according to the study itself, the dogs were "allowed to roam off-lead in a field, where sheep were present." So, they exactly replicated the real-life problem off-lead with free roaming sheep, and directly addressed the problem behaviour, unlike in the subsequent "main" study. In relation to the so-called "preliminary" study, the authors note that:

> "Six of the 8 dogs referred for sheep chasing only engaged in one or two approaches, and received *a single application* of the electric stimulus each time they approached sheep, which *led to cessation of approach*."[394] (emphasis added)

Of the other two dogs, one dog didn't attempt to approach the sheep at all. The final dog required five applications of the collar before it ceased chasing the sheep. In other words, the problem was solved mostly in one or two applications, or five at most, in *one single session*. This is perfectly consistent with my decades of experience dealing with these issues. But then, when they moved to their "main" study, they used different training methods, and never tested the dogs off lead or at a distance. And they failed to even mention these "preliminary" results in their conclusion, where they claim:

> The authors did a "preliminary" study which convincingly proved the great effectiveness of e-collar training – and then *completely ignored those results* and moved on to the "main" study using *completely different methods.*

> "…the finding that the use of an E-collar did not create a greater deterrent for disobedience we conclude that an E-collar is unnecessary for effective recall training."[395]

---

[393] Sidman, Murray. *Coercion and Its Fallout*. 2000. United States. Author's Cooperative.
[394] Cooper JJ, Cracknell N, Hardiman J, Wright H, Mills D (2014) *The Welfare Consequences and Efficacy of Training Pet Dogs with Remote Electronic Training Collars in Comparison to Reward Based Training*.
[395] China L, Mills DS and Cooper JJ (2020) *Efficacy of Dog Training With and Without Remote Electronic Collars vs. a Focus on Positive Reinforcement*. Front. Vet. Sci. 7:508. doi: 10.3389/fvets.2020.00508

CHAPTER 12 – Some real-life junk "science"

Professor Doug Elliffe notes that:

> "Such a result [curing predation by a single application of the e-collar] could certainly not be achieved by positive reinforcement—'reward-based learning'—alone."[396]

And:

> "There is well-established literature on the efficacy of e-collars at reducing serious undesirable behaviour by dogs..."[397]

So, what have we seen?

- A government agency, seeking "evidence" to justify their pre-existing promise to ban e-collars in order to appease activist "welfare" groups, funds a study.
- This funding is directed to biased researchers at a relatively low-ranking university, who had already been actively seeking to have e-collars banned.
- The authors did a "preliminary" study which convincingly proved the great effectiveness of e-collar training—and then *completely ignored those results,* and conveniently moved on to the "main" study using *completely different methods*.
- They failed to mention these preliminary results in their conclusion.
- In the main study they selectively allocated the easier dogs to the positive reinforcement group and the harder ones to the e-collar group.
- Their statistical analysis is so highly flawed that other scientists at a much more highly ranked university say that this could reverse the findings.
- They didn't even address the problem behaviours.

So, there you see yet another example of the type of *absolute junk pseudo-science* that is used by the reward-only cult to back up their ideology. When the "positive" ideologues claim they have "science" on their side, this is the type of "science" they're talking about.

> When the "positive" ideologues claim that they have "science" on their side, this is the type of "science" they are talking about.

---

[396] Elliffe, Dr. Doug. Commentary on China, L, Mills, D.S., & Cooper, J.J. (2020). *Efficacy of dog training with and without electronic collars vs. a focus on positive reinforcement.*
[397] As above.

Some real science

We've just seen two examples of oft-quoted studies (plus Skinner's), and what junk pseudo-science they truly are. And we could add example after example all along the same lines. In contrast to these types of pseudo-science surveys, flawed methodologies, and biased conclusions we see above, there is, in fact, some *real* science. Science that is based on repeatable experimentation rather than highly doubtful questionnaires replete with utterly flawed designs, flawed statistical analysis, and biased conclusions. Consider the real science I have repeatedly quoted from research (not a survey) done by Azrin and Holz:

> "One of the most dramatic characteristics of punishment is the *virtual irreversibility or permanence* of the response reduction once the behaviour has become completely suppressed. Investigators have noted that the punished response *does not recover* for a long period of time even after the punishment contingency has been removed... How quickly does punishment reduce behaviour? *Virtually all studies* of punishment have been in complete agreement that the *reduction of responses by punishment is immediate* if the punishment is at all effective. When the data has been presented in terms of number of responses per day, the responses have been *drastically reduced or eliminated on the very first day* in which the punishment was administered."[398]

You'll note that Herron never cited *that* study. Yet we saw these exact results in Cooper, Mills and China's *preliminary* study where the sheep chasing was fixed basically instantly. That is some real science right there, that accords perfectly with the *real world*. Or Marschark and Baenninger, in their 2002 study:

> "While positive reinforcement can be used exclusively for the training of certain behaviors, it is suggested that in the context of instinctive motor patterns, negative reinforcement and punishment may be desirable and necessary additions to positive reinforcement techniques."[399]

This is all in stark contrast to the "punishment doesn't work", "only makes things worse", "is never needed", etc., etc., etc., that the "positive" crowd continually parrots. Like Dunbar when he still insists on saying:

---

[398] Azrin, N. H., & Holz, W. C. (1966). *Punishment*. In W. K. Honig (Ed.), Operant Behavior: Areas of Research and Application (pp. 213-270). New York: Appleton-Century-Crofts.
[399] Marschark, Eve & Baenninger, Ronald. (2002). *Modification of instinctive herding dog behavior using reinforcement and punishment.*

> "Punishment-training is like the Myth of Sisyphus—an ***everlastingly laborious and theoretically impossible task***."[400] (emphasis added)

The outright ignorance or deceit (because it can only be one of the two) of these trainers, scientists and authors, is staggering. We also heard from Mazur in the 6th edition of Learning and Behaviour:

> "The question of whether punishment can be an effective way of controlling behaviour *has been settled*. This chapter has presented numerous studies, from both inside and outside the laboratory, that have demonstrated that punishment can change behaviour and, in many cases, *change it permanently*."[401] (emphasis added)

You cannot find *anything* that comes close to that effectiveness in positive reinforcement training studies. *Not even close.* You only have to consider the statements of most of the "positive" trainers themselves that we have seen earlier. Or, what about this research, again by Azrin and Holz:

> "When the effects of the four procedures [for reducing behaviour] were compared, punishment was found capable of producing a more immediate, complete and longer lasting response reduction than the others."[402]

Again, this relates to actual experimentation, not a survey surrounded by biased *commentary*.

I could go on and on (I know I already have), but you get the idea. Do you want to rely on this type of pseudo-science, or would you rather use your own logic and *real world* experience to see the *real world* reality? When used properly, punishment works highly effectively and very quickly, and we don't need to use much of it at all. It is far faster and easier, and therefore more practical, than so-called "positive" methods. The side-effects are all positive, and our dog's (or children's, or student's) standard of living (and safety) is much higher because we have been responsible enough to train them well. That, right there, is true animal (and human) welfare.

Unfortunately, governments, influenced by activist pressure and with the support of a misinformed public—including veterinarians and other professionals—are

---

[400] Dunbar, Ian. *The Good Little Dog Book*. 3rd ed. Berkeley, CA: James & Kenneth Publishers, 2003.
[401] Mazur, James E. *Learning and Behavior, 6th Edition.* New Jersey. Pearson Prentice Hall, 2006.
[402] Holz, W. C., & Azrin, N. H. (1963). *A comparison of several procedures for eliminating behavior.* Journal of the Experimental Analysis of Behavior, 6(3), 203-212. https://www.ncbi.nlm.nih.gov/pmc/articles/PMC1333086/

already legislating these options away. In some countries and jurisdictions extreme laws are already in place, and the restrictions are steadily increasing. (And guess what happens if you break these laws—they *punish you for using punishment!* So they hardly believe in the power of positive reinforcement.)

If we continue down this path, the freedom to make informed, effective training decisions will be taken entirely out of our hands.

13

LIMA, LIFE and PROTECT

Before leaving the subject of science, common training "frameworks" need addressing. LIMA —"Least Intrusive, Minimally Aversive"— is a popular framework introduced by Steven R. Lindsay, author of the three-volume *Handbook of Applied Dog Behavior and Training* (cited numerous times in this book).

In theory, LIMA seems like a reasonable approach: use the least intrusive, minimally aversive method that achieves the desired result. In practice, however, it often means that punishment is utilised only as a last resort. By that point, the behaviour has typically escalated, resulting in more serious issues that then require stronger corrective measures. As even Karen Pryor acknowledges:

> "Punishment has the best chance of halting a behavior in its tracks if the behavior is caught early, so that it has not become an established habit"[403]

So, as also discussed earlier, **very often punishment should be the *first resort*.** This is because the least amount is then required, habits have not been allowed to become established, and because the **positive reinforcement methods are generally doomed to fail anyway**.

Also, by trying to be "minimally aversive" it often has the effect of trainers using insufficient levels of punishment to be effective, at least initially. Apart from not getting the desired results, the dog becomes desensitized to the punishment. This

---

[403] Pryor, Karen. *Don't Shoot the Dog!: The New Art of Teaching and Training*. Rev. ed. New York: Bantam Books, 1999.

leads to more punishment of a higher intensity ultimately being required. Whereas, if a sufficient intensity of punishment is applied initially, this does not occur.

> Even though theoretically being "minimally aversive" is a worthy ideal, **what is actually minimally aversive is to use an effective level of punishment early.**

Going straight to an effective level of punishment early may seem harsh (just as a bitch does with its pups), but it actually means less punishment of a lower intensity is utilised overall—a better welfare outcome while achieving better results.

So, even though theoretically being "minimally aversive" is a worthy ideal, **what is actually minimally aversive in practice is to use the correct level of punishment right from the outset**. As Lindsay himself said:

> "Besides misrepresenting and confusing the facts, excessive moralizing about the use of punishment and other aversive training procedures may have a very undesirable effect on the dog-owning public, *making responsible dog owners feel guilty about exercising the necessary aversive prerogatives needed to establish constructive limits and boundaries over a dog's behaviour.* Many of the basic facts of life that all dogs must learn to accept (if they are to become successful and welcome companions) are won through the mediation of directive training, combining a balanced application of behaviour modification—not just positive reinforcement. Instead of grinding away on a very dull ax [positive reinforcement], *a dog's welfare is better served by teaching the owner when punishment is necessary and how to use it effectively and humanely*."[404] (emphasis added)

However, nowadays the fact that LIMA is endorsed by Dunbar's *Association of Professional Dog Trainers,* the related *Certification Council for Professional Dog Trainers*, and the *Karen Pryor Academy*, amongst others, proves the point that LIMA is used to promote ineffective "positive" methods. This fact alone should be enough to tell us to be very wary of it. Recently, in 2023, Eduardo Fernandez (a Senior Lecturer of Applied Animal Behaviour & Welfare in the School of Animal and Veterinary Sciences at the University of Adelaide here in Australia), noted exactly this:

> "LIMA has become one of the most commonly used ways *for force-free trainers to describe their training practices.*"[405] (emphasis added)

---

[404] Lindsay, Steven R. *Handbook of Applied Dog Behavior and Training, Volume 1: Adaptation and Learning*. Ames, IA: Blackwell Science, 2000.

[405] Fernandez, Eduardo J. 2023. *The Least Inhibitive, Functionally Effective (LIFE) Model: A New Framework for Ethical Animal Training Practices*. Journal of Veterinary Behavior.

CHAPTER 13 – LIMA, LIFE and PROTECT

In fact, the *Certification Council for Professional Dog Trainers* claims:

> "LIMA requires trainers/consultants *to work to increase the use of positive reinforcement and eliminate the use of punishment* when working with animal and human clients."[406] (emphasis added)

This idea that LIMA requires trainers to "eliminate the use of punishment" is not at all what Lindsay originally intended for LIMA to be, but unfortunately this is what it has largely become. It's been hijacked by the so-called "positive" training community, like we see above in this false statement from the CCPDT.

Because LIMA has been hijacked in this manner, it should not be endorsed by any sound trainers who actually desire to use effective methods that get optimal outcomes in terms of both *results* and *welfare* (which two are very closely correlated). Fernandez then moves on to admit the original intention of LIMA (which he, as a reward-only ideologue himself, argues against):

> "One of the biggest problems for the LIMA approach is the justification it has enabled for regularly using aversive stimuli or coercive training methods. Some of this is historical, which becomes more evident as we go back to Lindsay's original statements."[407]

(By the way, note the Sidmanesque reference to "coercive training methods".) Fernandez then quotes Lindsay as saying (correctly) that:

> "Aversive procedures are legitimate and valuable tools for controlling undesirable behaviour..."[408]

So, LIMA started out as a somewhat balanced way of using the least aversive methods necessary, but still using them when required. Although, in practice, as I noted, it actually had the effect of preventing aversive methods being used as early intervention, when they are actually most effective and required in the least amounts. Therefore, even then it often resulted in poorer welfare outcomes, but this has significantly worsened.

Fernadez promotes an alternative model he has developed, which he calls the LIFE model (least inhibitive, functionally effective). Unfortunately, it is a massive step backwards and doomed to certain failure. Following the LIFE model's guidelines

---

[406] *Least Intrusive, Minimally Aversive (LIMA) Effective Behavior Intervention Policy.* CCPDT, accessed July 8, 2024, https://www.ccpdt.org/about-us/least-intrusive-minimally-aversive-lima-effective-behavior-intervention-policy/.
[407] Fernandez, Eduardo J. 2023. *The Least Inhibitive, Functionally Effective (LIFE) Model: A New Framework for Ethical Animal Training Practices.* Journal of Veterinary Behavior.
[408] As above.

will simply fail to inhibit undesirable behaviour (being very appropriately named from that standpoint—"least inhibitive") and will result in very little that is "effective" in any sense, "functionally" or otherwise.

In the opening paragraph of his paper, Fernandez starts out by citing the thoroughly discredited study we just saw by Jonathan Cooper et. al. This alone immediately casts significant doubt on his subsequent conclusions (and indeed, on Fernandez's credibility). To even further erode his credibility, and that of his model, he also uncritically references Murray Sidman's fundamentally flawed book, *Coercion and Its Fallout*. Fernandez's entire thesis is grounded in this type of junk pseudo-science, with no *real world* evidence or experience anywhere to be seen.

> The LIFE model is quite simply just another dismal reward-only failure rooted in ideological blindness and pseudo-science.

The LIFE model is quite simply just another dismal, academic, reward-only failure rooted in poor science and Fernandez's ideological blindness.

PROTECT

For what it's worth, and for those who like frameworks (I don't), I have developed an alternative that actually works in the *real world* while achieving the best welfare outcomes. This simple acronym is superior to either of the failed LIMA or LIFE frameworks:

P.R.O.T.E.C.T.

Prioritize Results for Optimal Training to achieve EthiCal Transformation.

The three components are as follows:

Prioritize Results

Prioritizing results is by far the most critical component, and it should also be self-evident. However, in LIMA, and particularly in LIFE, *results* are relegated to a sort of forgotten, unimportant afterthought. After all, if training doesn't produce results (as we've seen with many so-called "positive" methods), can it even be called training? It's like debating whether it's better to cut a tree down with a butter knife or a chainsaw—one is inherently ineffective, making the entire argument pointless. You can argue all you like that the butter knife is lighter, cheaper, smaller, safer and so on, but if it can't do the job, of what relevance is any of it?

CHAPTER 13 – LIMA, LIFE and PROTECT

Similarly, arguing whether positive methods or balanced methods are superior is often a futile discussion, because the positive methods in many cases *simply don't work*. Therefore, frequently there isn't even a rational discussion to be had.

Consider this: if someone is having a heart attack, telling them to rest up and drink plenty of fluids is definitely a less intrusive or aversive intervention compared to open-heart surgery. But while it is gentle, non-intrusive, and non-aversive, it does nothing to fix the underlying issue—it's *ineffective* and ultimately *harmful* because it ignores *reality*. The same applies to training: **if a method doesn't work, it isn't truly ethical, no matter how gentle it may be.**

As we've heard from Jamie Dunbar:

> "When you use effective dog training techniques, your dog's behavior improves. If you've been trying to train your dog, and your dog's behavior hasn't improved, it's not your fault. **It means you've been using the wrong methods.** And that's not your fault either. **The world is full of self-proclaimed dog trainers who promote methods that sound reasonable, but they just don't work.** Again—**the proof is in the pudding**. If your dog's behavior improves, the training methods are good. If your dog's behavior doesn't improve, the methods are not good."[409] (emphasis added)

And from Ian Dunbar:

> "…it's **got to be easy** or else they [owners] can't do it. It's **got to be quick**, or else they probably won't… the sooner we as trainers get used to that… the quicker we can help people out. And it's **got to be effective**."[410] (emphasis added)

In this case, they're right—results are paramount. Unfortunately, most "positive" training approaches are the perfect examples of Jamie's "wrong methods". And they are rarely ever quick, easy, or effective.

Effective training, like a trip to the doctor or dentist, may involve tools or methods that are "more intrusive" or aversive, but in the *real world* they are vital for real, lasting change, and ultimately serve the welfare of the dog. **Ethics without effectiveness is simply virtue signalling.**

---

[409] Dunbar, Ian. *Six Simple Steps to Solve Your Dog's Behavior Problems*. Dunbar Academy. Accessed Feb 2024. https://www.dunbaracademy.com/courses/six-simple-steps.
[410] Balabanov, Ivan. *Training Without Conflict Podcast Episode Four: Dr. Ian Dunbar*. YouTube, April 30, 2021. Accessed Feb 2024. https://www.youtube.com/watch?v=EF-INPvEhA8

PART 5 – The scientific basis—fact and fallacy

Of course, the "positive" ideologues will hate this focus on *results,* given that this is the area they so regularly fail in (like attempting to cut a tree down with a butter knife). And because they fail in producing results, they therefore also fail in providing *ethical* training and optimum *welfare* outcomes. Why? **Due to the huge increase in rehoming, euthanasia, and the enormous reductions in freedom that occur when results aren't achieved.** *Without effective, practical results, true welfare is impossible.*

> "Positive" training is often *ineffective* and ultimately *harmful* because it ignores reality.

The problem with other frameworks, like Fernandez's LIFE model, is that they prioritize *ideology* over *results*. Therefore, they fail. LIMA also prioritizes using the least intrusive and minimally aversive approaches over results (whether this is explicit or simply the practical outcome), which results in people using far less effective methods. This simply wastes time and money, allowing behaviours to become more ingrained and therefore more difficult to fix, and resulting in more punishment being required long term to fix the problem.

> Good teaching results in fast learning.

So, results are paramount. If we use the most effective methods we will get the best and fastest results (good teaching results in fast learning). In the *real world*, these things matter. We must *Prioritize Results*. Even the Dunbars say the same.

Optimal Training

By Prioritizing Results, we get Optimal Training.

The use of the word *optimal* is deliberate. Even if a "positive" method achieved better results with less unwanted side-effects, but is so impractical that no-one in the *real world* could ever apply it (like the 150 people handling, grooming and training your pup in the first month after you get it home[411]) then it isn't *optimal*. It is a utopian fantasy that no one can ever actually apply.

Also, no training is perfect, in the sense that there are always trade-offs. Rocha and Hunziker (quoting Goldiamond) rightly noted:

> "Furthermore, assuming that "…the issue is *never coercion versus no coercion*… the issue is the amount and type of coercion we are willing to

---

[411] Dunbar, Ian, Dr. *Barking Up The Right Tree - The science and practice of positive dog training.* Novato, CA: New World Library, 2023.

accept, and the protections against abuse we set up" (Goldiamond, 1976, p. 23), behavior analysts *may better acknowledge the ubiquity, inevitability of coercion in society at large*, therefore putting forward *more pragmatic* and *less utopian* projects to social change."[412] (emphasis added)

We cannot eliminate "coercion" or "punishment" or "aversives" or "force" (even positive reinforcement methods contain these). Therefore, we simply have to weigh up the competing options and use the *optimal* approach. There is no *perfect* approach from this perspective. And as has been demonstrated throughout this book, "positive" training is often so far from "optimal" that it doesn't even get a look in, because it either simply doesn't work or is so impractical that it amounts to the same thing. Only balanced training, utilising correction and punishment when necessary (properly understood and properly applied) can ever be "optimal".

We *Prioritize Results for Optimal Training*.

Ethical Transformation

It just so happens that *results* and *welfare outcomes* are inextricably linked. So we *Prioritize Results* for *Optimal Training* in order to achieve *Ethical Transformation*.

Good training is indeed often transformational. A very common comment I receive from clients is that "he's like a new dog!" or "I can't believe my eyes!". They often can hardly believe the change they see in front of them in such a short space of time. And not only in behaviour, but also in emotion. The previously stressed and anxious dogs are now much calmer and more relaxed.

> "Hi Tully! Just wanted to thank you. Our dog has improved so much, and he's a completely different boy. He's very responsive, shuts up when asked, and comes immediately when called (even in full flight after a rabbit or kangaroo). We walk off lead all the time, and somehow he's become much more intuitive. Never explicitly taught heel or wait when off lead, but he does both. We still put him on the lead when there's a cyclist, but he's much better when they go past. We're even able to take him to markets and have random people pat him! He's much happier, as are we! Thanks so much." - *Maree*

---

[412] Rocha, C., & Hunziker, M. "*A Critical Assessment of Murray Sidman's Approach to Coercion.*" Brazilian Journal of Behavior Analysis 17, no. 2 (2021): 188-194.

Let's just take one example—lead reactivity. This is probably the most common problem I deal with. No matter the severity, it can always be fixed in a couple of sessions a week apart (much less, if I was doing the training myself). Prior to my intervention, walking such dogs was either a stressful chore or had even become completely untenable. In many cases these dogs were no longer walked at all, often after working with various "positive" trainers (and sometimes even sub-standard "balanced" trainers). Many other such dogs simply get rehomed or put down.

Yet, after a week's pretty simple training, this can all be solved. The dog is now back out walking regularly because all the reactivity is gone, and the owners are once again enjoying their walks. The dog gets walked more often, for longer. Also, the dog's exercise needs are now being better met, which also has other beneficial side effects. **The methods are truly transformational, and concurrently result in vastly improved welfare outcomes—this is truly ethical training.**

So, the equation is simple—effective training gets better welfare outcomes. Prioritizing Results for Optimal Training achieves Ethical Transformation. On the other hand, ineffective "positive" training methods—when ideology is prioritized over effectiveness—often simply result in the problem just not getting fixed. **For the greatest issues most dog owners face, so-called "positive" training nearly always results in poor welfare outcomes.**

Conclusion – LIMA, LIFE and PROTECT

Ultimately, the goal of any training framework should be to achieve real results that benefit both the owner and the dog. LIMA, and even more so LIFE, while they may sound well-intentioned, fall short because they prioritize (or have come to prioritize) ideology over practicality and effectiveness. They simply result in worse welfare outcomes.

PROTECT, on the other hand, is designed for *real world* success. By **Prioritizing Results for Optimal Training to achieve Ethical Transformation**, we ensure that both the effectiveness of the training and the welfare of the animal are achieved simultaneously.

Part 6

MYTHS ABOUT PRINCIPLES OF PUNISHMENT

Chapter 14 – Myths about principles of punishment

Chapter 15 – A model of obedience—balance scales

14

Myths about principles of punishment

In this section I'll examine some of the most common misconceptions about the application of punishment, held even by various trainers and researchers who are *not* against the use of, or in fact even actively support the use of, correction and punishment. There's a lot of misunderstanding and incorrect "science" about how to best use correction or punishment. To quote Dr. Dunbar yet again:

> "...punishment *must be consistent, immediate*, and *instructive*. Most people are rarely consistent or attentive 24/7, so it is next to impossible for people to punish effectively."[413] (emphasis added)

He also states:

> "the *exquisite timing* and *absolute consistency* that are required to discourage undesired behaviour with punishment."[414] (emphasis added)

Elsewhere Dunbar also states that a command should always be used before any punishment. So he claims that punishment *must* be:

---

[413] Dunbar, Ian, Dr. *Barking Up The Right Tree - The science and practice of positive dog training*. Novato, CA: New World Library, 2023.
[414] As above

1. Absolutely consistent 24/7
2. Immediate
3. Instructive
4. Have a command or warning beforehand.

Dr. Dunbar is wrong on all four points (as will be explained). However, even many advocates of using punishment (particularly academics) appear to have similar misconceptions.

Let's deal with each of these myths in turn.

Punishment must be consistent?

This is probably the biggest misconception of all. Dr. Dunbar likes to make the claim that *reward* does not have to be consistent, because it uses variable schedules of reinforcement, but that punishment does. As above, he makes the claim that because we can't be "absolutely consistent 24/7", that "it is next to impossible for people to punish effectively." Dunbar also claims:

> "For punishment to be effective, a dog must be immediately punished after *every* transgression. Every single one!"[415]

And:

> "…for punishment to be effective, the trainer needs to be present to punish the dog for an *infinite* number of improvisations, which of course requires an *infinite* amount of time."[416] (Dunbar's emphasis)

Wow. An "*infinite* number of improvisations" over an "*infinite* amount of time"! This is absurd beyond belief. In fact, it is the positive reinforcement approach that suffers most from this consistency criticism (and the huge amount of time and repetition required), as will be explained. But first, let's talk about why punishment often does *not* have to be consistent, and *why this is one of its greatest strengths*.

We only have to consider humans with speeding fines. *We don't have to get caught every time.* We know that just because we didn't get caught today, doesn't mean we won't get caught tomorrow. And so, we adjust our behaviour accordingly. In

---

[415] Dunbar, Ian, Dr. *Barking Up The Right Tree - The science and practice of positive dog training.* Novato, CA: New World Library, 2023.
[416] As above.

CHAPTER 14 – Myths about principles of punishment

fact, we don't even have to get caught *all that often*. So, the consequence can be completely inconsistent, and yet it still works.

So much for Dunbar's "it must be consistent 24/7 or it won't work" (or that the dog "must be immediately punished after *every* transgression. Every single one!"). Right there his claim (a claim which is supposedly backed up by "science") is already completely disproven. It doesn't. And that is one of its greatest strengths.

> Punishment often does *not* have to be consistent.
>
> *This is one of its greatest strengths.*

In fact, in my 2007 book *"Working Sheep Dogs – A practical guide to breeding, training and handling,"* I even talk about deliberately applying the concepts of "intermittent punishment" and "delayed punishment". Used correctly, the *deliberate use of inconsistent punishment* can actually be *more* effective. I won't go into details here, as it is a bit too advanced for a book such as this.

But say we have a dog that chases cars, maybe when visitors arrive or leave a property with a long driveway. A highly effective and permanent cure when done properly is the e-collar, set on a high-level, providing this occurs *while the dog is in the act*. With some more sensitive dogs or those in whom the habit isn't too firmly entrenched, one punishment will be enough (so much for Dunbar's farcical *infinite* number of applications and *infinite* amount of time). With more confirmed chasers, a repeat or two might be needed (remember Cooper et al.'s highly effective preliminary study on sheep chasers, which they completely ignored in their conclusion? Most dogs were cured with one or two applications, one dog required five).

Does this punishment have to happen *every* time the dog chases a car down the driveway? That is, two or three times *consecutively*? Does it have to be "perfectly consistent"? What if the owner can't find the remote the next time the behaviour occurs, or the battery is flat? Will that ruin their training because they weren't consistent, because the dog wasn't punished *every single time* it chased the car? Not at all. They press it the next time, or the one after that. Just like the speeding fines, the consequence doesn't have to occur every single time. It just has to occur *enough* that the dog knows it is a *possibility*, for the training to be 100% effective.

The same applies to a dog jumping on us. Do we have to give it a smack *every single time for all eternity* (Dunbar's "*infinite* number of improvisations" for "an *infinite* amount of time"), and if we fail *just once* we have ruined the training? Not at all. As long as it happens enough for the dog to realise it's a possibility (just like us with speeding fines), and it's firm enough to outweigh how much the dog enjoys jumping on us, it is highly effective very rapidly.

PART 5 – The scientific basis—fact and fallacy

By contrast, many "positive" methods certainly do have to be *absolutely* consistent to have *any* hope of success, and thus the reason you need "*perfect* control"[417] of the environment. This is a major reason why they're unsuccessful. Unlike the use of punishment, these positive reinforcement methods truly are like Dunbar's Myth of Sisyphus. In the ancient Greek myth, Zeus is condemned to a life of rolling a heavy boulder up a hill, only to have it roll back to the bottom again as it nears the top—and then he has to start all over again—a perfect analogy for a lot of "positive" methods. Zak George, when talking about teaching a dog not to jump on you, says:

> Dunbar's "Myth of Sisyphus" is actually positive reinforcement training, not punishment.

> "Of course, establishing proper jumping habits requires *relentless consistency*."[418] (emphasis added)

The problem, as Dunbar so rightly points out, is that:

> "Most people are rarely consistent or attentive 24/7…"[419]

And

> "People are simply not 100% consistent 100% of the time. Certainly people can concentrate for short periods, but not all the time. Even when people try their hardest to concentrate for limited periods, their attention often wanders."[420]

Take the example again of jumping on people. Because people are not consistent, the dog is rewarded for jumping on people. Whether that is guests, or kids, or people down the street who might give the dog attention and a pat when it jumps. And so we are back to competing rewards again. The "positive" trainer is attempting to reward an alternative behaviour, but the dog is still getting rewarded for the undesirable behaviour, and on a variable reinforcement schedule. So why on earth would it stop? As you've probably figured out by now, it won't. Unless everybody is perfectly 100% consistent always, for considerably extended periods of time, "positive" methods such as these simply will not work (as I discussed in

---

[417] George, Zak, and Dina Roth Port. *Zak George's Guide to a Well-Behaved Dog: Proven Solutions to the Most Common Training Problems for All Ages, Breeds, and Mixes*. New York: Ten Speed Press, 2019.
[418] As above.
[419] Dunbar, Ian, Dr. *Barking Up The Right Tree - The science and practice of positive dog training*. Novato, CA: New World Library, 2023.
[420] Dunbar, Ian. *The Good Little Dog Book*. 3rd ed. Berkeley, CA: James & Kenneth Publishers, 2003.

CHAPTER 14 – Myths about principles of punishment

Chapter Eight – Does Reward Work? – On extinction—and consistency). They are the epitome of Dunbar's Sisyphus Myth—you do all the reward-only work for months or years on end, almost make it to the top of the hill, and then someone slips up, once, and you are back to the start again (only worse—because you have increased the length of the variable reinforcement schedule). On the other hand, punishment, even if inconsistent, easily solves this problem. Why? Because it can *outweigh* those environmental rewards even if we don't have "perfect control" of the environment.

> Punishment, even if inconsistent, easily solves this problem, because it *outweighs* those environmental rewards even if we don't have "perfect control" of the environment.

So this is the first of Dunbar's myths busted about how punishment must be used—it does *not* have to be 100% perfectly consistent at all. This is one of its greatest strengths, and one of the greatest weaknesses of "positive" methods. Once again, the *exact opposite* to the "positive" idealogues claims.

Punishment must be immediate?

The second myth about the use of punishment—that it must always be immediate—is the closest to the truth of the four (at least for animals; for people it is an entirely different case). In fact, generally speaking, punishment should occur while the dog is *in the act* of the undesirable behaviour—not immediately after, but *during*. However, there's more to it than that.

This is one of the biggest things dog owners do wrong. If their dog is jumping on them, they push it off and *then* say "No". Or if the dog is chewing something, they pull the object away and *then* say "No". However, in both instances the *dog* didn't stop. It didn't make the *decision* or the *choice* to stop—the owner simply stopped the dog.

Instead, what the owner should do is *allow* the dog to jump on them or to chew the couch, quietly tell the dog "No" *while* it's chewing on the couch or jumping on them, and give it a *chance* to stop. If it doesn't, it gets a smack on the nose (or head area—just like the way dogs deal with each other—the old dog correcting the pup will bite the head). So all this occurs while the dog is *in the act*.

If the dog doesn't hop down, or doesn't stop chewing the couch, this process is repeated, only the smack is firmer this time. Consider speeding fines again. If the fine was $5, we would all ignore it. At some point, maybe $400 like it currently is here, we pay attention. The dog is the same—it will weigh up how rewarding jumping on you is, or chewing the couch is, versus the consequence. If the consequence is bigger than the environmental reward or the intrinsic reward, the dog will stop. It isn't silly.

That's how we outweigh the environmental reward, and why "positive" training fails—because the environmental rewards are bigger than the training rewards. The only thing we can find that outweighs the environmental rewards is a correction or punishment. The same as in nature, the same as dogs work with each other, the same way wolves work, the same as people. We're just weighing up the competing options.

> We are just weighing up the competing options.
>
> The same as in nature, the same as dogs work with each other, the same way wolves work, the same as people.

So once the smack is firm enough, the dog decides to stop doing whatever he is doing. He hops down, or he stops chewing the couch. We immediately praise him, "good boy" (don't pat him). Now the dog actually made the *choice* to stop, we didn't just push it down. And the consequence (the smack) occurred *while* the dog was *in the act*.

I always advise the vast majority of clients (who generally aren't highly experienced dog trainers) in the vast majority of situations to only punish *in the act*, never afterwards (and that this doesn't need to happen every time, so they don't need to stress if they aren't 100% consistent if they are too slow to catch the dog in the act—just wait until next time when you can catch it *in the act*).

However, we can indeed correct *after* the act, provided we meet a few requirements to ensure the dog connects the punishment to the behaviour. And we can do that by what is called bridging or linking.

Say the pup has just nipped the baby, nothing serious but more than just puppy mouthing which we certainly don't want to allow. (Dogs that bite kids or even adults often have a short life expectancy, so it's in the dog's interests (welfare), as well as the babies, that the behaviour never occurs again). However, we were a bit too far away when we saw it happen to do anything about it while the pup was actually in the act. But if we *yell* at the pup the instant we see it happen, and *keep yelling* at the pup as we run to it and give it a good smack (or maybe two or three if it's a tough character), it will certainly link the consequence back to the behaviour. We have that *unbroken chain* of correction (yelling at it) until we applied the consequence. This would be highly effective in preventing the pup from biting the baby ever again in the future. So punishment does not always have to be immediate.

CHAPTER 14 – Myths about principles of punishment

And what about Dunbar's statement that punishment requires "exquisite timing"?

> "the exquisite timing and absolute consistency that are required to discourage undesired behaviour with punishment."[421]

The timing is pretty simple. Punish while in the act. And if we can't, we bridge or link as described above. What exactly about that is "exquisite"? However, what *does* require "exquisite timing" is something like clicker training. Or, very often, the use of rewards in general. If you get the timing even slightly wrong you can be rewarding or shaping something you really don't want, which happens all the time.

> The timing is pretty simple. Punish while in the act.
>
> And if we can't, we bridge or link as described above.

Let's consider a common example of when reward is used with poor timing, even by some of it's most influential advocates. Consider Zak George when discussing curing chewing (and note that this isn't unique to him, this is very common advice from the majority of "positive" trainers):

> "…first, interrupt her [when she is chewing the object], maybe by *clapping your hands* or calling her name. Next, get her attention on you and quickly substitute the item she was chewing with an acceptable object such as a bone or chew toy. Make sure it's one your dog really likes! The new bone or chew toy can be *the reward* itself. If this is unsuccessful, you may want to grab one of your *nearby treats* to lure her away from the object she was chewing and then offer the desirable and acceptable chew toy. You'll need to be a master at redirecting your dog during this interim period. *It might take dozens of attempts* before your dog begins to make the connection…"[422] (emphasis added)

The only aspect of this entire process that has any chance of being even remotely successful is "clapping your hands" at the beginning. Although George claims he doesn't use punishment, for some sensitive dogs (like his thunder phobic dog Venus) this may be a big enough punishment that it outweighs how much it wants to chew the couch. Remember the dog hiding under a bed in response to a click from a clicker? A click on the clicker would have been a strong and effective punishment for that dog.

---

[421] Dunbar, Ian, Dr. *Barking Up The Right Tree - The science and practice of positive dog training*. Novato, CA: New World Library, 2023.

[422] George, Zak, and Dina Roth Port. *Zak George's Guide to a Well-Behaved Dog: Proven Solutions to the Most Common Training Problems for All Ages, Breeds, and Mixes*. New York: Ten Speed Press, 2019.

PART 5 – The scientific basis—fact and fallacy

If the clap or other loud noise works in altering future behaviour, then it was the *aversive* nature of the *punishment* that did the trick. However, for most dogs, clapping your hands is unlikely to be a sufficient punishment, and so the procedure will fail.

The rest of George's method is simply an exercise in rewarding the bad behaviour, as most pet owners instinctively realize when they hear this advice at puppy school, or see on YouTube. Every time the pup chews your couch, you go to it and play with it with its favourite toy—as George points out "one your dog really likes"—which he also rightly states is a "reward". But a reward *for what*? A reward for chewing the couch. Or, he advises you to use a treat to lure the dog away from the couch. So, your dog chews your couch, and out come the treats. When you use rewards at the wrong time like this, you are simply rewarding the behaviour you think you're fixing. Remember McDonald's used as positive reinforcement if the kids stop fighting? You are simply rewarding bad behaviour.

Compare the similarity of this approach to teaching a dog something using clicker training. What is the process of training with a clicker? Well, we click *while the dog is engaged in the desired behaviour* to *mark* that behaviour, and then the dog comes to us to get its treat (or we might throw it to the dog). The sound doesn't have to be a click—it can be any sound to mark the exact instant when the desired behaviour is occurring—marine trainers generally use a whistle, but any sound will do.

Isn't this *exactly the same* as the "positive" trainer's advice to stop the dog chewing the couch? Think about it—every time the dog chews the couch, George advises we "clap", which marks the behaviour (takes the place of the clicker), and then we give the dog its reward!

> Many "positive" approaches not only *fail*, but also *make things worse*.

So yet again we have the example of Zak George, fully endorsed by Dr. John Ciribassi of AVSAB (and many, many, others), advising strategies that are not only ineffective, but *counter-productive*. Reward-based approaches that not only won't fix the behaviours but will simply *make them worse*. Remember psychologist Jillian Enright:

"Punishing Unwanted Behaviour Just Makes it Worse."[423]

In reality, it is the "positive" approaches that regularly make behaviours worse, not punishment. The truth, once again, is the exact opposite to such claims.

---

[423] Neurodiversified. *Punishing Unwanted Behaviour Just Makes it Worse*. Medium, https://medium.com/neurodiversified/how-punishing-unwanted-behaviour-just-makes-it-worse-baf22793d07b.

CHAPTER 14 – Myths about principles of punishment

Another example of when timing is critical is when I use treats as part of a strategy when helping people with timid dogs to overcome their fear of strangers. This method is a combination of classical conditioning and positive reinforcement (or negative reinforcement, because it works better if the dog is hungry!). I might sit on a seat, and have some nice, tasty treats handy (yes, I do use treats with clients for some things, and even teach clicker training for some things—that just isn't the focus of this book). I'm careful not to feed the dog the treats while it's reactive or showing fear, as that would simply be rewarding that behaviour and that emotion (we can train emotions the same way as we train behaviours). But once the dog calms down, and *only then*, I will toss it a treat, someway between the dog and I. If it is really timid and scared, I'll toss the treat closer to the dog. If not so bad, closer to me. I want the dog to come *towards me* to get the treat, so that it's rewarded for *braving out* that which it's afraid of, and moving towards me. Then I progressively try to toss the treats closer and closer to me. And finally to have the dog take them out of my hand, etc. But at some point, the dog will often get a bit nervous and move backwards—if I toss the treat *then*, I will be rewarding it for being afraid and moving away. So timing is very important.

Or consider if you teach your pup to sit with treats. If you go to reward the pup with the treat when it sits, and it jumps up to get the treat, you've now rewarded it for jumping up. Or if it snatches the treat out of your hand, you've rewarded that (which is why treat training often ends up with pups and dogs nipping children's hands and snatching food from them).

> Treat training often ends up with pups and dogs nipping children's hands and snatching food from them.

So timing in reward-based training can be critical, and is often misused even by those whom the unsuspecting public assume know better, like Zak George endorsed by Dr. John Ciribassi.

So there is no reason to consider that punishment requires this "exquisite timing" as Dunbar claims, by comparison to what reward requires. In fact, I would say it is less exacting. If your dog is biting you, give it a good firm smack *while it is in the act* of biting you. Nothing particularly "exquisite" or even vaguely difficult about that. (Or, if you prefer, you could try Dunbar's approach and say, "ouch, that hurt you miserable worm" and burst out sobbing. Or George's approach and bring out the treats—just don't expect these approaches to work.)

So yes, punishment should generally be immediate, preferably while the dog is *in the act*, exactly the same as reward. However, that's not the whole story. There are times when we can, and might need to, do otherwise, and when that's a superior option.

PART 5 – The scientific basis—fact and fallacy

Punishment must be instructive (reward an alternative behaviour)?

Punishment does *not* need to be instructive in the sense these trainers mean (of telling the dog *what to do instead*) if we're simply using it to stop behaviour (something reward doesn't do). If we use an e-collar to stop a dog killing sheep, what about that needs to be instructive in this sense of telling the dog what to do instead? Absolutely nothing. We simply want the dog to associate something bad happening with attacking sheep, to punish that behaviour so that it doesn't do it again.

Certified Karen Pryor trainer Cindy Benson (like many others) regurgitates this type of nonsense when she says:

> "Teaching "don'ts" doesn't work. One of the problems I have with saying "no!" to a dog is that there is no information there for the dog regarding what you want him to do, rather than about what you don't want him to do"[424]

Dunbar likewise says:

> "A huge reason that aversive stimuli are largely ineffective is that they *lack instruction* and do not inform animals *which* specific behavior to stop or *what* to do instead." [425] (emphasis added)

Of course, Benson and Dunbar (and all the other "positive" ideologues) are completely and utterly incorrect. Teaching "No" actually works extremely well—very quickly and very effectively. We only have to look at nature to know this is true. When a dog growls at another dog it's usually just saying "No"—for example, "don't come near my food, or I will bite you." It doesn't instruct the other dog what to do instead. It doesn't say "Go and sit over there". There is absolutely no need to.

> Teaching "No" actually works extremely well – quickly and effectively.

I showed earlier that "aversive stimuli" certainly can be instructive in the sense of suggesting to the dog what we want it to do, just like "luring" (Dunbar's favourite) is instructive, and that Dunbar's and Benson's statements are simply false. But even when we are using punishment of whatever form to simply *stop* behaviours,

---

[424] Benson, Cindy. *Livestock Guardian Dog Training Manual*. Self-published, 2021.
[425] Dunbar, Ian, Dr. *Barking Up The Right Tree - The science and practice of positive dog training*. Novato, CA: New World Library, 2023.

CHAPTER 14 – Myths about principles of punishment

isn't that instructive? It tells the dog to *stop doing whatever it is doing at the time*. If I lean on an electric fence, the resultant shock instructs me very clearly to pull away from the fence, and to try not to touch it again. *It doesn't need to tell me to go and plant trees instead.*

Dunbar's statement that "aversive stimuli" also don't inform the animal "which specific behaviour to stop" is simply nonsense. The dog is jumping on us, it gets a smack while it is jumping on us, and with just a couple of repetitions it rapidly figures out to stop jumping. It knows *exactly* which behaviour to stop—the one it was engaged in at the time.

And what should it do instead? Anything else! We are not telling the dog, "you have to do *that*," we are simply saying "stop doing *this*". So, in many cases there is zero need for the punishment to be instructive in the sense of telling the dog *what to do instead*. Punishment is often used simply to tell the dog what to *stop* doing, and it does so very clearly and effectively.

Quite often, people who have heard this type of advice, like "An alternative behaviour should be made available which can be reinforced," simply make their training less effective. I had a good example recently. Another trainer sent me a video of him helping a client with a young dog. The dog was tied up while another dog was being trained, and was barking. The trainer moved towards the dog, telling it off (punishment). The dog stopped barking and lay down. The trainer then went immediately and patted the dog, "good girl". He was following this advice to "reward an alternative" (lying down).

The problem is that this is counterproductive in many circumstances. If a dog is attention barking and we correct it with pressure (vocal or positional—growling at it and moving towards it) or whatever we are using, and then it stops barking and we immediately give it a pat, it has still received positive attention as a result of the barking. Therefore, the correction will be far less effective, or in fact might even make the problem worse.

I think about balance scales—on one side of the scales is what the dog wants, and on the other side is what we apply. So the dog is trying to get attention by barking, and we correct it. But if we also then immediately pat it and give it attention, the dog is now factoring that in. Effectively, "yeah, I got told off, but I also got a pat!" So now we have a situation where, in order to be effective, the correction will have to be *stronger* to outweigh not only the barking itself, but *also the pat* the dog receives. So the process *might* still work, but it will be less effective than if we hadn't put reward on the end of it, and we will

> So now we have a situation where, in order to be effective, the correction will have to be *stronger* to outweigh not only the barking itself, but *also the pat* the dog receives.

PART 5 – The scientific basis—fact and fallacy

therefore need more and/or stronger punishment to outweigh it.

When we use punishment effectively, we ultimately end up using the least amount of it.

A study by Azrin and Holz agrees:

> "For example, one might be more disposed to supply reinforcement to an individual following administration of punishment, in a sense to try to "make up" for the punishment. If so, the punishment would be (inadvertently) paired with reinforcement and it could be expected to acquire a discriminative property. Thus, the severe punishment would be ineffective as a deterrent. On the other hand, one might be less disposed to supply reinforcing after punishing. A slap on the wrist, a frown, or a shout would appear only mildly aversive—if at all. However, if one were less inclined to reinforce after such mild punishment, then these events would become associated with extinction. In this instance, a trivial degree of punishment would be quite effective as a deterrent of behaviour."[426]

So, although the trainer who sent me the video wasn't rewarding the dog in an attempt to "make up" for the punishment, but was rather following the advice to "reinforce an alternative" (lying down), the effect is the same.

On the other hand, when we are using negative reinforcement to *teach* behaviour (rather than to *stop* behaviour) then rewarding that behaviour is definitely useful (though not essential) and will speed up training. E.g. "here"—we might put pressure of whatever form on the dog to come to us, then when it does we reward it with "good dog" and a pat. Or, if we are teaching a dog to "lie down", we put pressure on it until it does, and then reward it with "good dog" and a pat. But we generally should *not* do this when we're trying to *stop* behaviour.

So rewarding after correcting a behaviour we are trying to stop is very often counterproductive. What the trainer should have done in the above example is to correct the dog for barking, and then gone about their business. If the dog *remained* quiet, *then* they could have gone to it and given it a pat.

Consider two other common examples of where people often "reward an alternative" but simply make the behaviour worse. The first is in jumping up on people. When the dog jumps on them, people will often tell the dog to "sit" and *then* pat it or reward it with a treat. They think they are rewarding the sit (an "incompatible alternative" to jumping). However, we can reward "chains" of

---

[426] Holz, W. C., & Azrin, N. H. (1961). *Discriminative properties of punishment. Journal of the Experimental Analysis of Behavior*, 4(2), 225-230. https://onlinelibrary.wiley.com/doi/pdf/10.1901/jeab.1961.4-225

CHAPTER 14 – Myths about principles of punishment

behaviour—in this case a two-link chain, or a two-step process. (When complicated sequences of events are taught, such as obedience competition routines, trainers often do this deliberately.)

The dog learns that when it jumps on you to get your attention, you will ask it to sit, and pat it. It knows that jumping on you is what got your attention and instigated the process. You are simply rewarding a two-step process instead of rewarding the jumping directly.

Or, take a dog that is jumping on the door (or barking, whining, scratching, etc.,) wanting to be let inside. Obviously, if you open the door for it, you have rewarded that behaviour. But again, many people go to the door, ask the dog to sit, and then let it in. They think they're teaching the dog to sit to be let in. They think they're "rewarding an alternative" or "rewarding an incompatible behaviour". However, once again, they're actually just rewarding a "chain of behaviours". The dog knows full well that jumping on the door is what got you to come over and ask it to sit and let it in. So it does nothing to solve to the problem; in fact, again—as we're seeing for many "positive" approaches—it *makes it worse*.

When we're using punishment to stop behaviour, rewarding an alternative is often counterproductive. The punishment is generally more instructive and accurate on its own.

So we've now debunked Dunbar's Myth #3 about the requirements of using punishment. Punishment does *not* need to be instructive in the sense of telling the dog what to do instead. And, contrary to the "positive" trainer's claims that "don'ts don't work", *don'ts* in fact *do* work, highly effectively.

> Don'ts *do* work.
>
> Highly effectively.

Punishment must have a warning?

Myth #4 states that we should always *warn* our dog before applying punishment. Once again, Dunbar is completely wrong. As is Pryor, who likewise claims that punishment without a warning will have "no predictable effect on future behavior"[427]. In fact, used properly, it has *entirely predictable* and extremely useful effects on future behaviour.

---

[427] Pryor, Karen. *Don't Shoot the Dog!: The New Art of Teaching and Training*. Rev. ed. New York: Bantam Books, 1999.

PART 5 – The scientific basis—fact and fallacy

Certainly, at times, a warning or a command is necessary and/or useful. This is where I teach the use of a "No" command quite a lot to clients. If the dog jumps on them, initially they warn it with "No". This tells the dog to stop whatever it is doing, or there will be a consequence. If the dog doesn't respond, it gets a smack (while in the act of jumping). (Notice that we didn't say "No" and smack at the same time—the dog got *a chance not to get smacked* if it did as was asked.)

> "No" tells the dog to stop whatever it is doing, or there will be a consequence.

This is exactly the same as when a dog (or wolf, for those like Zak George or Dunbar or Stilwell who think dogs aren't like wolves!) growls at a pup—the growl says, "stop whatever you are doing, or you will get bitten". Or when a cat hisses at the pup, and if the pup doesn't desist he will get a swipe across the nose.

This prior warning, "No," is not actually necessary at all to teach a dog not to jump on us, so there is no *necessity* for it. If a pup gets a firm enough smack on the nose when he jumps, he will stop, regardless of whether he received a prior warning or not. In fact, it will actually be *more* effective than if a warning is given, as explained below. However, I like using the "No" because it then becomes trained and useful for a variety of other situations. Very quickly the pup learns to stop what it is doing when it hears the word "No"—whether that is jumping, chewing, digging, chasing the cat, etc., and so it avoids the consequence. It knows there will be a consequence if it doesn't listen.

> When we use punishment properly, we should *not* have to use it very much at all.
>
> Punishment is mentioned a lot in this book, in order to combat the reward-only craziness. And so it can seem like we are constantly punishing the dog.
>
> However, we actually use it *very little* when we use it properly. It does its job *quickly and easily*, and we can stop using it.

And so very quickly we no longer need to smack the pup—it now stops when we say "No". When we use punishment properly, we should *not* have to use it very much at all. Punishment is mentioned a lot in this book, simply because of the necessity of combating the many false claims of the "positive" ideologues. And so it can seem like we are constantly punishing the dog. That is not the case. The truth is, we actually use it *very little* when we use it properly. It does its job *quickly and easily*, and then we can stop using it.

On the other hand, however, by having given the pup this "No" warning, it quickly learns that it has *one chance* (provided we only tell the pup "No" *once* before the consequence, and don't nag it with repeated commands). So the pup learns that it can jump on us, or chew the couch, *until* we tell it "No", when it stops and so avoids the consequence. However, we don't want the pup to

CHAPTER 14 – Myths about principles of punishment

jump on people or chew the couch *at all*. So, after a few instances, once I'm happy the pup has made the connection with the behaviour, the warning disappears. Now if the pup jumps, there will just be an instant consequence, with no prior warning (although I still generally say "No" *as* the punishment happens—rather than before—as this increases the effectiveness of "No" as a conditioned punisher for use at other times). So now the pup cannot risk even jumping *the first time*, and so the behaviour can be completely eradicated. Because the behaviour is now *never* occurring, the *habit* can disappear. And now there is no longer a risk of grandma being knocked over by your big boisterous puppy and breaking her hip.

So, at times, a warning is useful. At other times it's counter-productive and should definitely *not* be used. Karen Pryor fails to understand these basic facts about training when she says that if we use punishment without a warning, for example when teaching a dog to walk on a lead, it will have:

> "…no predictable effect on future behaviour."[428]

Actually, when teaching a dog not to pull on the lead, punishment without warning has entirely predictable effects, and can be more effective than if we use commands (more on the actual methods in future volumes). This is due to much the same reason as we sooner or later need to stop warning if we want to fix a jumping problem as outlined above.

Another good example is in something like killing sheep (or wildlife, etc.). In this case we definitely should NOT give a warning. Otherwise, all we teach the dog is to not chase sheep *while we are around* to give the command. We want the dog to be reliable so that even if it escaped our backyard while we weren't home, it *still* wouldn't chase or kill sheep. So we deliberately make every effort to *completely disconnect ourselves* from the training. In that way, the dog associates the punishment with the sheep, not with us, and so the training is effective even in our absence. The same applies to snake avoidance training. (By the way, don't fall for the trap of those who claim to carry out "gentle" or "positive" avoidance training—it's another "positive" fairytale that *simply doesn't work*.)

Mazur agrees with *the real world reality* that a warning is *not* necessary, and so Dunbar and Pryor again are simply wrong. Mazur points out:

> "…animals can learn avoidance responses when there is no CS [conditioned stimulus—command or warning] to signal an upcoming shock."[429]

---

[428] Pryor, Karen. *Don't Shoot the Dog!: The New Art of Teaching and Training*. Rev. ed. New York: Bantam Books, 1999.
[429] Mazur, James E. *Learning and Behavior, 6th Edition*. New Jersey. Pearson Prentice Hall, 2006.

This is exactly what we are trying to create in the above scenarios—the dog avoids the snake or livestock without needing to be given a command.

Murray Sidman is known for developing what is called "Sidman Avoidance" (and he did this way back in 1953, so this is hardly "new science"), which also proves that animals can learn predictably without a prior command or warning (of course, good animal trainers have always understood this). Seligman's learned helplessness experiments prove the same thing.

So Dunbar and Pryor are incorrect and Myth #4 is debunked—a warning is *not* always required, and certainly should *not* be used in various situations. In others it is useful.

Conclusion - Myths about principles of punishment

In summary, these four points often cited about how punishment should be used (based on "science"!) are either completely incorrect, or only partially true, depending on which version we are looking at. According to Dunbar (the founder of APDT – the *Association of Pet/Professional Dog Trainers*), punishment must be:

1. **Absolutely consistent 24/7.** Completely untrue. Just like speeding fines, punishment can be inconsistent and still be highly effective. That is one of its great strengths.

2. **Immediate.** Mostly true—punishment is generally best applied *in the act*—but we can certainly move beyond this, and in various cases we must, provided certain rules are followed.

3. **Instructive – should reward an alternative and tell the dog what to do instead.** Completely untrue. Often we just want the dog to stop the behaviour, and rewarding an alternative is often counterproductive. "Don'ts" actually *do* work, extremely well.

4. **Have a command or warning beforehand.** Sometimes useful, often counter-productive.

Once again, the so-called "positive" idealogues are far from the truth.

15

A model of obedience—balance scales

As I've mentioned, this isn't a book going into detail about training methods and specifics (even so, there's already far more useful training advice in this book than every "positive" training manual ever written combined). I plan to write a number of further volumes explaining the actual methods in the future. Also, there is a general discussion of training principles in my book *Working Sheep Dogs* for those who are interested, although not specifically dealing with pets.

> There's far more useful training advice in this book than every "positive" training manual ever written combined.

I will, however, outline the model I generally use to explain gaining obedience and control of dogs in the presence of distractions. As we've learnt, treats can work in teaching the *meaning* of commands (although there are often better ways), but where they come up *very* short is in gaining *control*, or *obedience*, to those commands.

This is where Dunbar, George and others are quite misleading, when they incorrectly equate *understanding* with *obedience*. They use examples of school teachers teaching something and say, "you wouldn't punish someone for not understanding algebra, would you?" For example, Zak George says:

329

PART 5 – The scientific basis—fact and fallacy

> "Think about your favourite teacher when you were in school. Let's say you asked her for help with a particular concept or problem. What did she do? Did she yell at you?... Did she slap you on the wrist? No."[430]

Notice how he didn't use an example of a student *misbehaving* and *breaking rules* that they are crystal clear about, like throwing things in class, or punching another student, or just failing to be quiet when asked. Those would actually be applicable examples. This is known as a "straw man" argument, and is entirely misleading. Dunbar likewise says:

> "…it is not necessary to make *failure to learn* a more unpleasant experience than it already is. Maybe the teacher or the teaching contributed to the slow learning but regardless, who on earth would want to frighten or hurt children, pupils, employees and animals for *failing to learn*?"[431] (emphasis added)

No one is advocating punishing students or animals for failing to learn. Dunbar and George (and most other "positive" trainers) fail to distinguish the difference between *understanding* and *motivation*. They fail to distinguish between teaching the *meaning* of commands and getting *control* of them. For example, Dunbar claims:

> "Dog training is almost entirely about *communication*, specifically teaching dogs *our language*, that is, ESL."[432]

This is simply not true. Dog training is equally (indeed even more so) about *motivation* (and has exactly *zero* to do with teaching dogs our language). The dog can understand perfectly well what "sit" (or a whistle, or a hand signal) means, but *still choose not to sit*. That is not a "failure to learn", nor is it a lack of clear communication or a failure to teach our dog English. It is a failure to *obey*, which comes down to a lack of *motivation*. (In fact, motivation is not only fundamental to obtaining *obedience* but also in *learning*. A dog must be *motivated to learn* what we are teaching it—either to gain something

> The dog can understand perfectly what "sit" *means*, but still choose not to sit.
>
> That is not a "failure to *learn*" it is a failure to *obey*.

---

[430] George, Zak, and Dina Roth Port. *Dog Training Revolution: The Complete Guide to Raising the Perfect Pet with Love*. New York: Ten Speed Press, 2016.
[431] Dunbar, Ian. *Barking up the wrong tree – for 110 years?* Dog Star Daily, May 7, 2015. https://www.dogstardaily.com/blogs/dr-ian-dunbar/barking-wrong-tree-%E2%80%94-110-years.
[432] Dunbar, Ian, Dr. *Barking Up The Right Tree - The science and practice of positive dog training*. Novato, CA: New World Library, 2023.

CHAPTER 15 – A model of obedience—balance scales

or to avoid something. They are not "begging for an education" [433] as Dunbar would have us believe—we have to motivate them to want to learn and to want to understand.)

These trainers use this strawman argument because, while "positive" training can be highly effective (sometimes) at teaching the *meaning* of commands (or behaviours), it is singularly *ineffective* and completely impractical at gaining *obedience* to those commands, or in curing bad behaviour. Everyone is familiar with how easy it is to teach your dog to "sit" with treats, and also how that simply does *not* translate into any sort of *obedience* to the command. When we talk about *obedience*, the dog is simply weighing up competing motivations—what we apply (either reward or correction), versus what the environment or intrinsic motivation applies.

So instead of Skinner and his problematic "operant conditioning" theories, I prefer to consider "balance scales", and that dogs simply try to avoid things they don't like and gain things they do. They, like us, simply *weigh up* competing options. It is a simple, *real world* applicable model, that for most general dog training applications I believe works extremely well.

On one side of the balance scale is what the dog wants. And on the other side is what we apply in order to motivate the dog to do what we want.

Figure 18 - Balance scales—weighing up the options

---

[433] Dunbar, Ian, Dr. *Barking Up The Right Tree - The science and practice of positive dog training*. Novato, CA: New World Library, 2023.

PART 5 – The scientific basis—fact and fallacy

For example, the dog wants to chase a rabbit and we want it to come back when called. If we are treat training, there's a rabbit on one side of the balance scales and a treat on the other. The dog weighs it up, and which one wins? Obviously, for many dogs, the rabbit. *It far outweighs the treat.* The only way we can outweigh those environmental rewards is with some form of correction or consequence. We simply cannot carry around a big enough reward. Treat trainers add more weight to their side of the balance scales by utilising *deprivation* and *addiction* to make the treats worth more to the dog, and *stringent and extended restrictions on freedom* in a pointless attempt to remove the environmental rewards on the other side of the balance scales. (Or they falsely claim there is a cumulative effect of reward over time, such as "years of reinforcement history"—Susan Garrett incorrectly likens it to deposits into a bank account building up over time[434]). However, as we've learnt, these measures are rarely sufficient to outweigh the environmental rewards.

> Treat trainers attempt to tip the balance scales in their favour using *deprivation, addiction, stringent and extended restrictions* on their dog's freedom, and *high-value* treats.
>
> Even these measures are rarely sufficient.

Figure 19 – "Positive" attempts at outweighing environmental rewards

Consider, again, speeding fines. If the speeding fine was a paltry $5 most people would ignore it. As the speeding fine increases, at some point we start to pay

---

[434] Garrett, Susan. *Train My Dog to Come When Called.* Susan Garrett's Dog Agility Blog, April 3, 2019. https://susangarrettdogagility.com/2019/04/train-my-dog-to-come-when-called/.

CHAPTER 15 – A model of obedience—balance scales

attention. At $400 we decide we had better drive within the rules. However, some people might be wealthy enough that even a $400 penalty is not enough. So, "demerit" points were invented, where you also lose points off your driver's licence. If you lose enough points, you lose your driver's licence altogether. So the consequence must outweigh the behaviour. If it does, it will effectively control that behaviour. If it doesn't, it won't. It isn't complicated.

Like our misbehaving dance students we heard about earlier. The consequences of sitting out the rest of the dance lesson (and so getting behind and not dancing as well at the ball), along with the threat of being kicked out of the class altogether, were sufficient to immediately stop the behaviour. Did we do that to students who simply weren't understanding the dance steps (like in George's and Dunbar's strawman examples of school students)? Of course not—we would patiently help them understand. But for deliberate disobedience and misbehaviour? You bet. And the consequences were enough to outweigh whatever they were getting out of their misbehaviour.

Consider again the example of Karen Pryor using the spray bottle to cure her dog of getting into the rubbish bin. It worked because the dog disliked the spray in the face *more* than it wanted to get at the rubbish. However, for some dogs it *wouldn't* have worked. For them, the water spray might even have been a fun game, and so then it would have been rewarding the behaviour. What is punishment for one dog might not be for another, or might be an insufficient consequence (like the $5 speeding fine).

> What is punishment for one dog might not be for another, or might be an insufficient consequence (like the $5 speeding fine).

For Pryor's Border Terrier, the punishment outweighed the behaviour, just like the cap gun for Pat Miller's horse chasing Bull Terrier. For other dogs, they might not have. And so a smack with a rolled up newspaper might be necessary for another dog with the same bin diving behaviour, or an e-collar for a different horse-chasing Bull Terrier.

In simple terms, there are two important aspects to training—*understanding*, and *motivation*. **Understanding generally comes down to *timing plus repetition*, and *motivation* to the *level* of the consequence (either reward or punishment).** If the timing is good (generally during the act, either with reward or punishment) and repeated enough for the dog to see the pattern, then the dog will *understand*; and if the level of the consequence (either reward or punishment) outweighs the behaviour then the dog will be *motivated* to change.

> If the timing is good (generally during the act, either with reward or punishment) then the dog will *understand*, and if the level of the consequence (either reward or punishment) outweighs the behaviour, then the dog will be *motivated* to change.

PART 5 – The scientific basis—fact and fallacy

If you give your dog a smack two or three times *while* it is biting you, and that smack is firm enough, then the dog will both *understand* and be *motivated* to stop, and so it *will* stop. It isn't stupid. If it doesn't stop it is because you weren't firm enough; get firmer immediately—increase the level of the consequence.

Of course, Karen Pryor criticises the escalation of punishment—while using sufficient levels of it herself with her Border Terrier. She says:

> "Suppose we have punished a child, or a dog, or an employee, for some behavior, and the behavior occurs again. Do we say, "Hmm, punishment isn't working; let's try something else"? No. We escalate the punishment."[435]

She's saying that we shouldn't escalate the punishment, but that we should "try something else"—she means, of course, positive reinforcement. In fact, escalating the punishment (depending on the situation) is *exactly* what should happen (like speeding fines—if $5 doesn't work, make it $400). What would Pat Miller have done if the cap gun hadn't worked (after all her positive reinforcement approaches failed)? She would have needed to find a stronger punishment. An e-collar set to a high level would have been one effective option.

> The **ICEPIRP** negative reinforcement / punishment process:
>
> **Instant, Continuous, Escalating Pressure**
>
> Then
>
> **Immediate Release** and **Praise**.

In the ICEPIRP negative reinforcement / punishment process that I developed a few years ago for teaching sheepdog training, the E is a vital element—Escalation.

(The entire acronym stands for *Instant, Continuous, Escalating Pressure, then Immediate Release and Praise*. There are instructional videos on my sheepdog training website www.workingsheepdogtraining.com for those who would like to delve into this more deeply. Although it's aimed at sheepdog trainers, exactly the same principles apply to other dogs and animals, including horses.)

Most beginner trainers or owners are *simply not firm enough* and *fail to escalate quickly enough*—they keep using ineffective levels of punishment, and thereby simply desensitise their dog to that ineffective punishment. When punishment is done properly (during the act at an appropriate level for that dog in that situation), it works very quickly and effectively, and we can stop doing it. It's never

---

[435] Pryor, Karen. *Don't Shoot the Dog!: The New Art of Teaching and Training*. Rev. ed. New York: Bantam Books, 1999.

CHAPTER 15 – A model of obedience—balance scales

something we should be using for long—if we are, we're doing something wrong, and we need to change something.

One of the great things about e-collars is that the *level* of the consequence can be *very precisely* controlled. And so, just like speeding fines, we can find the level that works for *that* particular dog in *that* particular circumstance. For a sensitive dog they can be set very low, so low that I can't even feel them. For a tough dog, or a highly-driven dog that is fixated on killing livestock, they can be set much higher. So the consequence must simply *outweigh* how much the dog wants to indulge in the behaviour, just like us with speeding fines. (The other great thing about e-collars is that they make timing very simple—it is much easier to correct various behaviours during the act, particularly at a distance and so on.)

> Punishment is never something we should be using for long—if we are, we're doing something wrong, and we need to change something.

The difference between dogs and humans is that dogs don't understand *abstract* consequences. And so the punishment often has to be a direct physical one of some sort, just like the way dogs (or wolves) work with each other. For some really sensitive dogs, just telling them off in a stern voice can be an effective punishment. Or, possibly, for elephants, when one Karen Pryor was training hit her with it's trunk:

> "When I didn't fall for that, she whopped me on the arm. When Jim and I both yelled at her for that (a sign of disapproval, which elephants respect)..."[436]

Again, we see Pryor instantly resorting to the use of punishment (after her positive reinforcement training created an aggressive response). This is exactly the right thing to do, as I've mentioned, for example, when your pup growls at you when you go near its food, but which "positive" trainers tell us never to do!

So, to one dog, a stern reprimand might be a high enough level of consequence to be effective. And so, some people can rightly claim they never had to resort to *physical* punishment with their dog. However, they were still using punishment. To that dog the stern reprimand was a strong aversive, and therefore effective. But with other dogs, a stern word is like water off a duck's back and will be far from sufficient (unless, perhaps, we have first paired it with another sufficient punishment).

Just like Pryor's two dogs—one very timid, who she could let run free, and one robust, who at seventeen years old Pryor still couldn't let off the lead. She could

---

[436] Pryor, Karen. *Don't Shoot the Dog!: The New Art of Teaching and Training*. Rev. ed. New York: Bantam Books, 1999.

train dolphins to jump through hoops, but not her own dog to come back when called—it isn't a problem of *understanding*, but of *motivation*. The dolphins were in a captive environment where they had to perform in order to eat and had no competing environmental rewards—they were motivated through deprivation and extreme and extended restrictions on their freedom. **And yet Pryor couldn't motivate her own border terrier out in the *real world* even after seventeen years—it isn't a problem of *understanding*, but of *motivation*.**

To give another example of this idea of balance scales, consider the earlier example of a trainer trying to apply the idea of rewarding an alternative behaviour when using punishment. He growled at the dog to "be quiet" as he moved towards it, whereupon it lay down and stopped barking. At that point he immediately patted it. Now, on one side of the balance scales you have the dog's desire to bark, *plus* it gets a pat when it barks. On the other side you have the consequence (being told off). And so now the dog is effectively also factoring that in: "I want to bark, but I get told off, but then again I also get a pat…".

It would be no different than if we received a speeding fine and then got a partial refund. *Now the initial fine would have to increase in order to remain effective.* So, consider what you have on *each side* of the balance scales. If what we apply *outweighs* what the dog wants (and the dog understands what we want), we will achieve obedience. If not, we won't.

Reward good behaviour and correct or punish bad behaviour, with a simple, direct consequence the dog understands that is *sufficient to outweigh* what the dog got out of the undesired behaviour. Get the timing right either way, whether reward or punishment—mostly while the dog is in the act—so that it *connects the consequence with the behaviour*. And before you know it, you will have a well-trained dog.

> Reward good behaviour and correct or punish bad behaviour, with a simple, direct consequence the dog understands, that is *sufficient to outweigh* what the dog got out of the undesired behaviour.
>
> Get the timing right either way, whether reward or punishment – mostly while the dog is in the act – so that it *connects the consequence with the behaviour*.
>
> And before you know it, you will have a well-trained dog.

The principles aren't that difficult. Of course, we can get into a fair bit more nuance and detail on various points, and on their application. The best way to apply those principles for a variety of behaviours, and with different dogs and different owners in different situations, is where experience comes in. I plan to elaborate on the specifics in future volumes.

CHAPTER 15 – A model of obedience—balance scales

For now, just consider balance scales, what is on each side of those scales, and what outweighs what. It really is mostly just some common (or uncommon) sense.

PART 5 – The scientific basis—fact and fallacy

Part 7

THE MORALITY OF PUNISHMENT

Chapter 16 – The morality of punishment

16

The morality of punishment

When we consider that punishment is all around us, it becomes obvious that there is nothing intrinsically morally wrong with punishment. As described earlier, punishment occurs all the time in everyday life. The pup learns not to try to run across green lilies growing on a pond, not to crash into trees, not to jump off a high ledge. Dogs, and us, learn the rules of nature largely through punishment and correction, and also through reward. Is it morally wrong for a dog to punish another dog? Again, of course not. If we train the way nature trains, the way dogs (and wolves) train each other, then punishment will form part of that.

There are a number of pertinent questions to be answered when considering the morality of punishment:

1. Can we train without it?
2. Is there anything morally wrong with using it?
3. Is it really that bad?
4. Do its positives outweigh its negatives?

Most of these questions have already been answered throughout this book. However, I will quickly and directly summarize them again here.

PART 7 – The morality of punishment

Can we train without punishment?

As we've seen throughout this book, we simply cannot train without punishment. We've seen that many of the most vocal advocates of "positive" training grudgingly admit that punishment is sometimes necessary (often while they use it regularly while pretending not to). In fact, it's necessary *far* more than they are willing to admit to their naïve and inexperienced followers.

In the *real world* in a practical sense, punishment of whatever sort or level is *always* necessary. Even assuming you might be able to teach a good recall without it (and remember, Donaldson said you could get a *might/sometimes* recall after a LOT of work), who in the *real world* has time to even attempt that?

Zak George, while claiming there is no science that says punishment is ever necessary, admits his recall methods will not always succeed. I can tell you they will *rarely* succeed, like most of his methods and those of many others. As we've seen, the "positive" trainers have no effective methods for most of the common training requirements.

Everyone uses punishment at some point, even if they don't admit it. If you simply tell your dog off for something, you are using punishment (regardless if you try to rebrand it as a "social act" or "communication" [437] as Pryor does). If you ever correct it, whether vocally or with a lead, say for attempting to get in the wrong seat of the car, you've used punishment. Or, if you're "Doggy Dan" (Daniel Abdelnoor) and use leash corrections, or a front pull harness (as does Victoria Stilwell) or a head halter (like Donaldson or the Karen Pryor Academy), you are indeed using punishment and force.

So we simply cannot train without punishment or force, no matter how much the reward-only ideologues would like it to be so, or pretend it to be so. We live in the *real world*, and the reality is that punishment is all around us and is necessary if we want a well behaved dog that we can give a good standard of life to. The use of punishment is a vital ingredient in true animal welfare. As I said earlier: **Ethics without effectiveness is simply virtue signalling.**

---

[437] Pryor, Karen. *On my mind – Reflections on Animal Behavior and Learning.* Waltham, MA., USA. 2014.

CHAPTER 16 – The morality of punishment

Is there anything morally wrong with using punishment?

In reality, when we understand the facts, as I've mentioned in various places, it is actually morally wrong *not* to use punishment, like it is morally wrong not to discipline your child properly when necessary. They are the ones who end up suffering, as do you, and others around them, when you don't.

Your out of control, spoilt "treat brat" dog (or child) is a menace to itself, and to others. When your poorly trained dog jumps up on an old lady and knocks her to the ground, or gets off the lead and terrorizes someone walking their dog, or kills wildlife, or gets injured or killed itself from chasing cars, or kills your cat, or doesn't get walked or ever let off the lead, all because you were unwilling to use the necessary discipline to quickly and effectively train your dog properly, is that a good thing? Does that make you a good person? No, it does not. A good person is willing to do what is in the best interests of their dog (or child) even if that offends their sensibilities and they find it unpleasant.

> A good person is willing to do what is in the best interests of their dog (or child) even if that offends their sensibilities and they find it unpleasant.

If these things happen during the long year or two or more of labour and time intensive positive-reinforcement training that no one in *the real world* has time for, then these are the fault of the "positive" ideologues who convinced you to follow their failure of an ideology. If, during the time you are attempting ineffective "positive" methods, your dog kills some wildlife or livestock, that is on the "positive" crowd. Sure, you never smacked your dog, but it ripped another animal to pieces with its teeth while it was alive. If your dog attacks another dog during that time, when you could have had it trained properly a long time earlier, that is on you. If your dog gets killed or badly injured chasing a car, that is on you and those who convinced you not to use sound methods of training. If your dog gets bitten by a snake because you bought into the idea of "gentle" or "positive" avoidance training, then your dog's suffering and death is on you, and the reward-only ideologues. So much suffering of the dogs themselves, and other dogs and wildlife and people around them, is directly attributable to the "positive" "influencers".

Demonising the sound use of punishment is what is truly morally indefensible.

PART 7 – The morality of punishment

Is punishment really that bad?

An important question to ask is whether punishment is *really that bad?* The reward-only idealogues try to make out that punishment is the worst thing on earth. And nowadays most people aren't even game to say the word "punishment" in relation to dog (or human) training, due to all the negative connotations it has been given.

Is a few decent smacks on the nose with a rolled up newspaper to cure the aggressive dog that might bite a child on the face and put it in hospital, really *that* bad? I often give myself a good smack on the hand or leg with the newspaper to demonstrate this fact to clients. Sure, it stings a bit, but it's hardly the end of the world like the reward-only people make out.

I periodically touch an electric fence (which are far stronger than an e-collar) around the farm if I get a bit careless when I'm repairing it, and that hurts. But am I traumatised for life? No. It really isn't that bad. If we need to use something like that once or twice to cure a dog of ripping sheep to pieces, is it so terrible and evil? If that saves the lives and trauma of sheep in future being ripped apart and dying a slow, terrifying death, and means the dog can now be given so much more freedom and improved quality of life, is it so bad? And if it saves the dog from being euthanised? And given that I suffer that level (or higher) punishment every now and then from an electric fence on the farm, why the hysterics from the "positive" ideologues?

The same thing happens to the pup in its interactions with other dogs all the time. It pesters them, they growl at it, it doesn't stop, it gets bitten. *Not the end of the world*. The pup simply learns the rules and adjusts it's behaviour accordingly, to avoid getting bitten the next time. Is the pup terribly traumatised? No, not at all. (Not unless it is attacked *for no reason*—then we may have a problem depending on the severity of the attack and the nature of the pup. But then we're talking about *abuse*, not contingent punishment.)

To repeat an earlier example—a client's dog that wanted to kill their cat, and would have, given the chance. They had to be very careful to keep them separated, and both the dog and the cat were living semi-confined as a result. A simple bit of training with an e-collar that consisted of two or three applications, and a month or so later the client sent me a video of the cat and dog playing happily together. They were now great friends. Worth it? I'm sure the cat and the dog thought so. The cat was now safe from severe injury or death, both the dog and cat were now able to enjoy much more freedom, and they got to play with each other. And no "fallout" whatsoever (other than positive). Unfortunately, the "positive" crowd have done a great job with their propaganda of convincing people just how *terribly awful* any

CHAPTER 16 – The morality of punishment

sort of correction or punishment is (it isn't), and of the horrible "fallout" that is going to occur (it won't).

The simple fact is, it isn't really that bad at all. It's a simple and necessary part of life. Even Karen Pryor admits this:

> "Get Real. In real life you have to wade in with an aversive to stop something from happening. If you must yank a baby away from the light socket, stop a dog from grabbing the roast chicken off the table, so be it. Animals do reprimand (the official biologist's term) their young and each other. You'll have interrupted and stopped a dangerous event. Just don't kid yourself that you've taught or guaranteed any particular change for the future." [438]

Pryor is right about the first part—aversives *are* indeed just a part of life, and therefore we should "get real" and stop being so precious about them. However, she's *completely* wrong about the second part—if we use punishment properly it has *entirely* predictable and repeatable results. And, we actually get dogs (and children) trained and problems fixed, quickly and effectively.

Punishment, when used properly, really isn't that bad.

Do punishment's positives outweigh its negatives?

To answer this, we first need to remember there are no negatives when used properly (excluding the punishment itself—and "*get real*"—it really *isn't that bad*). It doesn't cause unwanted side-effects; in fact, the side-effects are all positive.

Given the facts that in the *real world* we simply cannot train without punishment (certainly not if we want to give our dogs *any* measure of freedom, and not just "manage" them like the *American Veterinary Society of Animal Behaviour* would have us do); and that it's far more effective, quicker, and easier than so-called "positive" training; that there are no unwanted side-effects when we use it properly; and that our well-trained dog is able to be given a much better quality of life, it's easy to see the answer. Dunbar says:

> "Why are speed and ease of training so important? Well, if dog training techniques are too complicated, requiring a masters in psychology or exacting leash dexterity, they will be well beyond the means and skillset of most owners. If dog training techniques are excessively time-consuming,

---

[438] Pryor, Karen. *On my mind – Reflections on Animal Behavior and Learning.* Waltham, MA., USA. 2014.

requiring the patience of Job, many people might not devote sufficient time."[439]

On this occasion he's dead right. Except, it's the "positive" methods that are incredibly convoluted and complicated, "requiring a masters in psychology" as we have seen, and that are "well beyond the means and skillset of most owners." And the "positive" methods are also the ones that are "excessively time-consuming, requiring the patience of Job," and therefore virtually no pet owner will ever have enough time to devote to them. And all that is assuming, of course, that even after all those criteria are met, that the "positive" methods actually get results (they rarely do).

As Dunbar says, speed and ease of training are highly important. For this reason alone, punishment is a far better option when training in many cases than is reward. We simply have to use the right tool for the job.

Punishment's positives *far* outweigh any negatives, real or imagined.

Conclusion – The morality of punishment

So when we consider the morality of punishment, we have to weigh up the pros and cons. We need to consider at least the following points:

1. So-called "positive" training fails practically in all the most important areas of dog training, for nearly all dog owners. So it's really a pointless discussion anyway.

2. Even if "positive" training actually appears to work with a tiny percentage of dogs in a tiny percentage of cases, it still requires great skill, time, and effort, none of which most dog owners have.

3. So-called "positive" training, to have any chance of being successful, must make use of the *Big Three Necessities of Positive Reinforcement Coercion*: Deprivation, Addiction, and Extended Stringent Restriction of Freedom, *for life* with most dogs.

4. Punishment is far faster, easier, requires less skill, and is far more effective.

---

[439] Dunbar, Ian, Dr. *Barking Up The Right Tree - The science and practice of positive dog training*. Novato, CA: New World Library, 2023.

CHAPTER 16 – The morality of punishment

5. Punishment, when used correctly, has no long-term undesirable side-effects, and many positive side-effects.

6. Punishment, when used correctly, is not used very much at all.

7. A well-trained dog, trained utilising correction-based methods when appropriate, can be given a *lot* more freedom a *lot* sooner; is much *safer* to itself and others; is more likely to be taken on *more walks*, more often; is less likely to be rehomed or euthanised, and so on.

8. The welfare of livestock, wildlife, other dogs, other pets, the owners, and other people also need to be taken into consideration—these are all drastically improved with a well-trained dog.

Given all these facts, the answer is simple. It's the reward-only ideologues who have caused, and are increasingly causing, a huge amount of suffering in the dog world (as they have done and are increasingly doing in the human world). Responsible dog owners (or parents, or school teachers) will make the effort to understand and use correction and punishment as a necessary component of training, and avoid reward where it is inappropriate or counter-productive.

Only the irresponsible, ignorant, inexperienced, or ideologically driven will demonise the use of punishment.

Avoiding punishment for some misguided, utopian fantasy, based on some junk pseudo-science (or nice sounding marketing strategies), is what is morally repugnant and irresponsible.

PART 7 – The morality of punishment

CONCLUSION

REAL WORLD DOG TRAINING –

Why **CORRECTION SUCCEEDS** and **"POSITIVE" TRAINING FAILS**

Conclusion

Real World Dog Training—

Why correction succeeds and "positive" training fails

In final conclusion, I've attempted throughout this book to debunk the many lies and myths surrounding the use of punishment, and on the other hand to expose the ineffectiveness and downsides of the extreme "positive" dog training cult. For the dog's (and their owner's) sake (and children's, parent's, student's and teacher's), we *must* debunk these myths, and learn the truth.

We need to put an end to the dark ages of so-called "positive" training.

> We need to put an end to the dark ages of so-called "positive" training.

We've seen that the common myths, lies, and propaganda about punishment are completely false. We've seen that not only does punishment work, but that it's far faster, easier, and works far more effectively in *the real world* than "positive" methods ever can.

Much of the evidence for this has come from the very mouths of some of the top, most influential figures in the "positive" cult. People like Dr. Ian Dunbar, Karen Pryor, Jean Donaldson, Zak George, Victoria Stilwell, Pat Miller, and others. We've seen their own statements on the effectiveness of punishment like these:

CONCLUSION

> "My parents punished me exactly twice in my whole upbringing... both behaviours stopped instantly."[440]

> "I gritted my teeth and sprayed the dog in the face... She stayed away from the wastebaskets from then on."[441]

> "Just as he ducked under the wire into the pasture, I fired... It was the first time we had succeeded in stopping him in mid-charge... Twice was enough to convince Happy to leave horses alone forever."[442]

The evidence also comes from all around us every day, in the interactions of dogs with other dogs (or wolves with other wolves). A growl means "no". If the other dog ignores that, there is a consequence—they will get bitten. It is how dogs and wolves have worked for who knows how long, and all without negative side-effects. There is *absolutely nothing natural* about so-called "positive" training.

We see the evidence in the *real* science, and in a huge amount of *real world* experience.

> "The question of whether punishment can be an effective way of controlling behaviour *has been settled*. This chapter has presented numerous studies, from both inside and outside the laboratory, that have demonstrated that punishment can change behaviour and, in many cases, *change it permanently*."[443] (emphasis added)

We've seen that punishment, when used properly, does *not* cause any negative side effects. In fact, there is a *much greater likelihood of positive results*.

> "The **lack of undesirable side effects** associated with the use of punishment has also been noted in the applied literature (e.g., Brantner & Doherty, 1983; Harris, 1985; Johnston, 1972; van Oorsouw et al., 2008). Indeed, the use of punishment-based interventions typically has been *related to* **increases in positive behavior** (e.g., Bostow & Bailey, 1969; Firestone, 1976; van Oorsouw et al., 2008; Risley, 1968). For example, Matson and Taras (1989) reviewed **382 applied studies** employing different punishment procedures during interventions with individuals with developmental disabilities and concluded that the results reviewed *did not*

---

[440] Pryor, Karen. *Don't Shoot the Dog!: The New Art of Teaching and Training*. Rev. ed. New York: Bantam Books, 1999.
[441] As above.
[442] Miller, Pat. *The Power of Positive Dog Training*. 2nd ed. Hoboken, NJ: Howell Book House, 2008.
[443] Mazur, James E. *Learning and Behavior, 6th Edition*. New Jersey. Pearson Prentice Hall, 2006.

CONCLUSION

*provide evidence supporting the occurrence of undesirable side effects. Instead, **the majority (93%) reported positive side effects** during punishment interventions, such as increases in social behavior and responsiveness to the environment."*[444] (emphasis added)

You'll have a calmer, less anxious dog (or child) that bonds more strongly to you (despite the claims to the contrary). Punishment can even be used highly effectively to quickly cure fear and anxiety (when done properly), and result in a calmer and much more confident animal.

We've also seen that punishment doesn't cause aggression, again contrary to the claims (unless badly misused). In fact, it is often a *very fast and effective cure for it*.

> "Aversive shocks are known to produce aggression when the shocks *are not dependant on behaviour* and to suppress behaviour when the shocks are arranged as a dependant punisher...*these results show that aggression is eliminated by direct punishment of the aggression...*"[445]

On the other hand, aggression and learned helplessness were side-effects that even Karen Pryor created on numerous occasions, while sometimes reacting to the aggression immediately (and correctly) with punishment, while advising against its use and claiming it "almost never really works."

Sure, if you *abuse* your dog (or child) you will create problems. But if you follow a few simple rules (only correct while the dog is *in the act*, and being sufficiently *firm enough*, being the main ones), so that the dog connects the behaviour to the consequence (a contingent punisher), then no problems will ever arise. The dog has complete power over the outcome, and the "side-effects" will be all positive. There will be no "learned helplessness" (contrary to Zak George's and other's false claims), unless the punishment is continuous and *non*-contingent, or, in other words, unless it isn't connected to (or contingent upon) behaviour.

For us, we know "don't speed, and you won't get a fine". It's our choice. For the dog, "don't steal food off the kitchen bench, and you won't get a smack". It's their choice. Yes, non-contingent (random) punishment (abuse) certainly will cause problems. That is not what any good trainer is

> Yes, non-contingent (random) punishment (abuse) certainly will cause problems. That is not what any good trainer is advocating.

---

[444] Fontes, R. M., & Shahan, T. A. (2021). *Punishment and its putative fallout: A reappraisal.* Journal of the Experimental Analysis of Behavior, 115(1), 185–203.
[445] Azrin, N. H., Ulrich, R., Wolfe, M., & Dulaney, S. (1969). *Punishment of shock-induced aggression.* Journal of Experimental Analysis of Behavior, 12(6), 1009–1015.

CONCLUSION

advocating. But the "positive" influencers like to twist and distort the truth, or they simply do not understand basic psychology.

And keep in mind that people are far more likely to abuse their dogs (or their children) when they get frustrated and angry, after all the "positive" methods they've been told to use by the "experts" *simply haven't worked*. Far better to teach owners (and parents) to use a little bit of correction and punishment early, when they can apply it calmly and deliberately, than to let things get out of hand due to a lack of some simple discipline.

> "Besides misrepresenting and confusing the facts, excessive moralizing about the use of punishment and other aversive training procedures may have a very undesirable effect on the dog-owning public, **making responsible dog owners feel guilty about exercising the necessary aversive prerogatives needed to establish constructive limits and boundaries over a dog's behaviour.** Many of the basic facts of life that all dogs must learn to accept (if they are to become successful and welcome companions) are won through the mediation of directive training, combining a balanced application of behaviour modification—not just positive reinforcement. Instead of grinding away on a very dull ax [positive reinforcement], *a dog's welfare is better served by teaching the owner when punishment is necessary and how to use it effectively and humanely*."[446] (emphasis added)

We've seen, that despite claims to the contrary, punishment does *not* have to be "100% consistent 24/7" as Dunbar and others falsely claim. It is some of the intrinsic aspects of "positive" methods such as "extinction" or having *"perfect control of the environment"* that have to be 100% consistent 24/7, 365 days a year—and not only from you, but also from your kids, your guests, and random people down the street—something that Dunbar rightly points out is completely impossible in the *real world*, and this is why such methods *simply do not work*.

Indeed, we've seen the fact that punishment does *not* have to be consistent (just remember speeding fines) is actually one of its greatest strengths.

We've seen that while reward can be useful for teaching the *meaning* of commands, in most cases it fails to get *obedience* to those commands under distraction. Sure, your dog will sit while you have treats and there are no distractions, but when the guests arrive, or any other type of distraction, it looks very different. Just like Pryor's terrier who, at 17 years of age, she still couldn't let off lead. Donaldson, for all her vitriol aimed at trainers who use negative

---

[446] Lindsay, Steven R. *Handbook of Applied Dog Behavior and Training, Volume 1: Adaptation and Learning.* Ames, IA: Blackwell Science, 2000.

CONCLUSION

reinforcement or punishment, likewise admitted that positive reinforcement could well fail, and that she herself would, in fact, use negative reinforcement.

And we've seen *why* "positive" training fails—because the environment (and all this entails) is also training our dog very effectively. As Jean Donaldson said:

> "The environment is training the dog all the time, expertly."[447]

The competing "environmental rewards" are rewarding all of our dog's undesirable behaviours and disobedience. So reward training is working—just in the wrong direction! And the environment is using far *higher value* rewards than we can ever carry around in our pocket, on a variable schedule of reinforcement of "incredible power"[448]. Your dog is simply *weighing up* "treat" versus "chase the cat", "play with other dogs", "attack other dogs", "chase cars", and so on and so on. We just have *competing* rewards.

We also seen the many downsides of "positive" training (apart from its ineffectiveness and complete impracticality). We learnt about what I call the *Big Three Necessities of Positive Reinforcement Coercion*: Deprivation, Addiction, and Extended Stringent Restriction of Freedom.

We've learnt about the strategy known as "deprivation" or "closing the economy", where you simply stop feeding your dog meals, thereby keeping it in a constant state of hunger, which is *necessary* when relying on positive reinforcement training. We learnt that this means you are *not* using positive reinforcement at all—you are using negative reinforcement. The aversive component (hunger) is simply out of sight. You also heard from Dunbar—in a bizarre and feeble attempt to justify this—that if you feed your dog an actual meal then you are, in fact, causing your dog terrible psychological abuse!

Addiction was another *unavoidable* and *necessary* downside to reward. Variable reinforcement schedules are at the very core of reward-based training—and they're the very psychology of addiction. **Get your dog addicted enough, dependant enough, and you might just be able to control it.**

Excessive, almost absolute, constraint on your dog's freedom is another significant *necessary* and *unavoidable* downside. We saw that if you want "positive" methods to work, you have to control *every aspect of your dog's life and environment* for an extended period of time—probably for its entire life. As Pat Miller said:

---

[447] Donaldson, Jean. *The Culture Clash*. 2nd ed. Berkeley, CA: James & Kenneth Publishers, 2005.
[448] Pryor, Karen. *On my mind – Reflections on Animal Behavior and Learning.* Waltham, MA., USA. 2014.

CONCLUSION

> "All you have to do is figure out how to *prevent your dog from being rewarded for the behaviours you don't want* and rewarded consistently and generously for the behaviours you do want." (emphasis added)[449]

As Zak George suggests, you need to keep your dog on a lead not only outside, but also *inside* the house, or put it in a crate. If you're a competition dog trainer (or a YouTube influencer) whose entire life revolves around training your dog, maybe you can even make this work (with some dogs). For everyone else, everyone with a job, kids, spouse, hobbies, and a life, it is never going to happen. Not in *the real world*.

> If you want "positive" training to work, you will HAVE to rely on these
>
> *Big Three Necessities of Positive Reinforcement Coercion* –
>
> Deprivation, Addiction, and Extended Stringent Restriction on Freedom.

So, if you want "positive" training to have *any* chance of working, you will HAVE to rely on these **Big Three Necessities of Positive Reinforcement Coercion – Deprivation, Addiction, and Extended Stringent Restriction of Freedom**.

And, instead of direct, clear communication, "don't do that", the "positive" trainers have to invent very devious, roundabout, "cunning" protocols of psychological manipulation in order to get what they want. While saying "we don't force our dogs to obey", they simply go about *attempting* to gain obedience in much more manipulative, time consuming, ineffective ways. Especially for kids and people this is incredibly dishonest and leads to various problems.

And we've seen from the mouths of the best known "positive" trainers that training with positive reinforcement is a *very* long, slow, tedious process.

> "Leash training usually takes extensive follow-through for a solid year at least or you risk regression!"[450]

> "Are you beginning to get a feel for the enormity of reinforcement that is necessary to build a recall?"[451]

---

[449] Miller, Pat. *The Power of Positive Dog Training*. 2nd ed. Hoboken, NJ: Howell Book House, 2008.

[450] George, Zak, and Dina Roth Port. *Zak George's Guide to a Well-Behaved Dog: Proven Solutions to the Most Common Training Problems for All Ages, Breeds, and Mixes*. New York: Ten Speed Press, 2019.

[451] Killion, Jane. *When Pigs Fly!: Training Success with Impossible Dogs*. Wenatchee, WA: Dogwise Publishing, 2007.

> "If you are going to take your dog to a park or trial and let him off leash, *you may need literally years of substantial reinforcement history of coming when called."* [452] (emphasis added)

> "Is it all too much? Sometimes a dog owner will be *horrified at what is actually required* to proof their training effectively, wondering how they can possibly be expected to go to all that trouble." [453] (emphasis added)

And as Ian Dunbar stated:

> "…if dog training techniques are too complicated, requiring a masters in psychology or exacting leash dexterity, they will be well beyond the means and skillset of most owners. If dog training techniques are excessively time-consuming, requiring the patience of Job, many people might not devote sufficient time."[454]

He was trying to suggest, of course, that his "positive" methods are faster and easier. Nothing could be further from the truth.

And we've seen that after all this extended time and effort following complicated training protocols, many of the techniques may simply *not work*, as they admit.

> "then [after all this work] there's *some chance* it [the recall] might kick in, in a *real situation*. Mastering this exercise *doesn't produce* a squirrel-proof recall on every dog. But it does *sometimes*." (emphasis added) [455]

And what happens if, during all these long, drawn-out processes, your dog gets killed because you didn't train it effectively? Or it attacks another dog? Or you can't walk it for a year or more while you are trying to train it? Or it kills the cat? All for the sake of a simple bit of discipline, that, as demonstrated, *really isn't that bad*. Is a smack on the nose really the end of the world? Dogs interact with each other like that on a regular basis, only with teeth! I can tell you from experience that a smack is less painful.

---

[452] Killion, Jane. *When Pigs Fly!: Training Success with Impossible Dogs*. Wenatchee, WA: Dogwise Publishing, 2007.
[453] Mattinson, Pippa. *Total Recall: Perfect Response Training for Puppies and Adult Dogs*. Shrewsbury: Quiller Publishing, 2013 reprint.
[454] Dunbar, Ian, Dr. *Barking Up The Right Tree - The science and practice of positive dog training*. Novato, CA: New World Library, 2023.
[455] Donaldson, Jean. *The Culture Clash*. 2nd ed. Berkeley, CA: James & Kenneth Publishers, 2005.

CONCLUSION

Even higher-level punishment for livestock killers or high-level aggression—is it really that terrible? I get zapped by electric fences while repairing them on the farm occasionally. Sure, it's not pleasant, but it really isn't that bad.

People in modern times have become incredibly soft and sensitive, but the end result is the negative welfare outcomes of dogs having a much poorer standard of living, or being euthanised.

(Indeed, some people seem to prefer that—they would rather get their dog put down than use effective discipline. I suppose at least then they can claim "I've never smacked a dog!")

If you've been attempting to train with "positive" methods, trainers will often try to make you feel like it's *your* fault—*you* haven't been consistent enough, *you* don't put enough time into it, *you* haven't controlled your dog's environment enough, etc.. The fact is, it isn't your fault. The methods simply don't work, or *at best* are completely impractical.

> The fact is, it isn't your fault. The methods simply don't work, or *at best* are completely impractical.

Karen Pryor, one of the world's most influential "positive reinforcement" trainers, *could not teach her dog to recall in the presence of distractions.* **Why on earth would anyone else even bother attempting it?**

It really is a great scam—the vast majority of pet owners are going to give up on the methods long before the years of work they are advised it will take are up, and so they will assume that they simply didn't stick to it long enough. And "positive" trainers who know no better will assume the same. But it can make long-term clients!

There is a reason for the following statement from Zak George:

> "It's important to find an authentic positive trainer. Almost all dog trainers claim to be positive trainers, but most of them are not. Their general game plan may be to pay lip service to positive training methods, but if they don't get instant results, they'll often view this as justification to use an overabundance of force in lieu of real teaching."[456]

He's correct to a point. Most (though certainly not "almost all") dog trainers nowadays do follow the fashion, and claim to be positive trainers. But then, when they try the methods they learnt in their "positive" certified dog training course, or read in a "positive" dog training book, or learnt on YouTube, they find that they

---

[456] George, Zak, and Dina Roth Port. *Dog Training Revolution: The Complete Guide to Raising the Perfect Pet with Love.* New York: Ten Speed Press, 2016.

CONCLUSION

simply don't work in *the real world*. And so, not surprisingly, they try things that *do* actually work. Remember the definition of insanity—"doing the same thing over and over again expecting a different result"?

They might feel guilty about it, they possibly think that they are doing something wrong and that it's them that are at fault, not the methods. But it isn't. *Stop feeling guilty.* It's the whole *ideology* which is at fault.

So-called "positive" training simply *does not* work in the real world.

If these "qualified" dog trainers can't make these methods work, maybe there actually is something wrong with the methods! They either knowingly or unknowingly just keep peddling the same "positive" myths, failing to get results even after extended periods of helping dog owners, and taking their money.

> *Stop feeling guilty* about using punishment.
>
> It's the entire "positive" ideology that is actually at fault.

As Jamie Dunbar noted:

> "When you use effective dog training techniques, your dog's behavior improves. If you've been trying to train your dog, and your dog's behavior hasn't improved, it's not your fault. **It means you've been using the wrong methods.** And that's not your fault either. **The world is full of self-proclaimed dog trainers who promote methods that sound reasonable, but they just don't work. Again—the proof is in the pudding.** If your dog's behavior improves, the training methods are good. If your dog's behavior doesn't improve, the methods are not good."[457]

He's dead right. Of course, he's claiming that it's the *"positive"* methods that are effective and get results. If they did, we would all use them, and those claiming to be "positive" trainers wouldn't feel the need to resort to any other methods. And Karen Pryor would be able to let her terrier off the lead, and Pat Miller wouldn't have needed to resort to a cap gun to train Happy, and Jean Donaldson and Zak George and Pippa Mattinson and Jane Killion wouldn't have recognized that even with all of their elaborate, laborious techniques your dog still might not come back when called.

So, contrary to George's statement, if you want results (ie. you actually want your dog trained and its behavioural problems solved), you certainly should *not* find a

---

[457] Dunbar, Ian. *Six Simple Steps to Solve Your Dog's Behavior Problems*. Dunbar Academy. Accessed Feb 2024. https://www.dunbaracademy.com/courses/six-simple-steps.

CONCLUSION

"positive" trainer, and definitely not a "veterinary behaviourist" (who will likely also have no effective solutions and will just want to drug your dog into zombie land). Rather, you should find a trainer who actually understands and is experienced in the use of all dog training principles—*especially* correction and punishment.

> Find a trainer who *understands* and is *experienced* in the use of all dog training principles—including *especially* correction and punishment.

We've also seen that because of the ineffectiveness and impracticality of "positive" training methods, it simply results in far more dogs being surrendered or euthanised. Or of having a much poorer quality of life because their owners can't walk them, or ever let them off lead, or take them anywhere.

On the other hand, responsible dog owners will *train their dogs properly*. I have saved countless dogs from these fates by quickly and effectively—and with only positive side effects—curing their behavioural problems.

I challenge any "positive" or "force free" "trainer" to back up their claims in a real world test. Whether that is Zak George, who claims he has a faster and better way; or Susan Garrett, who claims her methods are the fastest and most effective; or Karen Pryor, who claims punishment hardly ever works and likewise claims her methods are "more effective in every meaningful dimension"; or Ian Dunbar, who claims rewards are so fast and easy they should be mandatory, and that "lure-reward training is always so much more effective, especially for eliminating noncompliance and misbehaviour"[458], and who, when it comes to punishment says "I never recommend its use"[459]; or Jamie Dunbar, who tells us "if it works the methods are good"; or any other "positive" trainer on the planet. And whether that test is recall, or lead aggression/reactivity, or predatory aggression, or play biting, or stealing food off the bench, and so on. And I will challenge them to a very large wager. And I want all the authors of the studies claiming that "positive reinforcement is better" to be present, and to apologise to all the dog owners and dog trainers they have misled, and to all the dead dogs their methods and claims are responsible for.

For the rest of us, *a judicious mix of reward and punishment* is the responsible, effective option.

Pippa Mattinson described it well when she criticized "the old days":

---

[458] Dunbar, Ian, Dr. *Barking Up The Right Tree - The science and practice of positive dog training*. Novato, CA: New World Library, 2023.
[459] As above.

CONCLUSION

"In the old days, training methods were fairly straightforward. The dog got a pat on the head when he was good, and a clip round the ear when he was bad."[460]

As we've seen throughout this book, this is *still* basically the best approach. Reward good behaviour and punish bad behaviour. The old trainers knew what they were doing. Here's looking forward to a future with a lot more happy, well-trained dogs, and happy owners. And looking forward to an end to the insanity of the "positive" ideological cult.

We need to push back against this "positive" madness before legislation spreads further in forcing the use of methods that *simply don't work,* and outlaws those that do. We need to end the *dark ages* of "positive" training, in both animals and humans.

> We need to push back against this "positive" madness before legislation spreads further in forcing the use of methods that *simply don't work.*
>
> We need to end the dark ages of reward-only training.

THE POSITIVE REINFORCEMENT EXPERIMENT IS OVER.

IT HAS FAILED.

---

[460] Mattinson, Pippa. *Total Recall: Perfect Response Training for Puppies and Adult Dogs.* Shrewsbury: Quiller Publishing, 2013 reprint.

APPENDIX – Some client testimonials

APPENDIX

SOME CLIENT TESTIMONIALS

...his rates are peanuts compared to what he delivers and what's out there!

After paying hundreds of dollars for another trainer to come out to my home for 1 hour, I had no improvement whatsoever. Then 1 visit to Tully & not a minute later all our troubles with pulling and walking on the lead were over! It's been weeks now and my dog has not faltered. Tully has given me the tools and knowledge to enjoy my dog as a pet, and now we are working on sheep so he will soon become a trusty work mate too. Thank you Tully for giving Buddy & I the chance to understand one another and develop a great partnership. I highly recommend Tully to help you train your dog (for work or as a pet). His rates are peanuts compared to what he delivers and what's out there! Thanks again.

Anita

Wow... I am gobsmacked!! Thankyou!

Wow, I went home on my lunch break and went outside to see Boof and usually he is so excited and jumps all over me and he just stood in front of me! I am gobsmacked!! Thankyou!

Bronwyn Irving

...couldn't believe my eyes

So grateful! After months of training with positive reinforcement with another trainer without any result, Max our one year old puppy walked on the lead within 2 minutes without any pulling thanks to Bendigo Dog Training. Couldn't believe my eyes. He is a different puppy on the lead. Our walks are now enjoyable instead of frustrating. Worth every dollar!!! Thanks.

Christa Goodall

APPENDIX – Some client testimonials

...a wonderful experience

What a wonderful experience we had with Tully and our puppy Daisy! Tully taught us in a respectful and knowledgeable way how to train our pup. Tully clearly has a wealth of experience and his strategies for the most simple commands through to more complicated tasks were spot on. We now have a delightful puppy—we couldn't have done it without him! Thank you so much Tully.

Adrian & Erica Carr

Honestly amazing! I cannot recommend Tully enough.

My partner and I honestly can't thank Tully enough for his expertise, knowledge and support in our private dog training lessons. So far we have had 2 private lessons for our 14 month old purebred Labrador, Argo. Argo's largest problem was pulling severely on the lead. We honestly cannot believe the incredible improvement Argo has made since seeing Tully. Today (after just 2 weeks) I was able to take Argo for a walk with nearly NO PULLING! This is honestly amazing! Tully, we cannot thank you enough for all that you've done already and we cannot wait to see what future sessions entail. Thank you thank you. For anyone looking for dog training I cannot recommend Tully enough. His knowledge, tips/tricks & experience is exceptional!

Jemma K. Flanagan

...learnt so much in such a short time!

Hi Tully, thank you so much for our lesson last Thursday. I thoroughly enjoyed it and learnt so much in such a short time!

Melissa Harper

...delighted with the results

Bendigo Dog Training came highly recommended to us and we were delighted with the results. The techniques that Tully taught us for handling our dog were invaluable.

Sophia & Justin White

...like watching a miracle unfold

Hi Tully, I just wanted to thank you for demonstrating how to properly walk/lead a dog (especially a big eager heeler pup) on a lead. It was like watching a miracle unfold. It has given me the confidence to give it a really good go... Thanks again.

Simone

APPENDIX – Some client testimonials

...can't speak highly enough

Belinda & I enlisted the help of Bendigo Dog Training for our 12 month old Hungarian Viszla. After the very first visit Memphis was a completely different dog, so calm and relaxed. Tully's training methods are very calm but assertive to ensure the dog knows exactly what is wanted when told. He also instills in the owners that we can do it. With time, patience & practice it will come. We can't speak highly enough of Tully & Bendigo Dog Training.

David Ruffell

...practical, simple and easy

A friend recommended Tully to us after they had a "breakthrough" with their dog. I was keen to see what help Tully could provide us—we didn't have any "major" problems—but we knew there were a few areas where our dog could improve. Tully's advice was very practical and simple & easy enough for us to follow through with. We noticed a change in our dog straight away—walking her on the lead soon became a breeze! Thanks Tully.

Bec

...very clear with his instructions

I am thrilled to be having lessons with Tully. I have made progress with my Border Collie in a matter of months, after having years of frustration. Tully is very clear with his instructions, ensuring I understand the methods/exercises correctly. I finally feel that being able to work sheep effectively with my dog is achievable.

Louise

Simply brilliant!

At the start of 2016, I decided that I wanted to have a try at sheepdog work. Both sporting and practical. I had literally no idea where to begin. I purchased a dog, and was extremely lucky when the dog's seller recommended Tully for some lessons. From having no idea where to start, to managing to muster my sheep into a yard in 4 lessons was astounding! I have found my lessons with Tully to be fun, precise, and he has the ability to find a way to explain exactly what the handler needs to understand.   Simply brilliant!

Melissa Harper

Highly recommended

I highly recommend Tully Williams if you are having trouble with your working dogs, or are just wanting to improve your own training skills. Tully seems to have an exceptional talent when it comes to training dogs, and when training people to train dogs. A great setup which makes learning easy and fun.

Tarsha Walters

APPENDIX – Some client testimonials

...cannot recommend Tully's services highly enough!!

We cannot thank Tully enough for the help he has given us with our 1 year old Great Dane. We had a lot of problems with her constantly barking at our neighbours kids and at us, and after just 2 sessions with Tully, her barking has completely stopped. We are blown away by this result and cannot recommend Tully's services highly enough!!

Kat Young

...he's worked absolute wonders

Tully is wonderful, he's worked absolute wonders with our boy & us. We couldn't have come this far without his help & advice. Thank you once again Tully, we would recommend you to anyone

Carolyn Harrington

I highly recommend!!

He has helped me a lot. And will be getting Tully Williams back to do more. I highly recommend!!

Amy Evans

Don't put it off...you won't be disappointed

Can't thank Tully enough! We decided to try the New Puppy Consult & highly recommend anyone with a new puppy to give it a go! In 1 hour Tully had stopped her from trying to get in the house, shown us how to walk her properly as she wasn't keen on the lead (now loves her walk) stopped her from jumping. I was amazed at the difference it made to our gorgeous girl in just 1 hour. Don't put it off give it a go as you won't be disappointed.

Jodi Lewis

Still can't quite get over the change his training has made to her and us!

We adopted a very boisterous Staghound cross, and as first time dog owners with two indoor cats we were completely out of our depth. After just a few sessions with Tully our 'girl' is toilet trained, no longer chases the cats, walks on a lead, responds to commands and is generally a well-disciplined dog. We still can't quite get over the change his training has made to her and us! Tully has a relaxed approach and gives very clear and easy to understand instruction. If you follow his advice you can't go wrong. Our home has gone from a place of chaos to peace and calm again. We highly recommend Tully and the service he provides.

Helen Kitching

APPENDIX – Some client testimonials

We would recommend you to anyone

A huge thank you to Tully for helping us with our little boy. He has turned a little mischievous German Shepherd into a more calmer boy. Tully has a lot of patience especially when it comes to showing us owners how to be more confident in training. Tully has helped us out a lot & we must admit we really needed his help. Thank you once again Tully we would have been lost without all your help, we would recommend you to anyone.

Carolyn Harrington

2 sessions…and the fighting has stopped

2 sessions and our 2 dogs are so much improved. The fighting has stopped & the humans in the house are much more relaxed and in control. Simple & effective. Thanks so much Tully!

Fleur Hastings

…amazing knowledge and experience…we strongly recommend you to anyone

Thank you so much for sharing your amazing knowledge and experience Tully. We really appreciate you taking the time to train us so we can successfully train our dog. We are also really grateful that you travelled out to Heathcote. We strongly recommend you to anyone who needs help with their fur baby! 10/10

Meaghan Lee

Best money you will ever spend!

It's the best money you will ever spend!

Renee Dawson

Highly recommend Tully's service

Highly recommend Tully's service. We have a French Bulldog that was continually bringing rocks inside to chew and we were worried he'd swallow one and we'd end up with a very sick dog and a huge vet bill. We sought the services of Tully and after two sessions we haven't had a rock inside since.

Carl Schubert

…Tully is the man to get

Tully is the man to get. Highly recommended.

Lavena Derham

APPENDIX – Some client testimonials

Allows you to make an informed choice of training method

Great range of info and training techniques. No hard sells or push in any direction. Allows you to make an informed choice of training method.

Elizabeth McGovan

...he was awesome!

We used Tully for a little refresher when we adopted our 9 month old boxer Otis—he was awesome!

Lesley Shubring

...best thing we ever did

Tully is wonderful, he comes to your home and teaches you how to teach your furbaby in your own environment... best thing we ever did.

Michelle Lougoon

...attention to detail is second to none

A very special thanks to Tully for teaching "us" how to stop our special doggie Lola to stop pulling on the lead when we walk her, and also to stop jumping on people. His attention to detail was second to none, and we now have a doggie that is very well behaved. I would highly recommend him to anyone.

Michael Foord

...blown away by the difference in our dogs

We cannot express enough the gratitude we feel for Tully. Our two dogs were noisy, anxious and disruptive. Essentially real ankle biters. After only a few visits, they are now quiet, calm and respectful. Our household is now more peaceful and relaxed. We no longer fear the sound of the doorbell because of the barking and aggression that occurred. We also no longer fear letting them outside as the constant barking caused problems with the neighbours. Now they are quiet. It was just such an incredible change in such a short time, and we are just so blown away by the difference in our dogs and the mood of the household. Thank you so much Tully.

Wendy Richards

...shocked and thrilled how well it is working

Hi Tully, I wanted to say thank you so much for your visit last night, and the wealth of knowledge you shared. Tim and I learnt so much of what we were seeking, as well as a lot we hadn't considered. We have starting implementing you advice wholly and I'm shocked and thrilled at how well it is working, for the most part.

Riva Bee

APPENDIX – Some client testimonials

Awesome bloke, would recommend his training.

We had Tully out to our place to visit our 3 Jack Russells and in one hour of training us humans our puppy is coming along very well. The older boys need a bit more training. Awesome bloke, would recommend his training.

Louise Beggs

Such practical advice!

Thankyou Tully! Such practical, and easy to put into practice, advice...now the training really starts!

Elizabeth Maw

Zoom consult—simple and worked extremely well

Highly recommend Tully's services. Just used Zoom for a consult for our new puppy which was very simple and worked extremely well. Thank you Tully

George Elyse

Tully is a dog whisperer

OMG—Tully is a dog whisperer. We've got a new pup. Thought we were doing the right thing. But encountering mischievous issues and toileting challenges. We were discipline like we would a child. But oh boy were we mistaken. Tully explained so much in simple language and made it seem so easy. In one session are set with enough techniques to get us going on the right track.

Debra Wakefield

...truly astounding!

Thanks so much Tully. We really appreciate all you've done for Benson. From a dog we were at our wits end with to the dog he is today, is truly astounding!

Jessica Templeton

I strongly recommend Tully and his invaluable service.

Tully is amazing! We recently contacted Tully as we needed some assistance with walking our Border Collie pup. He instantly recognised the problem and corrected the behaviour. Within 2 days of using the technique shown by Tully, our beautiful boy was walking perfectly. Tully also provided us with some very valuable dietary advice which has resulted in avoiding our older girl (Border Collie) undergoing a second bout of surgery. I strongly recommend Tully and his invaluable service.

Gail Sherwood

APPENDIX – Some client testimonials

Honestly cannot thank Tully enough...

Cannot recommend Tully's service highly enough. My parents have a male adult border collie who inexplicably developed acute anxiety about a year ago. Tully was able to reassure my parents that this is a common occurrence and can be fixed. He provided them with the tools and advice on how to manage their dog's anxiety and after only a few days they are starting to see progress. After so many months of stress it is wonderful to see how the whole household's mood has improved. Honestly cannot thank Tully enough for his help.

Eloise Beyer

If you want it done right, call Tully and get the best!

Tully is worth his weight in gold. Our dog has come leaps and bounds thanks to the wealth of knowledge Tully has given us. Problems we thought were too big are now a non-issue. If you want it done right, call Tully and get the best!

Atlanta Armstrong

...an overwhelmingly positive outcome

Highly Recommend! Training with my two dogs went beautifully with instant results after the first session. Second session was an overwhelmingly positive outcome that showed that a perfect walk exists. We are excited for more training with Tully.

Matt Richardson

He's much happier, as are we!

Hi Tully! Just wanted to thank you. Our dog has improved so much, and he's a completely different boy. He's very responsive, shuts up when asked, and comes immediately when called (even in full flight after a rabbit or kangaroo). We walk off lead all the time, and somehow he's become much more intuitive. Never explicitly taught heel or wait when off lead, but he does both. We still put him on the lead when there's a cyclist, but he's much better when they go past. We're even able to take him to markets and have random people pat him! He's much happier, as are we! Thanks so much.

Maree

...amazing outcome

I had problems with my dog Kia jumping on everyone she saw whilst out walking. Since my lessons with Tully we now walk up the street with no jumping, no chasing cats or other dogs, and best of all NO LEAD! Thankyou Tully—an amazing outcome.

Jenny Round

APPENDIX – Some client testimonials

Amazing... tried other avenues—they didn't work.

Amazing. Our staffy x has stopped jumping, nipping, rushing through doors and we have had calm walks that everyone enjoyed. Tully explained what we should be doing clearly and the reasons why. Tully was always prompt and easy to organise sessions with. I'd recommend him 100%, we had tried other avenues—they didn't work. Thanks Tully—one happy dog and very happy humans

Lucy Schepisi

...a completely different dog since working with you

Hi Tully. I just wanted to thank you for coming out the other day! Jed has been a completely different dog since working with you. No jumping and no scratching at the door, and now we are using the same techniques for other little issues that arise! He was even super calm and gentle with my friends 8 month old baby! Cannot thank you enough!

Tanika Love

Can't thank Tully enough...

We can't thank Tully enough, we were on the verge of having to re-home our family pet, and after three sessions we have a dog who will calmly interact with our family and is quickly learning to do everything he's told.

Nikki Anthony

...cured in 45 minutes

Fantastic and fast dog training. Our Ridgeback Sam had a phobia of bridges (refused to cross the Mitchell St train bridge, or any others), and was hard to get into the car. Cured in 45 Minutes. Highly Recommended.

Andrew Wilkinson

APPENDIX – Some client testimonials

Index

A

Abdelnoor, Daniel, 29, 86, 87, 230, 342
abstract concepts, 101, 102
abuse
    aggression, 112, 121, 218
    anger and abuse, 240
    Coercion and Its Fallout, 273, 275, 276
    does punishment cause undesirable side-effects, 111, 145
    downsides of positive training, 214, 240, 243
    fear and anxiety, 128
    hunger, 229
    learned helplessness, 133, 134, 135
    LIMA, LIFE and PROTECT, 309
    morality of punishment, 344
    stringent restriction of freedom, 237, 238
    why correction succeeds, 353, 354, 355
Academy for Animal Training and Behaviour, 253
addiction
    continuous or variable schedules of reinforcement, 42
    downsides of positive training, 214, 232–35, 232, 248, 249
    hunger, 226, 232
    model of obedience, 332
    morality of punishment, 346
    observational evidence vs. scientific studies, 263
    stringent restriction of freedom, 235, 237, 238
    why correction succeeds, 355, 356
affection, 230, 231, 232, 234, 235, 248
aggression, 85, 289, 295, 296, 297
    answering case against punishment, 152
    can punishment create behaviors?, 146, 150
    continuous or variable schedules of reinforcement, 41
    demonisation of punishment, 23
    does punishment cause undesirable side-effects, 107, 108, 109, 111–21, 118, 145
    does reward work?, 176
    downsides of positive training, 214, 218–20, 248
    effect of punishment on relationship, 144
    effectiveness of punishment with dogs, 66
    fear and anxiety, 124, 126
    learned helplessness, 131, 132, 134, 135, 221
    negative reinforcement, 46
    positive reinforcement, 44
    punishment in nature, 67, 87
    real life junk science, 284, 285, 286, 287
    rehoming and euthanasia, 246
    reward-only training, 34
    stringent restriction of freedom, 237
    survey, 283
    why correction succeeds, 353, 358, 360
alpha roll, 86
American Veterinary Society of Animal Behavior, 24, 290
    demonisation of punishment, 23, 24, 28, 29
    does reward work?, 180, 181, 191
    effect of punishment on relationship, 140
    learned helplessness, 133
    morality of punishment, 345
    myths about principles of punishment, 320
    observational evidence vs. scientific studies, 262
    recall, 163, 167, 170, 173
Amichien Bonding, 230
anger, 240
animal kingdom, 167, 168
animal welfare, 65, 177, 342
Anthony, Nikki, 9, 371
anthropomorphism, 76
anxiety
    does punishment cause undesirable side-effects, 110, 121–31, 145
    downsides of positive training, 214, 222–26, 248
    learned helplessness, 131, 132
    rehoming and euthanasia, 247
    why correction succeeds, 353
APDT, 198, 253, 328, *See* Association of Professional Dog Trainers
Applied Animal Behaviour Science, 113
Armstrong, Atlanta, 370
Association of Pet Dog Trainers, 23, 29, 59, 160, 172, 197
Association of Professional Dog Trainers, 23, 198, 253, 289, 304
asymptote, 123
autoshaping, 204
aversive control, 214, 273
aversive stimulus
    aggression survey, 290
    can punishment create behaviors?, 146, 147, 148, 149
    Coercion and Its Fallout, 271, 273
    does punishment cause undesirable side-effects, 108, 109
    downsides of positive training, 214, 215, 217
    effect of punishment on relationship, 140
    effectiveness of punishment with dogs, 56, 57
    effectiveness of punishment with people, 102
    fear and anxiety, 130, 131

Index

hunger, 227
junk science vs. real-world training, 284, 288, 289, 295
learned helplessness, 133, 134
LIMA, LIFE and PROTECT, 303, 304, 305, 307, 308, 309
model of obedience, 335
morality of punishment, 345
myths about principles of punishment, 320, 322, 323, 324
positive reinforcement, 44, 45, 46
problems with operant conditioning, 49, 50
punishment in nature, 69, 76
recall, 161, 163, 167, 173
rehoming and euthanasia, 243, 244
stringent restriction of freedom, 237, 238
why correction succeeds, 354, 355
withdrawal of, 97
avoidance
  aggression, 115
  Coercion and Its Fallout, 272
  demonisation of punishment, 21, 28
  does punishment cause undesirable side-effects, 106, 109, 110
  does reward work?, 176
  fear and anxiety, 123, 124, 127, 224
  hunger, 227
  morality of punishment, 343
  myths about principles of punishment, 327
  positive reinforcement, 45
  snakes, 327
AVSAB. *See* American Veterinary Society of Animal Behavior
Azrin, N.H., 68, 112, 113, 121, 134, 202, 267, 285, 286, 300, 301, 324

B

Baenninger, Ronald, 168, 300
Balabanov, Ivan, 84
balanced approach, 138, 145, 248, 256, 281
balanced trainers, 34, 97, 310
barking, 323
  anger and abuse, 240
  continuous or variable schedules of reinforcement, 42
  does punishment cause undesirable side-effects, 132
  does reward work?, 188, 199, 202, 209
  learned helplessness, 132
  model of obedience, 336
  myths about principles of punishment, 323, 324, 325
  punishment in nature, 66, 75
  rewarding alternative behaviours, 193
Beckedorff, Barbara, 116
Beggs, Louise, 369
behavioural frequency, 151

behaviours
  alternative, 100, 191, 194, 198
  good, 21, 87, 100, 138, 168, 198, 199, 202, 219, 240, 336, 361
  ignoring bad, 191
  incompatible, 191
  management, 99, 102, 106, 140, 162, 167, 173, 192, 243
  modification, 38, 244, 304, 354
  undesirable, 87, 100, 185, 211, 236, 287, 299, 305, 306, 316, 317
Bekoff, Marc, 76, 77
Benson, Cindy, 35, 128, 225, 322
Berman, Jean, 116
Beyer, Eloise, 370
bias, 46, 122, 259, 260, 284, 286, 287, 288, 292
Big Three Necessities of Positive Reinforcement
  Coercion, 226, 232, 233, 235, 238, 346, 355, 356
bin diving, 90, 189, 333
biting
  Coercion and Its Fallout, 279
  does punishment cause undesirable side-effects, 122
  effect of punishment on being operant, 136
  effectiveness of punishment with people, 99
  model of obedience, 333
  myths about principles of punishment, 318, 321
  punishment in nature, 70, 77, 78
  rewarding alternative behaviours, 192, 193, 194, 195
  why correction succeeds, 360
Blackfish, 27, 229
Blackwell, Emily, 288, 295
boredom, 202
boundaries, 67, 137, 140, 144, 244, 247, 304, 354
breeders, 70, 72, 122
breeds, 78, 85, 112
  Amstaff, 166
  Border Collie, 223, 234
  Border Terrier, 58, 189, 295, 333, 334, 352
  Bull Mastiff, 120
  Campaspe sheepdogs, 71
  Cavoodle, 166
  Chihuaha, 78
  Chihuahua, 222
  Cocker Spaniels, 114, 143, 144
  dangerous, 114
  Doberman Pinscher, 140, 141, 142, 147
  German Shepherd, 209
  Golden Retriever, 208
  Great Danes, 62, 195
  Jack Russell, 142, 227, 228
  Kelpies, 202
  Labrador, 114, 119, 167, 170
  Maltese, 223
  Maremma, 71, 207, 208
  Pit Bull, 59, 62, 159, 176
  Pit Bull Terrier, 78

Poodle, 162, 171, 208
Ridgebacks, 120
Sheepdog, 234
Terriers, 202
Turkish Kangals, 71, 119, 207, 208
Breland, K, 38, 79, 200
Breland, Marian, 38, 79
Brinker, Ellen, 116
Brooks, Alan, 4, 64

C

California Veterinary Medical Association, 23
Campaspe Working Dogs, 63
canine consent, 76, 80
Carr, Adrian & Erica, 364
Cataudo-Williams, Silvia, 94
cats, 57, 58, 67, 160, 186, 192, 292
Cavanagh, Rod, 144
Certification Council of Professional Dog Trainers (CCPDT), 23, 289, 305
Certified Trainer, 108, 116, 176
check chains. *See* choke chains
chewing, 40, 67, 68, 72, 87, 188, 189, 201, 202, 209, 297, 317, 318, 319, 320, 326
child rearing, 21, 127
child training, 94
China, Lucy, 291, 295
choke chains, 126, 167, 175, 176, 177
Ciribassi, Dr. John, 29, 133, 134, 180, 191, 320, 321
classical conditioning, 36, 321
Click to Calm – Healing the Aggressive Dog, 116
clicker training, 65, 125, 211, 217, 231, 319, 320, 321
close the economy, 228, 229, 230, 232, 248
coercion, 29, 270, 271, 274
   addiction, 233
   Coercion and Its Fallout, 270, 272, 273, 274, 275, 276, 277, 281
   hunger, 226, 230, 232
   LIMA, LIFE and PROTECT, 308, 309
   manipulation and dishonesty, 239
   morality of punishment, 346
   Murray Sidman, 265
   observational evidence vs. scientific studies, 251
   punishment in nature, 74
   rehoming and euthanasia, 247
   stringent restriction of freedom, 235, 237, 238
   why correction succeeds, 355, 356
Coercion and Its Fallout, 29, 265, 281, 306
Coffey, Thérèse, 292, 293
communication, 23, 76, 171, 240, 330, 342, 356
competing rewards, 159, 170, 177, 182–88, 182, 186, 187, 189, 192, 316, 355
Conant, Susan, 116
conditioned punishers, 36, 278, 279, 280, 327
Conditioned stimulus (CS), 327
conditioning. *See* classical conditioning; operant conditioning; counter-conditioning

confidence, 85, 98, 114, 121, 122, 128, 131, 258
confinement, 102, 203
confirmation, 44, 98, 121, 122, 216, 260, 288
consequences, 36
   aggression, 112
   anger and abuse, 240
   can punishment create behaviors?, 149, 150
   Coercion and Its Fallout, 269, 272, 277
      Coercion and Its Fallout, 272
   correction, 36
   does punishment cause undesirable side-effects, 111
   does reward work?, 175, 188, 191
   effect of punishment on being operant, 136, 137
   effect of punishment on relationship, 141, 144
   effectiveness of punishment with dogs, 88
   effectiveness of punishment with people, 93, 95, 100, 101, 102
   fear and anxiety, 121, 123, 125, 128
   gaining obedience to commands, 159
   learned helplessness, 134
   manipulation and dishonesty, 240
   model of obedience, 332, 333, 334, 335, 336
   myths about principles of punishment, 315, 317, 318, 326, 327
   negative punishment, 47
   punishment in nature, 67, 68, 80, 86
   recall, 170
   rehoming and euthanasia, 247
   rewarding alternative behaviours, 194
   teaching the meaning of commands, 157
   why correction succeeds, 352, 353
consistency, 164, 182, 185, 198, 199, 200, 213, 313, 314, 316, 317, 319
Cooper, Jonathan, 291, 306
correction, 296
   addiction, 235
   aggression, 220
   anger and abuse, 241, 242, 243
   Coercion and Its Fallout, 278
   competing rewards, 186
   demonisation of punishment, 21, 22, 23
   does punishment cause undesirable side-effects, 105, 108
   does reward work?, 176, 190, 205
   effect of punishment on relationship, 140, 145
   effectiveness of punishment with dogs, 55, 58, 59, 63, 87, 88, 89
   explained, 36
   fear and anxiety, 124, 125, 128, 223, 225
   hunger, 230, 231, 232
   junk science vs. real-world training, 285, 287
   learned helplessness, 133, 222
   LIMA, LIFE and PROTECT, 309
   model of obedience, 331, 332
   morality of punishment, 341, 345, 347
   myths about principles of punishment, 313, 318, 323

Index

observational evidence vs. scientific studies, 256
punishment in nature, 67, 68, 69, 72, 74
recall, 164, 166, 170
rehoming and euthanasia, 245
reward, 33
reward-only training, 34
why it succeeds, 351, 354, 360
correction collars. *See* choke chains
correlational studies, 261
corruption, 254, 260, 292
counter-conditioning, 223
Cracknell, Nina, 291, 293, 295
Critical assessment of Murray Sidman's approach to coercion, 271
Culture Clash, The, 29, 161, 197

D

Dawson, Renee, 367
DEFRA (Department for Environment, Food and Rural Affairs, 292, 293
DEFRA (Department for Environment, Food and Rural Affairs), 292
Denenberg, Dr. Sagi, 147
depression, 109, 135, 221
deprivation
    Coercion and Its Fallout, 274, 275, 276
    continuous or variable schedules of reinforcement, 42
    downsides of positive training, 214, 248, 249
    hunger, 228, 229, 230, 231, 232
    model of obedience, 332, 336
    morality of punishment, 346
    observational evidence vs. scientific studies, 263
    positive reinforcement IS negative reinforcement, 276, 277
    reward, 34
    stringent restriction of freedom, 235, 236, 237, 238
    teaching the meaning of commands, 156
    why correction succeeds, 355, 356
Derham, Lavena, 367
desensitization, 115, 223, 224
directive training, 244, 304, 354
discipline, 137, 243, 343, 354, 357, 358
    aggression, 115
    effect of punishment on being operant, 137
    effect of punishment on relationship, 138, 140, 143
    fear and anxiety, 131
    lack of, 128
    rehoming and euthanasia, 247
dishonesty, 92, 239
distraction-proofing, 161, 186
distractions, 158, 159, 182, 329, 354, 358
dog bites, 40, 136, 194, 219, 245, 246, 247, 258
Dog Psychology, 24
dog to dog interactions, 72

Doggy Dan. *See* Abdelnoor, Daniel
Dogstar Daily, 23
dominance, 220
    aggression, 111, 120
    effect of punishment on relationship, 139
    effectiveness of punishment with dogs, 80–87
    hunger, 230
dominant dog, 81, 85, 220
Don't Shoot the Dog!, 27, 49, 276
Donaldson, Jean, 294
    addiction, 233
    aggression, 218
    anger and abuse, 242
    Coercion and Its Fallout, 276
    competing rewards, 182, 183, 186, 187
    demonisation of punishment, 29
    does reward work?, 174, 175, 177, 181, 190, 205, 211
    effectiveness of punishment with dogs, 89
    hunger, 228, 229, 230, 231, 232
    morality of punishment, 342
    observational evidence vs. scientific studies, 263
    recall, 161, 163, 164, 167, 168
    rewarding alternative behaviours, 197
    stringent restriction of freedom, 236
    why correction succeeds, 351, 354, 355, 359
Duke, James B., 254
Dunbar, Dr. Ian, 23–27, *Passim*
Dunbar, Jamie, 60, 61, 62, 89, 142, 165, 169, 208, 307, 359, 360
Dunbar, Kelly, 117, 238

E

e-collars, 46, 59, 110, 114, 125, 149, 175, 206, 208, 256, 263, 291, 292, 296, 298, 299, 315, 322, 333, 334, 335, 344
    junk science vs. real-world training, 291, 292, 293, 298
electric fences, 48, 110, 151, 323, 344
Elliffe, Doug, 4, 296, 299
Elyse, George, 369
English as a Second Language. *See* ESL
Enright, Jillian, 35, 56, 60, 63, 67, 94, 98, 99, 320
environment
    Coercion and Its Fallout, 271, 274, 278
    competing rewards, 182, 184, 185, 187, 188
    demonisation of punishment, 27
    does punishment cause undesirable side-effects, 107, 108, 145
    does reward work?, 182, 189, 200, 205, 207, 208, 211
    downsides of positive training, 213, 217
    fear and anxiety, 122, 123, 130
    junk science vs. real-world training, 292
    model of obedience, 331, 336
    myths about principles of punishment, 316, 317
    punishment in nature, 69, 70

recall, 168, 169, 170, 173
rewarding alternative behaviours, 194, 195
why correction succeeds, 352, 353, 354, 355, 358
why correction succeedst, 355
epidemiology, 261
errorless learning, 274
escalation, 113, 114, 334
ESL, 75, 76, 78, 80, 100, 195, 238, 330
ethical transformation, 309, 310
ethics, 307, 342
euthanasia, 214, 219, 243, 245, 246, 247, 258, 288, 289, 308
exercise, 73, 86, 97, 110, 161, 178, 186, 196, 202, 225, 310, 320, 357
explosive behavior, 108, 110, 124, 176
Extended Stringent Restriction of Freedom, 232, 238, 346, 355, 356
extinction, 41, 123, 124, 198, 199, 202, 203, 219, 221, 267, 268, 317, 324, 354

F

fear
    aggression, 112, 119, 120, 220
    anger and abuse, 242
    answering case against punishment, 152
    can punishment create behaviors?, 146
    Coercion and Its Fallout, 278, 279, 280
    demonisation of punishment, 22
    does punishment cause undesirable side-effects, 107, 108, 109, 110, 111, 121–31, 145
    does reward work?, 176
    downsides of positive training, 214, 222–26, 248
    learned helplessness, 131, 135
    myths about principles of punishment, 321
    recall, 163
    why correction succeeds, 353
Fennell, Jan, 230
Fernandez, Eduardo, 304
Flanagan, Jemma K., 364
Fontes, R, 106, 168, 279
food lures, 24, 25, 164, 238
Foord, Michael, 368
force free, 29, 34, 47, 55, 88, 127, 175, 205, 217, 304, 360
frustration, 42, 241, 242, 243

G

Garrett, Susan, 29, 74, 93, 105, 107, 110, 135, 187, 205, 220, 221, 222, 241, 360
genetics, 201, 226
Gentle Leader, 217
George, Zak, 28–29, *Passim*
Goldiamond, 274, 275, 308, 309
Goodall, Christa, 9, 363
Gopnik, Alison, 224

growling, 70, 85, 102, 112, 143, 323
Grubis, Roo, 116
guide dogs, 63, 64, 65, 90, 106
Guide to a well-behaved dog, 28

H

habits, 115, 182, 201, 303, 316
Handbook of Applied Dog Behavior and Training, 44, 227, 303
Harper, Melissa, 364, 365
Harrington, Carolyn, 366, 367
Hastings, Fleur, 113, 367
hat trick rule, 97, 98, 102, 121, 143
Herron, Meghan, 113, 120, 131, 132, 134, 283, 286, 288, 289, 295, 297, 300
    aggression, 114, 119
    Coercion and Its Fallout, 268
    does punishment cause undesirable side-effects, 113, 134, 135
    does reward work?, 202
    junk science vs. real-world training, 283, 290
hierarchy, 82, 83, 84, 86, 87
Holz, W.C., 68, 202, 267, 300, 301, 324
hunger, 39, 146, 217, 226–32, 232, 235, 355
hyperactivity, 108, 110, 124, 176

I

ICEPIRP, 334
improvisations, 314, 315
in the act, 102, 111, 194, 203, 315, 317, 318, 319, 321, 326, 328, 336, 353
intelligence, 78, 109
International Guide Dog Federation, 64
International Society for Applied Ethology, 23
intrinsic motivation, 331
intrinsic rewards, 202
irritability, 72, 108, 110, 124, 176
Irving, Bronwyn, 363

J

Jones, Erin, 76
Journal of Applied Animal Welfare Science, 181, 244
Journal of the Experimental Analysis of Behavior, 89
jumping
    aggression, 115
    can punishment create behaviors?, 151
    Coercion and Its Fallout, 267
    competing rewards, 185, 187
    does reward work?, 188, 189, 199, 202, 203, 209
    effect of punishment on being operant, 136
    effectiveness of punishment with people, 96
    learned helplessness, 133, 134
    myths about principles of punishment, 315, 316, 317, 321, 323, 324, 325, 326, 327

Index

negative punishment, 40, 47
problems with operant conditioning, 49
rewarding alternative behaviours, 192

K

Ken Lord Award, 64
Killion, Jane, 44, 136, 165, 166, 215, 234, 294, 359
Kitching, Helen, 137, 366

L

Landsberg, Dr. Gary M., 147
language, 51, 75, 76, 78, 100, 101, 102, 134, 161, 330
Lawrence, Kay, 21, 40, 44, 217
learned helplessness, 221, 222
    answering case against punishment, 152
    can punishment create behaviors?, 146
    does punishment cause undesirable side-effects, 107, 108, 109, 111, 131–35, 145
    downsides of positive training, 214, 221, 248
    effect of punishment on relationship, 141, 144
    learned helplessness, 220, 221
    myths about principles of punishment, 328
    punishment in nature, 68
    why correction succeeds, 353
Learning About Dogs, 40
Learning and Behavior, 98
leash checks, 44, 216
leash training, 178, 356
Least Intrusive Minimally Aversive. See LIMA, *See* LIMA
Lee, Meaghan, 367
Leist, Daniel & Morgan, 10
Lewis, Jodi, 244, 366
LIFE framework, 115, 243, 251, 257, 303, 305, 306, 308, 310
life rewards, 34, 158, 172, 185, 188
LIMA, 115, 243, 251, 257, 303, 304, 305, 306, 308, 310
Lindsay, Steven, 44, 45, 97, 227, 303
livestock chasing, 291, 295, 297
Lougoon, Michelle, 368
Love, Tanika, 9, 371
luring, 148, 156, 322

M

Management Magic, 174
management tool, 102
Mandell, RoseAnne, 116
manipulation, 92, 239, 268, 280, 356
Marschark, Eve, 168, 300
Massa, Louis, 116
maternal punishing activity, 72
Mattinson, Pippa, 162, 163, 164, 240, 359, 360

Maw, Elizabeth, 369
Maxwell, Dr Deborah, 3
Mazur, James, 89, 98, 103, 123, 256, 286, 301, 327
McGovan, Elizabeth, 368
McGreevy, Paul, 76
McLean, Ian, 296, 297
Mech, Dr. L. David, 83, 84, 86
Medical News Today, 49
Meir, Israel, 116
Miller, Pat, 29
    anger and abuse, 242
    competing rewards, 183
    effectiveness of punishment with dogs, 59, 60, 61, 89, 90
    fear and anxiety, 125
    model of obedience, 333, 334
    observational evidence vs. scientific studies, 263
    recall, 159, 160, 172, 173
    stringent restriction of freedom, 235
    why correction succeeds, 351, 355, 359
Mills, Daniel, 286, 287, 290, 291, 293, 295
misbehaviour, 90, 189, 333, 360
Misbehaviour of Organisms, The, 38, 79
morality of punishment, 339, 341, 346
mother with her pups, 69–72
muzzling, 114
myths
    about principles of punishment, 311, 313, 314, 316, 317, 325, 328
    aggression, 120
    Coercion and Its Fallout, 265
    continuous or variable schedules of reinforcement, 42
    demonisation of punishment, 22, 23, 29
    does punishment cause undesirable side-effects, 145
    does reward work?, 188, 199
    downsides of positive training, 249
    effectiveness of punishment with dogs, 88, 90
    effectiveness of punishment with people, 103
    fear and anxiety, 122
    punishment, 36
    punishment in nature, 66, 70
    why correction succeeds, 351, 359
Myths
    about principles of punishmet, 311
    junk science vs. real-world training, 301

N

Natural Method of Dog Training, The, 24, 226
Negative Effects of Positive Reinforcement, 213
Nelson, Leslie, 60, 160, 172, 173, 205, 294
nipping, 188, 196, 209, 240, 267, 321
noncoercive techniques, 139, 192, 229, 231, 276
noncompliance, 189, 360

Index

O

obedience
    addiction, 233, 234
    aggression, 115
    anger and abuse, 242
    can punishment create behaviors?, 149, 150
    competing rewards, 187
    does reward work?, 155, 175, 176, 177, 207
    downsides of positive training, 216, 248
    effect of punishment on relationship, 140
    fear and anxiety, 124, 125
    gaining obedience to commands, 159
    hunger, 232
    model of, 329
    model of obedience, 330, 331, 336
    myths about principles of punishment, 325
    observational evidence vs. scientific studies, 257
    recall, 166, 174
    rehoming and euthanasia, 245
    teaching the meaning of commands, 157, 158
    to commands, 158
    why correction succeeds, 354, 356
Obedience
    model of, 311
observational evidence, 257, 259, 266
off leash, 25, 165, 166, 169, 171, 243, 294, 295, 357
O'Keefe, Patrick, 164, 263
operant conditioning, 31, 33, 36, 37, 40
    can punishment create behaviors?, 151
    continuous or variable schedules of reinforcement, 41, 42
    does reward work?, 200
    downsides of positive training, 215
    explained, 37, 38
    hunger, 227
    model of obedience, 331
    problems with, 43–53, 43, 48, 49, 51
    problems with operant conditioning, 43
    punishment in nature, 79
Optimal Training, 306, 308, 309, 310
Overcoming Your Dog's Leash Reactivity, 116

P

pack, 82, 84, 85, 182, 197
Parsons, Emma, 108, 110, 116, 118, 124, 176, 180, 187, 201
Parsons, Virginia, 116
Patel, Chirag, 76
Pavlov, Ivan, 36
Peaceable Paws Dog & Puppy Training Centre, 160
Perone, Dr. Michael, 44, 45, 102, 213, 214, 215, 216, 217, 239, 273
phobias, 131, 222, 223
physical restraints, 99

Pigs Might Fly! Training success with impossible dogs, 136
pinch collars, 177
play-biting, 77, 195
pleasurable recoil, 97
Plonsky, Mark, 48
Port, Dina Roth, 28, 133
positive behavior, 107, 145, 271, 352
positive ideologues, 283
positive reinforcement training, 34, 295
    addiction, 232
    Coercion and Its Fallout, 276
    competing rewards, 182, 185
    does reward work?, 208, 210, 212
    downsides of positive training, 214
    junk science vs. real-world training, 291, 294, 301
    learned helplessness, 222
    model of obedience, 335
    problems with operant conditioning, 46
    recall, 161, 162
positive training. *See also reward-only training*
    addiction, 235
    aggression, 115, 118, 121, 218, 219, 220
    anger and abuse, 241, 242, 243
    can punishment create behaviors?, 146, 149, 151
    Coercion and Its Fallout, 273, 274, 276
    competing rewards, 183, 184, 186, 188
    confinement, 204
    continuous or variable schedules of reinforcement, 42
    demonisation of punishment, 22, 24, 28
    does punishment cause undesirable side-effects, 105, 106, 145
    does reward work?, 155, 176, 178, 188, 189, 190, 202, 204, 205, 206, 208, 209, 210, 211, 212
    downsides and side effects of, 213
    downsides and side-effects of, 213, 214, 248
    downsides of, 153
    downsides of positive training, 213, 217, 249
    effect of punishment on relationship, 139, 143, 144
    effectiveness of punishment with dogs, 56, 61, 62, 66, 87, 88
    effectiveness of punishment with people, 100, 103
    failure of, 95
    fear and anxiety, 126, 130, 224, 225, 226
    gaining obedience to commands, 159
    hunger, 226, 228, 229, 230, 231, 232
    junk science vs. real-world training, 284, 289, 290, 294, 301
    learned helplessness, 135, 221
    LIMA, LIFE and PROTECT, 305, 307, 309, 310
    model of obedience, 329, 330, 331, 335
    morality of punishment, 342, 345, 346
    myths about principles of punishment, 318, 319

Index

negative punishment, 40
negative reinforcement, 47
observational evidence vs. scientific studies, 254, 256, 257, 260, 262, 263
positive reinforcement, 45, 46
punishment, 35
punishment in nature, 69, 79, 87
recall, 160, 163, 165, 166, 167, 170, 172, 173, 174
rehoming and euthanasia, 243, 244, 245, 247
rewarding alternative behaviours, 192, 193, 194, 195, 197, 198
reward-only training, 34
stringent restriction of freedom, 238
teaching the meaning of commands, 156, 157, 158
why correction succeeds, 351, 352, 355, 356, 358, 359, 360, 361
possessive-aggressive behavior, 114, 143
Power of Positive Dog Training, The, 29, 59, 159
Power of Positive Training and *Proof Positive, The*, 174
Premack Principle, 172, 185, 186
pressure/release training, 46, 92
Prioritizing Results for Optimal Training. *See* PROTECT
prong collars. *See* pinch collars
PROTECT, 115, 243, 251, 257, 303, 306, 310
Pryor, Karen, 27–28, *Passim*
pseudo-science, 120, 130, 169, 202, 257, 260, 263, 270, 289, 299, 300, 301, 306, 347
Punishing Unwanted Behaviour Just Makes it Worse, 35, 56, 94, 98, 255, 320
punishment
    answering the case against, 53–55, 151
    can it create behaviors?, 146–53
    can we train without it, 342
    demonisation of, 19, 21, 22, 29, 51, 245, 247, 265
    does punishment cause undesirable side-effects, 105–46
    does reward work?, 207
    effectiveness of, 55–91, 89, 105, 144, 238, 249, 319, 335
    escalating, 94
    explained, 35–36
    in nature, 66–69
    in training people, 103
    in traning people, **Error! Not a valid bookmark in entry on page** 91
    is it that bad?, 344–45
    morality of, 343
    must be consistent, 314–17
    must be immediate, 317–21
    must be instructive, 322–25
    must have a warning, 325–28
    negative, 21, 34, 37, 39, 40, 41, 43, 47, 48, 99, 186
    non-contingent, 220
    physical, 35, 94, 98, 167, 168, 170, 171, 335
    positive, 33, 34, 36, 40, 43, 47, 48, 50, 59, 62, 159, 288
    positives vs negatives, 345–46
    principles of, 311, 313, 328
    problems with operant conditioning, 51
Punishment and Its Putative Fallout: A Reappraisal, 106, 279
puppy schools, 100, 198

Q

quick fix, 89, 126, 132, 142, 144, 157

R

Rand, Ayn, 88
Ray, Carol, 139, 235
reactivity, 115, 116, 209, 237, 288, 310, 360
real world
    aggression, 112, 114
    anger and abuse, 240
    Coercion and Its Fallout, 266, 268
    competing rewards, 184, 185, 188
    continuous or variable schedules of reinforcement, 41
    demonisation of punishment, 23, 25, 26, 27, 28
    does punishment cause undesirable side-effects, 145
    does reward work?, 155, 156, 180, 199, 200, 210, 211
    downsides of positive training, 215
    effectiveness of punishment with dogs, 56, 66, 87, 88, 89
    effectiveness of punishment with people, 101, 102
    extinction and consistency, 203
    fear and anxiety, 122, 127, 128, 223, 224
    gaining obedience to commands, 158, 159
    junk science vs. real-world training, 291, 300, 301
    LIMA, LIFE and PROTECT, 306, 307, 308, 310
    model of obedience, 331, 336
    morality of punishment, 342, 343, 345
    myths about principles of punishment, 327
    observational evidence vs. scientific studies, 254, 257, 258, 259, 260
    optimal training, 308
    punishment in nature, 68, 69, 70, 73, 79, 80
    recall, 161, 166, 167, 168
    stringent restriction of freedom, 236, 238
    teaching the meaning of commands, 157
    why correction succeeds, 351, 352, 354, 356, 359, 360
Real World Dog Training website, 128

Index

Really Reliable Recall, 60, 160, 164, 172, 173, 174, 205, 294
recall, 296
    anger and abuse, 243
    Coercion and Its Fallout, 280
    competing rewards, 182, 183, 185, 186, 188
    does reward work, 159–72
    does reward work?, 181, 188, 206, 210
    effectiveness of punishment with dogs, 62
    junk science vs. real-world training, 291, 294, 295, 298
    morality of punishment, 342
    observational evidence vs. scientific studies, 263
    recall, 172, 173, 174
    why correction succeeds, 356, 357, 358, 360
rehoming, 214, 219, 243, 246, 247, 258, 308
reinforcement history, 165, 172, 294, 357
reinforcement schedule
    addiction, 235
    aggression, 219
    continuous, 41
    continuous or variable schedules of reinforcement, 41
    downsides of positive training, 217
    hunger, 232
    myths about principles of punishment, 314
    variable, 40, 41, 42, 45, 158, 187, 200, 201, 203, 204, 211, 235, 355
reinforcement, negative, 272
    addiction, 234, 235
    aggression, 118
    can punishment create behaviors?, 146, 148, 149
    Coercion and Its Fallout, 266, 269, 270, 272, 273, 274, 275, 276, 277, 279, 281
    does punishment cause undesirable side-effects, 131
    does reward work?, 188
    downsides of positive training, 214, 215, 217, 226, 248
    effect of punishment on relationship, 140, 141
    effectiveness of punishment with dogs, 62
    effectiveness of punishment with people, 91, 92, 93
    explained, 39, 40
    fear and anxiety, 125, 128, 130, 225
    hunger, 227, 229
    junk science vs. real-world training, 295, 300
    manipulation and dishonesty, 239
    model of obedience, 334
    myths about principles of punishment, 321, 324
    operant conditioning, 37
    problems with operant conditioning, 43, 45, 46, 47, 48, 49
    recall, 166, 167, 168
    rehoming and euthanasia, 247
    reward-only training, 34
    teaching the meaning of commands, 158
    why correction succeeds, 355

why correction succeedss, 355
reinforcement, positive, 296
    addiction, 232, 233, 234, 235
    aggression, 116, 121, 218, 219
    Coercion and Its Fallout, 266, 267, 269, 270, 271, 273, 274, 275, 276, 277, 278, 279, 280, 281
    competing rewards, 182, 183, 185
    confinement, 204
    does reward work?, 155, 188, 206, 208, 209, 210, 211, 212
    downsides of positive training, 214, 215, 216, 217, 218, 248, 249
    effect of punishment on relationship, 139
    effectiveness of punishment with dogs, 59, 60, 61, 66
    effectiveness of punishment with people, 91, 92, 97, 98, 99, 100
    explained, 38
    fear and anxiety, 123, 125, 126
    gaining obedience to commands, 159
    hunger, 227, 228, 232
    junk science vs. real-world training, 287, 289, 290, 291, 294, 295, 296, 299, 300, 301
    learned helplessness, 135, 220, 221, 222
    LIMA, LIFE and PROTECT, 303, 304, 305, 309
    lure-based, 156
    manipulation and dishonesty, 239
    methods contain punishment, 215–18, 215
    model of obedience, 334, 335
    myths about principles of punishment, 314, 316, 320, 321
    negative punishment, 40
    observational evidence vs. scientific studies, 253, 256, 259, 260, 262
    problems with operant conditioning, 43, 44, 45, 46, 47, 50
    punishment in nature, 87
    recall, 161, 162, 163, 165, 167, 168, 169, 171
    rehoming and euthanasia, 244
    reward, 33
    rewarding alternative behaviours, 191
    reward-only training, 34, 35
    stringent restriction of freedom, 238, 239
    teaching the meaning of commands, 158
    why correction succeeds, 354, 355, 356, 358, 360
    why correction succeedss, 355
reinforcer junkie, 233, 234, 235, 248
Reisner, Ilana R, 283, 286, 288
Reisner, Ilana R., 283
repetitions, 58, 90, 97, 111, 122, 177, 195, 225, 267, 279, 298, 314, 323, 333
reprimands, 190
respect, 139, 218, 278, 335
response-cost contingency, 95
reward
    alternative behaviours, 191–98
    competing, 211
    explained, 33–34

381

Index

non-contingent, 222
schedule. *See* reinforcement schedule
stopping unwanted behaviour with, 188–89
reward-only training
   aggression, 111, 115, 220
   anger and abuse, 241
   can punishment create behaviors?, 146
   Coercion and Its Fallout, 265, 266, 268, 269, 270, 277
   competing rewards, 185
   demonisation of punishment, 21, 22, 24, 25, 28, 29
   does punishment cause undesirable side-effects, 105, 107
   does reward work?, 175, 205
   downsides of positive training, 248, 249
   effect of punishment on being operant, 137
   effect of punishment on relationship, 138, 141
   effectiveness of punishment with dogs, 55
   effectiveness of punishment with people, 93, 97, 100
   explained, 34
   fear and anxiety, 122, 130, 131
   hunger, 228
   junk science vs. real-world training, 283, 286, 287, 288, 289, 290, 291, 295, 297, 299
   learned helplessness, 135
   LIMA, LIFE and PROTECT, 305, 306
   morality of punishment, 342, 343, 344, 347
   myths about principles of punishment, 317
   negative reinforcement, 39
   observational evidence vs. scientific studies, 256, 257, 258, 259, 262
   operant conditioning, 40
   problems with operant conditioning, 48, 51
   punishment in nature, 66, 69, 72, 78, 79, 80
   recall, 169, 172
   rehoming and euthanasia, 246, 247
   rewarding alternative behaviours, 191
   stringent restriction of freedom, 237
Richards, Wendy, 132, 368
Richardson, Matt, 61, 370
Rocha and Hunziker, 267, 271, 273, 274, 276, 308
Round, Jenny, 10, 370
Royal College of Veterinary Surgeons, 23
Ruffell, David, 365
rule breaking, 100, 136, 137, 170, 175

S

Sargisson, Rebecca, 296, 297
satiation, 202, 226
Schepisi, Lucy, 10, 371
Schubert, Carl, 367
Schultz, Penny, 116
science, 303, 306
   Coercion and Its Fallout, 265, 266, 270
   competing rewards, 182, 184

   continuous or variable schedules of reinforcement, 42
   demonisation of punishment, 22, 26, 27
   does punishment cause undesirable side-effects, 145
   does reward work?, 202
   downsides of positive training, 249
   effectiveness of punishment with dogs, 56
   effectiveness of punishment with people, 103
   fear and anxiety, 131
   hunger, 226, 228
   junk, 251, 283–303
   morality of punishment, 342
   myths about principles of punishment, 313, 315, 328
   observational, 257
   observational evidence vs. scientific studies, 253–65
   problems with operant conditioning, 43, 49
   punishment in nature, 68, 79
   recall, 168, 169
   stringent restriction of freedom, 238
   why correction succeeds, 352
Science in an Age of Unreason, 254
Scientific American, The, 262
scratching, 34, 199, 202, 325
self-injurious behaviour, 98, 99
Seligman, Martin, 132, 133, 134, 141, 147, 328
Sessa, Alex, 116
Seton, Ernest Thompson, 75
Shanan, T, 106, 168, 279
Shea, Cathy, 116
Sherwood, Gail, 369
Shofer, Frances, 283, 286, 288
Shubring, Lesley, 368
side-effects, 301
   anger and abuse, 243
   answering case against punishment, 53, 152
   can punishment create behaviors?, 146, 150
   Coercion and Its Fallout, 265, 269, 270, 271, 272, 274, 275, 278, 280, 281
   continuous or variable schedules of reinforcement, 41, 42
   demonisation of punishment, 27
   does reward work?, 155, 212
   downsides of positive training, 153
   effectiveness of punishment with dogs, 55
   effectiveness of punishment with people, 92, 99, 100, 103
   LIMA, LIFE and PROTECT, 308, 310
   morality of punishment, 345, 347
   observational evidence vs. scientific studies, 257
   of positive training, 213
   of potisitive training, 249
   positive, 118, 145, 257, 271, 353
   positive reinforcement, 45, 46
   punishment, 35, 36
   punishment in nature, 67, 68, 74

Index

recall, 164, 166
rehoming and euthanasia, 244, 248
rewarding alternative behaviours, 195
stringent restriction of freedom, 237
teaching the meaning of commands, 156
undesirable, 105–46, 105, 106, 145
why correction succeeds, 352, 353, 360
Sidman Avoidance, 272, 328
Sidman, Murray, 74, 306
    can punishment create behaviors?, 146
    Coercion and Its Fallout, 265–83
    demonisation of punishment, 29
    junk science vs. real-world training, 283
    myths about principles of punishment, 328
    observational evidence vs. scientific studies, 251, 257, 258, 263
    punishment in nature, 69
    rehoming and euthanasia, 247
Sierra Veterinary Medical Association, 23
SIRIUS Puppy Training, 197
Sisyphus Myth, 317
Skinner box, 38, 200
Skinner, B. F., 3, 29, 33, 37, 38, 40, 42, 43, 50, 79, 80, 184, 200, 202, 214, 231, 232, 258, 265, 266, 267, 268, 270, 275, 283, 285, 300, 331
smacks, 114, 149, 150, 344
socialisation, 26, 115, 117, 127, 197
sound methods, 27, 163, 166, 242, 343
speeding fines, 35, 100, 314, 315, 317, 328, 332, 334, 335, 354
spontaneous recovery, 202
spray bottle, 58, 61, 125, 133, 189, 206, 333, 352
Staddon, John, 3, 254, 261, 293
Standard Training for Effective Partnerships. *See* STEP program
STEP program, 63, 64, 65, 184
Stern, Geoff, 116
Stilwell, Garret, 141
Stilwell, Victoria, 51
    addiction, 233
    aggression, 120
    anger and abuse, 242
    demonisation of punishment, 29
    does reward work?, 175, 177, 205
    effectiveness of punishment with dogs, 61, 87, 89
    extinction and consistency, 203
    fear and anxiety, 223, 224
    junk science vs. real-world training, 283
    learned helplessness, 131
    morality of punishment, 342
    myths about principles of punishment, 326
    negative punishment, 40
    observational evidence vs. scientific studies, 255
    punishment in nature, 81, 82, 83
    rehoming and euthanasia, 245, 246, 247, 248
    rewarding alternative behaviours, 196, 198
    stringent restriction of freedom, 236
    why correction succeeds, 351
stop/go method, 174, 175, 177, 178, 180, 181
strangle collars, 175, 177
stress, 130, 199, 216, 237, 289, 318, 370
stringent and extended restriction of freedom, 235, 237
Stubbs, Margaret, 116
students, 38, 75, 330, 333
    Coercion and Its Fallout, 278, 280
    demonisation of punishment, 27
    downsides of positive training, 215, 249
    effect of punishment on relationship, 138, 143, 144
    effectiveness of punishment with dogs, 87, 88
    effectiveness of punishment with people, 93, 95, 97, 103
    fear and anxiety, 127, 224
    misbehaving, 93
    operant conditioning, 38
    positive reinforcement, 45
    problems with operant conditioning, 49
    punishment in nature, 79
    rehoming and euthanasia, 247
    rewarding alternative behaviours, 192
submissive dog, 82, 85, 220
surveys, 135, 257, 260, 261, 262, 263, 285, 289, 300
Suterko-Cory Award, 64

T

tail wagging, 124, 143, 176
The Scientific Method – How Science Works, Fails to Work, and Pretends to Work, 255
Tilikum, 27, 46, 139, 219, 229, 289
time outs, 48, 73, 93, 101, 102
timing, 101, 102, 109, 111, 313, 319, 321, 333, 335, 336
Todd, Zazie, 130, 131, 202, 295
Total Recall – Perfect Response Training for Puppies and Adult Dogs, 162, 163
toxic dependency, 235
Trainers from the Dark Side, 24
training. *See also* treat training; positive training; positive reinforcement training:fear and anxiety
    balanced, 74, 244, 291, 296, 309
    competition, 95–98
    effective, 126, 144, 237, 243, 302, 310
    punishment-based, 62, 119, 216, 231, 245, 288
training methods
    Coercion and Its Fallout, 273
    demonisation of punishment, 29
    does punishment cause undesirable side-effects, 118, 125, 133
    does reward work?, 208
    downsides of positive training, 239, 241, 243
    effectiveness of punishment with dogs, 61, 62, 90

Index

junk science vs. real-world training, 283, 285, 293, 298
LIMA, LIFE and PROTECT, 305, 307, 310
model of obedience, 329
problems with operant conditioning, 45
punishment in nature, 79, 81
why correction succeeds, 358, 359, 360, 361
treat training
competing rewards, 182
does reward work?, 157, 181, 206, 207, 208
effectiveness of punishment with dogs, 62
fear and anxiety, 125
gaining obedience to commands, 159
hunger, 226, 228, 229
model of obedience, 332
myths about principles of punishment, 321
positive reinforcement, 46
punishment, 217
recall, 172
rewarding alternative behaviours, 195
teaching the meaning of commands, 157
Trigger warnings, 127

U

unconditioned punishers, 279
unconditioned reinforcer, 280
undesirable behaviours, 66, 155, 180, 188, 210, 285, 355
University of Sydney here in Australia, 26
unwanted behaviours. *See* undesirable behaviours
US Department of Health and Human Services, 246

V

van Oorsouw et al, 107, 145, 271, 352
verbal feedback, 195, 198
virtue signalling, 307, 342

W

wait and reward, 181
Wakefield, Debra, 369
walking on the lead, 61, 174, 175, 176, 180, 181, 182, 187, 210
Walters, Tarsha, 365
water bottle. *See* spray bottle
whining, 199, 325
White, Sophia & Justin, 364
Whitney, Dr. Leon, 24, 25, 28, 69, 70, 71, 226, 227, 228, 232
Whole Dog Journal, 160
Wilkinson, Andrew, 9, 131, 371
Williams, Emma, 288, 295
withdrawal, 44, 97, 109, 110, 124, 176
Wittenberg, Anya, 116
Wolff, Joel, 116
wolves, 66, 70, 74–80, 81, 82, 83, 84, 86, 194, 220, 318, 326, 335, 341, 352
addiction, 235
Wood, Rick, 116
Working Sheep Dogs – A Practical Guide to Breeding, Training and Handling, 257
Wright, Hannah, 291, 293, 295

Y

Young, Kat, 366
YouTube
aggression, 219
anger and abuse, 241
demonisation of punishment, 28
does reward work?, 199, 210
effectiveness of punishment with dogs, 62, 63
fear and anxiety, 125
myths about principles of punishment, 320
observational evidence vs. scientific studies, 253
why correction succeeds, 356, 358

www.ingramcontent.com/pod-product-compliance
Lightning Source LLC
Chambersburg PA
CBHW082149070526
44585CB00020B/2146